Undergraduate Texts in Mathematics

Readings in Mathematics

Editors
S. Axler
K.A. Ribet

Graduate Texts in Mathematics
Readings in Mathematics

Ebbinghaus/Hermes/Hirzebruch/Koecher/Mainzer/Neukirch/Prestel/Remmert: *Numbers*
Fulton/Harris: *Representation Theory: A First Course*
Murty: *Problems in Analytic Number Theory*
Remmert: *Theory of Complex Functions*
Walter: *Ordinary Differential Equations*

Undergraduate Texts in Mathematics
Readings in Mathematics

Anglin: *Mathematics: A Concise History and Philosophy*
Anglin/Lambek: *The Heritage of Thales*
Bressoud: *Second Year Calculus*
Hairer/Wanner: *Analysis by Its History*
Hämmerlin/Hoffmann: *Numerical Mathematics*
Isaac: *The Pleasures of Probability*
Knoebel/Laubenbacher/Lodder/Pengelley: *Mathematical Masterpieces: Further Chronicles by the Explorers*
Laubenbacher/Pengelley: *Mathematical Expeditions: Chronicles by the Explorers*
Samuel: *Projective Geometry*
Stillwell: *Numbers and Geometry*
Toth: *Glimpses of Algebra and Geometry,* Second Edition

Arthur Knoebel
Reinhard Laubenbacher
Jerry Lodder
David Pengelley

Mathematical Masterpieces

Further Chronicles by the Explorers

 Springer

Arthur Knoebel
Albuquerque, NM
USA

Jerry Lodder
Department of Mathematical Sciences
New Mexico State University
Las Cruces, NM 88003-8001
USA
jlodder@nmsu.edu

Reinhard Laubenbacher
Virginia Bioinformatics Institute
Virginia Polytechnic Institute and
    State University
Blacksburg, VA 24061
USA
reinhard@vbi.vt.edu

David Pengelley
Department of Mathematical Sciences
New Mexico State University
Las Cruces, NM 88003-8001
USA
davidp@nmsu.edu

ISBN 978-0-387-33060-0 (hardcover)
ISBN 978-0-387-33061-7 (softcover)
e-ISBN 978-0-387-33062-4

Library of Congress Control Number: 2006940178

Mathematics Subject Classification (2000): 01-01, 11-01, 40-01, 53-01, 65-01

Cover illustration (clockwise from top left): Jakob Bernoulli, Christiaan Huygens' construction
of the evolute in *Horologium oscillatorium* (*The Pendulum Clock*), Isaac Newton, Christiaan
Huygens, the *Shu Shu Jiu Zhang* by Qin Jiu-Shao, and Gotthold Eisenstein.

Printed on acid-free paper.

9 8 7 6 5 4 3 2 1

springer.com

Preface

In introducing his essays on the study and understanding of nature and evolution, biologist Stephen J. Gould writes:

> [W]e acquire a surprising source of rich and apparently limitless novelty from the primary documents of great thinkers throughout our history. But why should any nuggets, or even flakes, be left for intellectual miners in such terrain? Hasn't the *Origin of Species* been read untold millions of times? Hasn't every paragraph been subjected to overt scholarly scrutiny and exegesis?
>
> Let me share a secret rooted in general human foibles. . . . Very few people, including authors willing to commit to paper, ever really read primary sources—certainly not in necessary depth and completion, and often not at all. . . .
>
> I can attest that all major documents of science remain chock-full of distinctive and illuminating novelty, if only people will study them—in full and in the original editions. Why would anyone *not* yearn to read these works; not hunger for the opportunity? [99, p. 6f]

It is in the spirit of Gould's insights on an approach to science based on primary texts that we offer the present book of annotated mathematical sources, from which our undergraduate students have been learning for more than a decade. Although teaching and learning with primary historical sources require a commitment of study, the investment yields the rewards of a deeper understanding of the subject, an appreciation of its details, and a glimpse into the direction research has taken.

Our students read sequences of primary sources. These provide authentic motivation for seminal problems, and trace the creation of new concepts and techniques for their solution through the centuries. The broader mathematical and social context provided by primary historical sources allows technical elements to appear in their proper place, understood and appreciated as by the creators themselves. Students will even find themselves asking many of the same questions the pioneers did and answering these for themselves within the historical path of human discovery, thereby engendering a sense of adventure and immediacy, along with deeper motivation and a real grasp of the scope of each subject.

Primary sources also inject students directly into the process of mathematical research. They become active participants at the cutting edge of their own knowledge, experiencing actual research through grappling with the writings of great thinkers of the past. This creative immersion into the challenges of the past helps students better understand the problems of today. Finally, students gain a more profound technical comprehension, since complexity is introduced gradually and naturally.

Here, we present four independent chapters, each a story anchored around a sequence of selected primary sources showcasing a masterpiece of mathematical achievement. Our stories in brief are these:

1. The dynamic interplay between the discrete and continuous in mathematics stretches from Zeno's paradoxes and Pythagorean geometric number theory to the present, aiming to quantify exactly how separated, distinct, and finite objects blend with connected, homogeneous, and infinite spaces. Today, the bridge between the continuous and discrete is more important than ever, with digital technology increasingly emulating continuous phenomena.

2. A similarly ancient history underlies the development of algorithms for finding numerical solutions of equations. This evolution has gone hand in hand with multiple expansions of our notion of number itself, and today, questions of algorithmic robustness and rates of convergence are vital for modern science, exemplified in the appearance of fractal phenomena.

3. In contrast, our contemporary understanding of curvature began more recently, relying on the emerging calculus of the seventeenth century. Impetus for comprehending curvature has ranged from attempts to develop accurate maps and clocks for navigating the world to our present efforts to understand the geometric nature and dimensionality, large and small, of the physical universe we live in.

4. Finally, number theory has been driven over several centuries by the mysterious yet crucial nature of prime numbers. Their behavior and patterns remain ever enticing and mysterious, yet they obey a few beautiful fundamental laws. Recently, prime numbers have emerged into a broader limelight, their elusive properties increasingly important to the security of modern electronic communication.

Our goal is to tell these stories by guiding readers through the words of the masters themselves.

The present work is similar in format to our earlier book *Mathematical Expeditions* [150], which chronicled the development of five mathematical topics at the beginning undergraduate level. However, the current endeavor encompasses different topics and at a higher level, and is for advanced undergraduates who know at least a year of calculus and have some maturity with mathematics at the upper-division level. The book has emerged from a course at New Mexico State University taken by juniors and seniors majoring in mathematics, secondary education, engineering, and the sciences. While our focus is on the mathematics itself through the words of the masters, the richly historical nature of the presentation has encouraged professors at some colleges to use these materials for teaching the history of mathematics as well.

The book is quite flexible. The chapters are entirely independent of each other, except for minor biographical cross-referencing, so they can be read in any combination and order, or used individually to supplement another course. Moreover, the introduction to each chapter is an extensive free-standing summary of the relevant mathematics and its history. Within the chapter introduction, the reader is referred to the subsequent sections of annotated original sources. The individual sections can be read independently as well, preferably in conjunction with the introduction. In our own one-semester course, we usually focus on just one or two chapters; there is plenty of material in the book for at least two semesters. In the classroom, we often work through the introduction together with students, jumping to the later sections as the sources are mentioned, asking students to read and write their own reactions and questions in advance of classroom exegesis of the primary source. The annotation after each source is there to help with sticky points, but is used sparingly in class. We have included many exercises throughout based on the original sources, and we provide extensive references for further reading, as well as some internet resources [144].

During the past fifteen years, discussion and use of history in teaching mathematics have expanded significantly, including the approach we take based on primary sources [30, 52, 71, 122, 132, 145, 146, 147, 160, 187, 215, 233]. And there are now increasingly many resource materials available to support the use of history [40, 53, 144, 150, 234]. Our own approach is to have students read primary sources directly, keeping the original notation as much as possible, translating only the words into English. We strongly encourage the reader to go beyond this book to explore the rich and rewarding world of primary sources. There are substantial collections of original sources available in English, which we have endeavored to compile in a Web bibliography for using history in teaching mathematics [144]. Collected works of mathematicians are also a great resource [196].

This book has been ten years in the making, and we are grateful for the help of many people and institutions. Directors Tom Hoeksema and Bill Eamon of our university's Honors College provided extensive support and encouragement for the course from which this book grew. Our department heads Carol Walker and Doug Kurtz believed enough in this approach to help us make it a permanent part of our university curriculum. A grant from the Division of Undergraduate Education at the National Science Foundation (NSF) provided extensive resources, including assistance and apprentice teaching by graduate student Karen Schlauch. Our outside NSF advisory consultants, John McCleary and Victor Katz, generously provided expert and extremely helpful advice, including diligent reading and editorial suggestions on several drafts. We also owe great thanks to the help of our libraries, particularly interlibrary loan.

Others have also provided invaluable special assistance and encouragement. Our colleague Mai Gehrke has taught the course with drafts of two chapters, and we are most grateful for her helpful suggestions. Harold (Ed) Edwards read and gave extremely valuable suggestions for our "bridge" chapter, as did Manfred Kolster and Jens Funke for the chapter on primes. We received assistance with French translation from Mai Gehrke, and with Latin from Danny Otero, Joe Ball, Jens Funke, and Marty Flashman, to whom we are very grateful. Keith Dennis always tells us how to find things, from sources to portraits, and we appreciate Andrea Bréard's help with Chinese sources. The special and generous technical assistance with file recovery offered by Ron Logan in a time of crisis went way beyond the call of duty. We also offer great thanks to Sterling Trantham for superb photography.

John Fauvel's tremendous enthusiasm, encouragement, and generous detailed suggestions over the years will never be forgotten. We are sad he is no longer alive to continue to hold us to the highest standards; we must aspire to them on our own and can only hope that the final form of this book would meet with his approval.

The greatest credit for this book must go to our students. Without them, it would surely never have been written. We have used many versions of the manuscript with students at New Mexico State University, as well as at Vanderbilt University, and Hélène Barcelo has taught with some of our materials at Arizona State University. Our students' enthusiasm and accomplishments have convinced us that teaching with primary sources is invaluable to them, and their feedback greatly improved the book.

We are ever grateful to Ina Lindemann, from Springer, who showed great interest in our project, supported us with just the right mixture of patience and prodding, and whose enthusiasm provided much encouragement. And we thank David Kramer for very thoughtful copyediting, and Mark Spencer for his interest and shepherding us through final production of the book.

The first author appreciates that his wife, Patricia, provided a peaceful and productive setting in which to write, and for relaxation, planned lively backpacking trips to England and the canyons of southern Utah. The second author thanks his wife, Maria Elena, for her unwavering love and support while this work was done, even though it led to many canceled weekend motorcycle rides together. The third author would like to thank the NSF for its generous support from both the Division of Undergraduate Education and the Division of Mathematical Sciences, permitting a fruitful excursion into differential geometry that united the author's research and teaching. The fourth author thanks his wife, Pat Penfield, for her enduring love, encouragement, and support for this endeavor; for her excellent ideas and incisive critiques of several chapter drafts; and for showing us Stephen Jay Gould's essay quoted above. And he remembers his parents, Daphne and Ted, for their constant love, support, and inspirational role models for integrating history with science.

Las Cruces, New Mexico   *Arthur Knoebel*
April 2006   *Reinhard Laubenbacher*
                                                            Jerry Lodder
                                                         David Pengelley

Contents

Preface V

1   The Bridge Between Continuous and Discrete   1
    1.1   Introduction   1
    1.2   Archimedes Sums Squares to Find the Area Inside
          a Spiral   18
    1.3   Fermat and Pascal Use Figurate Numbers,
          Binomials, and the Arithmetical Triangle to
          Calculate Sums of Powers   26
    1.4   Jakob Bernoulli Finds a Pattern   41
    1.5   Euler's Summation Formula and the Solution for
          Sums of Powers   50
    1.6   Euler Solves the Basel Problem   70

2   Solving Equations Numerically: Finding Our Roots   83
    2.1   Introduction   83
    2.2   Qin Solves a Fourth-Degree Equation by
          Completing Powers 110
    2.3   Newton's Proportional Method 125
    2.4   Simpson's Fluxional Method 132
    2.5   Smale Solves Simpson 140

3   Curvature and the Notion of Space 159
    3.1   Introduction 159
    3.2   Huygens Discovers the Isochrone 167
    3.3   Newton Derives the Radius of Curvature 181
    3.4   Euler Studies the Curvature of Surfaces 187
    3.5   Gauss Defines an Independent Notion of Curvature   196
    3.6   Riemann Explores Higher-Dimensional Space 214

4   Patterns in Prime Numbers:
The Quadratic Reciprocity Law 229
   4.1   Introduction 229
   4.2   Euler Discovers Patterns for Prime Divisors of
        Quadratic Forms 251
   4.3   Lagrange Develops a Theory of Quadratic Forms
        and Divisors 261
   4.4   Legendre Asserts the Quadratic Reciprocity Law 279
   4.5   Gauss Proves the "Fundamental Theorem" 286
   4.6   Eisenstein's Geometric Proof 292
   4.7   Gauss Composes Quadratic Forms: The Class Group 301
   4.8   Appendix on Congruence Arithmetic 306

References 311

Credits 323

Name Index 325

Subject Index 329

1

The Bridge Between Continuous and Discrete

1.1 Introduction

In the early 1730s, Leonhard Euler (1707–1783) astonished his contemporaries by solving one of the most burning mathematical puzzles of his era: to find the exact sum of the infinite series $\frac{1}{1} + \frac{1}{4} + \frac{1}{9} + \frac{1}{16} + \frac{1}{25} + \cdots$, whose terms are the reciprocal squares of the natural numbers. This dramatic success began his rise to dominance over much of eighteenth-century mathematics. In the process of solving this then famous problem, Euler invented a formula that simultaneously completed another great quest: the two-thousand-year search for closed expressions for sums of numerical powers. We shall see how Euler's success with both these problems created a bridge connecting continuous and discrete summations.

Sums for geometric series, such as $\frac{1}{1} + \frac{1}{2} + \frac{1}{4} + \frac{1}{8} + \cdots = 2$, had been known since antiquity. But mathematicians of the late seventeenth century were captivated by the computation of the sum of a series with a completely different type of pattern to its terms, one that was far from geometric. In the late 1660s and early 1670s, Isaac Newton (1642–1727) and James Gregory (1638–1675) each deduced the power series for the arctangent,[1] $\arctan t = t - \frac{t^3}{3} + \frac{t^5}{5} - \cdots$, which produces, when evaluated at $t = 1$, the sum $\frac{\pi}{4}$ for the alternating series of reciprocal odd numbers $1 - \frac{1}{3} + \frac{1}{5} - \frac{1}{7} + \frac{1}{9} - \cdots$ [133, pp. 492–494], [135, pp. 436–439]. And in 1674, Gottfried Wilhelm Leibniz (1646–1716), one of the creators of the differential and integral calculus, used his new calculus of infinitesimal differentials and their summation (what we now call *integration*) to obtain the same value, $\frac{\pi}{4}$, for this sum by analyzing the quadrature, i.e., the area, of a quarter of a unit circle [133, pp. 524–527].

Leibniz and the Bernoulli brothers Jakob (1654–1705) and Johann (1667–1748), from Basel, were tantalized by this utterly unexpected connection

---

[1] This power series had also been discovered in southern India around two hundred years earlier, where it was likely derived for astronomical purposes, and written in verse [125], [133, pp. 494–496].

between the special number π from geometry and the sum of such a simple and seemingly unrelated series as the alternating reciprocal odd numbers. What could the connection be? They began considering similar series, and it is not surprising that they came to view the sum of the reciprocal squares, first mentioned in 1650 by Pietro Mengoli (1626–1686), as a challenge. Despite hard work on the problem, success eluded the Bernoullis for decades, and Jakob wrote, "If someone should succeed in finding what till now withstood our efforts and communicate it to us, we shall be much obliged to him" [258, p. 345]. The puzzle was so prominent that it became known as the "Basel problem."

Around 1730, Euler, a student of Johann Bernoulli's, took a completely fresh approach to the Basel problem by placing it in a broader context. He decided to explore the general discrete summation $\sum_{i=1}^{n} f(i)$ of the values of an arbitrary function $f(x)$ at a sequence of natural numbers, where n may be either finite or infinite. The Basel problem, to find $\sum_{i=1}^{\infty} \frac{1}{i^2}$, fits into this new context, since the sum can be written as $\sum_{i=1}^{\infty} g(i)$ for the function $g(x) = \frac{1}{x^2}$. Euler's broader approach also encompassed an age-old question, that of finding formulas for sums of numerical powers, as we will now explain.

By the sixth century B.C.E., the Pythagoreans already knew how to find a sum of consecutive natural numbers, which we write as

$$1 + 2 + 3 + \cdots + n = \sum_{i=1}^{n} i^1 = \frac{n(n+1)}{2} = \frac{n^2}{2} + \frac{n}{2}.$$

Archimedes of Syracuse (c. 287–212 B.C.E.), the greatest mathematician of antiquity, also discovered how to calculate a sum of squares. Translated into contemporary symbolism, his work shows that

$$1^2 + 2^2 + 3^2 + \cdots + n^2 = \sum_{i=1}^{n} i^2 = \frac{n(n+1)(2n+1)}{6} = \frac{n^3}{3} + \frac{n^2}{2} + \frac{n}{6}.$$

Throughout the next two millennia, the search for general formulas for $\sum_{i=1}^{n} i^k$, a sum of consecutive kth powers for any fixed natural number k, became a recurring theme of study, primarily because such sums could be used to find areas and volumes. All these previous efforts also fit within Euler's general context, since they are simply $\sum_{i=1}^{n} f_k(i)$ for the functions $f_k(x) = x^k$. While the function g for the Basel problem is very different from the functions f_k that produce sums of powers, Euler's bold vision was to create a general approach to any sum of function values at consecutive natural numbers.

Euler's aim was to use calculus to relate the discrete summation $\sum_{i=1}^{n} f(i)$ (with n possibly infinity) to a continuous phenomenon, the antiderivative, i.e., the integral $\int_0^n f(x)dx$. We know that these two provide first approximations to each other, since the sum can be interpreted as the total area of rectangles with tops forming a staircase along the curve $y = f(x)$, while the antiderivative, appropriately evaluated between limits, can be interpreted as the area

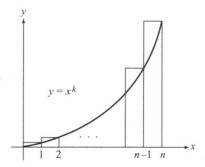

Fig. 1.1. Comparing a discrete sum with a continuous area.

under the curve itself (Figure 1.1). It is precisely the delicate difference, in both numerical value and in concept, between such discrete sums and continuous areas that mathematicians had in fact been exploring for so long when trying to find formulas for sums of powers.

For such powers the discrete sum is

$$\sum_{i=1}^{n} f_k(i) = \sum_{i=1}^{n} i^k = 1^k + 2^k + 3^k + \cdots + n^k,$$

while the continuous quantity for comparison is

$$\int_0^n f_k(x)dx = \int_0^n x^k dx = \frac{n^{k+1}}{k+1}.$$

Notice that the latter provides the first term in each of the polynomial summation formulas displayed above from the Pythagoreans and Archimedes. Understanding the dynamic between discrete and continuous amounts to quantifying exactly how separated, distinct, and finite objects blend with connected, homogeneous, and infinite spaces. Scholars as far back as Zeno, in classical Greece, grappled with this tension. Out of the fog of using discrete sums to approximate areas emerged the discovery of the differential and integral calculus in the seventeenth century. We shall see that Euler then turned the tables around in the eighteenth century by applying calculus to solve problems of the discrete.

Euler reconciled the difference between a discrete sum and a continuous integral via a striking formula using a corresponding antiderivative $\int_0^n f(x)dx$ as the first approximation to the summation $\sum_{i=1}^{n} f(i)$, with additional terms utilizing the iterated derivatives of f to make the necessary adjustments from continuous to discrete. Today we call this the *Euler–Maclaurin summation formula*. Euler applied it to obtain incredibly accurate approximations to the sum of the reciprocal squares, for solving the Basel problem, and these successes likely enabled him to guess that the infinite sum was exactly $\frac{\pi^2}{6}$. Armed with this guess, it was not long before he found a proof, and announced a solution of the Basel problem to the mathematical world.

Euler's correspondents were greatly impressed. Johann Bernoulli wrote, "And so is satisfied the burning desire of my brother [Jakob], who, realizing that the investigation of the sum was more difficult than anyone would have thought, openly confessed that all his zeal had been mocked. If only my brother were alive now" [258, p. 345].

Euler also used his summation formula to provide closure to the long search for closed formulas for sums of powers. By now this thread had wound its way from classical Greek mathematics through the medieval Indian and Islamic worlds and into the Renaissance. Finally, during the Enlightenment, Jakob Bernoulli discovered that the problem revealed a special sequence of numbers, today called the *Bernoulli numbers*. These numbers became a key feature of Euler's summation formula and of modern mathematics, since, as we shall soon see, they capture the essence of converting between the continuous and the discrete. We will trace this thread through original sources from Archimedes to Euler, ending with Euler's exposition of how his general summation formula reveals formulas for sums of powers as well as a way to tackle the Basel problem. That Euler used his summation formula to resolve these two seemingly very different problems is a fine illustration of how generalization and abstraction can lead to the combined solution of seemingly independent problems.

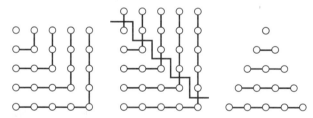

Fig. 1.2. Square, rectangular, and triangular numbers.

We return now to the very beginning of our story, which revolves around the relationship between areas and formulas for discrete sums of powers, such as the closed formulas above for the sums of the first n natural numbers and the first n squares. For the natural numbers it is not hard both to discover and to verify the formula oneself, but the Pythagoreans would not have written it as we do. For them, number was the substance of all things. Numbers were probably first represented by dots in the sand, or pebbles. From this, patterns in planar configurations of dots began to be recognized, and these were related to areas of planar regions, as in Figure 1.2 [18, p. 54f], [113], [133, p. 48ff], [135, p. 28ff], [258, p. 74ff].

In the figure, the arrangement and number of dots in each configuration suggests general closed formulas for various types of sums, illustrated by the three types $1+3+5+7+9 = 5^2$, $2+4+6+8+10 = 5 \cdot 6$, and $1+2+3+4+5 = (5 \cdot 6)/2 = 15$. The reader may easily conjecture and prove general summation formulas with n terms for each of these.

For the third type, the total number of dots in the triangular pattern is clearly half of that in the rectangular pattern, which can be verified in general either algebraically, or geometrically from Figure 1.2. Thus we have deduced the closed Pythagorean formula above for the sum of natural numbers,[2] and we also see why the numbers $\frac{n(n+1)}{2}$ (i.e., 1, 3, 6, 10, 15, ...) deserve to be called *triangular numbers*. Notice that each of the three types of sums of dots has for its terms an *arithmetic progression*, i.e., a sequence of numbers with a fixed difference between each term and its successor. The first and third types always begin with the number one; the Pythagoreans realized that such sums produce *polygonal numbers*, i.e., those with dot patterns modeled on triangles, squares, pentagons, etc. (Exercises 1.1, 1.2).

The closed formula for a sum of squares, which we pulled from thin air earlier, is implicit in the work of Archimedes. At first sight it may seem unexpected that such a discrete sum should even have a closed formula. Once guessed, though, one can easily verify it by mathematical induction (Exercise 1.3). The formula arises in two of Archimedes' books [7]. In *Conoids and Spheroids* Archimedes develops and uses it as a tool for finding volumes of paraboloids, ellipsoids, and hyperboloids of revolution. In *Spirals* he applies it to obtain a remarkable result on the area enclosed by a spiral, stated thus in his preface:

> If a straight line of which one extremity remains fixed be made to revolve at a uniform rate in a plane until it returns to the position from which it started, and if, at the same time as the straight line revolves, a point move at a uniform rate along the straight line, starting from the fixed extremity, the point will describe a spiral in the plane. I say then that the area bounded by the spiral and the straight line which has returned to the position from which it started is a third part of the circle described with the fixed point as the centre and with radius the length traversed by the point along the straight line during the one revolution.

---

[2] Another way of obtaining this formula occurs in a story about the developing genius Carl F. Gauss (1777–1855). When Gauss was nine, his mathematics teacher, J. G. Büttner, gave his class of 100 pupils the task of summing the first 100 integers. Gauss almost immediately wrote 5050 on his slate and placed it on his teacher's desk. Gauss had noticed that adding the numbers first in the corresponding pairs 1 and 100, 2 and 99, 3 and 98,..., produced the sum 101 exactly 50 times, and then he simply multiplied 101 by 50 in his head. Fortunately, Büttner recognized Gauss's genius, and arranged for special tutoring for him. Gauss became the greatest mathematican of the nineteenth century [133, p. 654].

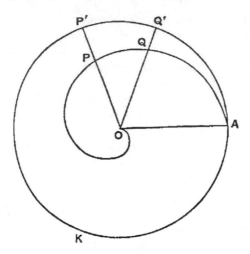

Fig. 1.3. Archimedes' area of a spiral.

Figure 1.3 illustrates Archimedes' claim that the area $OPQAO$ within the spiral is exactly one-third the area $AKP'Q'A$ of the "first circle."

Our original source will focus on Archimedes' expression for a sum of squares, and the resulting theorem on the area of the spiral, using the classical Greek *method of exhaustion*. Here we will see an early historical link between the discrete, in the form of the sum of squares formula, and the continuous, namely the area bounded by a continuous curve.

We will see also that Archimedes does not actually need an exact sum of squares formula to find the area in his spiral, but rather only the inequalities

$$\frac{n^3}{3} < \sum_{i=1}^{n} i^2 < \frac{(n+1)^3}{3},$$

which are highly suggestive of a more general pattern related to antidifferentiation of the kth-power functions $f_k(x) = x^k$ (Exercises 1.4, 1.5).

Our mathematical forebears were extremely interested in formulas for sums of higher powers $\sum_{i=1}^{n} i^k$, since they could use these to compute other areas and volumes. Let us pause to review from modern calculus how sums of powers are explicitly involved in the interpretation of the area under the curve $y = x^k$, for $0 \leq x \leq 1$, as the definite integral $\int_0^1 x^k \, dx$. Recall that to calculate this area from its modern definition as a limit of Riemann sums, we can subdivide the interval into n equal subintervals, each of width $1/n$, and consider the sum of areas of the rectangles built upwards to the curve from, say, the right endpoints of these subintervals, obtaining $\sum_{i=1}^{n} \frac{1}{n} \cdot \left(\frac{i}{n}\right)^k = \frac{1}{n^{k+1}} \sum_{i=1}^{n} i^k$ as an approximation to the area under the curve. The exact area is then the limit of this expression as n approaches infinity, since increasing n refines the accuracy of the approximation. Thus it is clear why having a closed formula for $\sum_{i=1}^{n} i^k$ (or perhaps just inequalities analogous to those of Archimedes above) is key for carrying this calculation to completion. While this modern formulation streamlines the verbal and geometric versions of our ancestors, still the algebraic steps were essentially these.

As we continue to powers higher than $k = 2$, a formula for $\sum_{i=1}^{n} i^3$ jumps off the page once we compute a few values and compare them with our previous work. (The reader who wishes to guess the formula before we introduce it may consult Exercise 1.6 now.) It seems likely, from the work of the neo-Pythagorean Nicomachus of Gerasa in the first century C.E., that the mathematicians of ancient Greece knew this too; while it is not explicit in extant work, it is implicit in a fact about sums of odd numbers and cubic numbers found in Nicomachus's *Introductio Arithmetica* [19], [113, p. 68f] (Exercise 1.7).

The general formula for a sum of cubes first appears explicitly in the *Āryabhaṭīya*, from India [133, p. 212f], a book of stanzas perhaps intended as a short manual for memorization, which Āryabhaṭa wrote in 499 C.E., when he was 23 years old. Without any proof or justification, and in the completely verbal style of ancient algebra, he wrote:

> The sixth part of the product of three quantities consisting of the number of terms, the number of terms plus one, and twice the number of terms plus one is the sum of the squares. The square of the sum of the (original) series is the sum of the cubes.

The earliest proof we have of the sum of cubes formula is by the Islamic mathematician Abū Bakr al-Karajī (c. 1000 C.E.), one of a group who began to develop algebra, in particular generalizing the arithmetic of numbers, centered around the House of Wisdom established in Baghdad in the ninth century [133, p. 251ff]. Al-Karajī's argument is noteworthy for its use of the method of "generalizable example" [113, p. 68f], [133, p. 255].

The idea of a generalizable example is to prove the claim for a particular number, but in a way that clearly shows that it works for any number. This was a common method of proof for centuries, in part because there was no notation adequate to handle the general case, and in particular no way of using indexing as we do today to deal with sums of arbitrarily many terms. Al-Karajī proves that $(1 + 2 + 3 + \cdots + 10)^2 = 1^3 + 2^3 + 3^3 + \cdots + 10^3$ in a way that clearly generalizes: He considered the square $ABCD$ with side $1 + 2 + \cdots + 10$ (Figure 1.4), subdivided into gnomons (L-shaped pieces) as shown, with the largest gnomon having ends $BB' = DD' = 10$. The area of the largest gnomon is 10^3 (the reader should carry this "calculation" out in a way that is convincing of "generalizibility"). By the same generalized reasoning the area of the next-smaller gnomon is 9^3, and so on for all the smaller gnomons, with only a square of side 1 left over.

Now one can think of the area of the large square in two ways. As the sum of gnomons it has area $1 + 2^3 + 3^3 + \cdots + 10^3$. On the other hand, as a square it has area $(1 + 2 + 3 + \cdots + 10)^2$. Today we would be inclined to use an algebraic proof by mathematical induction here; but it appears unnecessary if one sees how

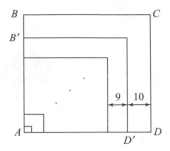

Fig. 1.4. Gnomons for the sum of cubes (not to scale).

to break the square up into gnomons, each identifiable numerically as a cube. This could all be done algebraically, although it would be excruciating, which is what leads us to use mathematical induction if we are invoking algebra rather than geometry (Exercise 1.8).

At this point we can be optimistic that for each fixed natural number k there is a polynomial in n for $1^k + 2^k + 3^k + \cdots + n^k$. Based on our examples and the analogy to integration of x^k, the reader should try to guess the degree of the polynomial, the leading coefficient, and inequalities that might bound the polynomial like those of Archimedes. On the other hand, no general pattern is yet emerging for the details of the formula for various values of k, and worse, all the formulas we obtained emerged from ad hoc methods, each demanding separate verification.

The work of the Egyptian mathematician Abū ʿAlī al-Ḥasan ibn al-Haytham (965–1039) gives us the first steps along a path toward understanding these formulas in general [133, p. 255f]. He needed a sum of fourth powers in order to find the volume of a general paraboloid of revolution (in contemporary terms this involves integrating x^4). At that time, Islamic mathematicians were studying, rediscovering, and extending the work of Archimedes and others on volumes by the method of exhaustion. Ibn al-Haytham's specific expression for fourth powers came from his equation (expressed here in modern symbolism) connecting sums of powers for different exponents:

$$(n+1)\sum_{i=1}^{n} i^k = \sum_{i=1}^{n} i^{k+1} + \sum_{p=1}^{n}\left(\sum_{i=1}^{p} i^k\right).$$

Although ibn al-Haytham did not state a completely general result, rather only for $n = 4$ and $k = 1, 2, 3$, his proof, like al-Karajī's, clearly generalizes for all n and k from his example, and uses a kind of mathematical induction (Exercise 1.9). In fact we can also prove his equation by interchanging the order of the double summation (Exercise 1.10). Letting $k = 3$, one can now obtain a formula for $\sum_{i=1}^{n} i^4$, as did ibn al-Haytham, by solving for it in his equation, first substituting the known formulas for smaller exponents. He did this, again by generalizable example. This is not quite as easy as we have made it sound, though, since in the process the double summation will actually give rise to the very thing one is solving for again, in addition to its already stated occurrence. The reader may see how this actually works out in practice in Exercise 1.11.

Having followed in ibn al-Haytham's footsteps, we should now be reasonably convinced that in principle we could calculate a polynomial formula in terms of n for the sum $\sum_{i=1}^{n} i^k$ for any particular k. But we imagine that this will quickly become increasingly tedious and complicated with increasing k, and with no discernible pattern in the final formulas for different values of k. As our story unfolds, we will gradually uncover intricate patterns in these formulas reflecting the subtle connections between integration and discrete summation.

In the seventeenth century, the European creation of the calculus became a driving force in the development of formulas for sums of powers. In the second quarter of the century, a number of brilliant mathematicians had great success at *squaring* heretofore intractable regions (i.e., finding their areas), in particular the regions under the curves we write as $y = x^k$, which they called *higher parabolas*. Their successes, and especially the increasing use of *indivisible* methods, were the immediate precursors to the emergence of calculus later in the century. For instance, on September 22, 1636, Pierre de Fermat (1601–1665), of Toulouse, wrote to Gilles Persone de Roberval (1602–1675) that he could "square infinitely many figures composed of curved lines" [133, p. 481ff], including the higher parabolas. Roberval replied that he, too, could square all the higher parabolas using the inequalities

$$\frac{n^{k+1}}{k+1} < \sum_{i=1}^{n} i^k < \frac{(n+1)^{k+1}}{k+1}.$$

The reader is invited to confirm that these inequalities suffice for computing $\int_0^a x^k dx$ using our modern definitions (Exercise 1.12), and also, conversely, that Roberval's inequalities follow easily if we already know modern calculus (Exercise 1.13).

In reply to Roberval, Fermat claimed more, that he could solve "what is perhaps the most beautiful problem of all arithmetic" [19], namely finding the precise sum of powers in an arithmetic progression, no matter what the power. Fermat, apparently unaware of the works of al-Haytham, thought that the problem had been solved only up to $k = 3$, and stated that he had reached his results on sums of powers by using the following theorem on the *figurate numbers*[3] derived from "natural progressions":

The last number multiplied by the next larger number is double the collateral triangle;
the last number multiplied by the triangle of the next larger is three times the collateral pyramid;

---

[3] Fermat was likely also unaware of the work of Johann Faulhaber (1580–1635), who managed to develop explicit polynomials for $\sum_{i=1}^{n} i^k$ for all k up to 17. Faulhaber's interest and methods were also related to figurate numbers, but his work did not yield any general insight into the larger picture for all k [19].

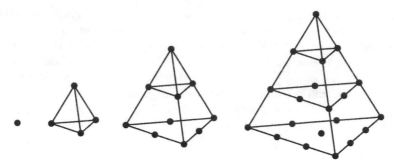

Fig. 1.5. Pyramidal numbers.

the last number multiplied by the pyramid of the next larger is four
times the collateral triangulo-triangle;
and so on indefinitely in this same manner [19].

By a "natural progression" Fermat simply means an arithmetic progression
$1, 2, \ldots, n$, whose "last number" is n. By the "collateral triangle" he means the
triangular number (Figure 1.2) on a side with n dots. The figurate numbers
then generalize this by counting dots in analogous higher-dimensional figures.
For instance, by the "collateral pyramid" Fermat means to count the dots in
a three-dimensional triangular pyramid on a side with n dots (Figure 1.5).

Fermat, typically, did not reveal his methods. But we can fill in the details
of his claims by studying the figurate numbers and discovering their agreement
with the numbers in the "arithmetical triangle"[4] (Figure 1.6) of his contemporary and correspondent Blaise Pascal (1623–1662). This we will explore in
our section on the work of Fermat and Pascal.

Fermat's results on figurate numbers, and his derivation therefrom of formulas for sums of powers, could indeed be carried on indefinitely, but the
process quickly becomes cumbersome and seemingly lacks insight. Despite
Fermat's enthusiasm for the problem, it appears at first that his procedure
yields not much more than ibn al-Haytham's. But what it did introduce was
a major role for the figurate numbers that appear in the arithmetical triangle.
And since the numbers in the arithmetical triangle have yet other important
properties and patterns, namely in their roles as combination numbers and
binomial coefficients, Fermat helped pave the way for future developments.

Blaise Pascal, in his *Treatise on the Arithmetical Triangle* [100, v. 30],
made a systematic study of the numbers in his triangle, simultaneously encom-
passing their figurate, combinatorial, and binomial roles. Although these numbers had emerged in the mathematics of several cultures over many centuries
[133], Pascal was the first to connect binomial coefficients with combinatorial
coefficients in probability.

---

[4] Today called Pascal's triangle.

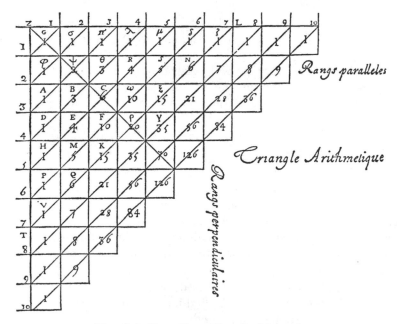

Fig. 1.6. Pascal's *arithmetical triangle*.

A major motivation for Pascal's treatise was a question from the beginnings of probability theory, about the equitable division of stakes in an interrupted game of chance. The question had been posed to Pascal around 1652 by Antoine Gombaud, the Chevalier de Méré, who wanted to improve his chances at gambling: Suppose two players are playing a fair game, to continue until one player wins a certain number of rounds, but the game is interrupted before either player reaches the winning number. How should the stakes be divided equitably, based on the number of rounds each player has won [133, p. 431, 451ff]? The solution requires the combinatorial properties inherent in the numbers in the arithmetical triangle, as Pascal demonstrated in his *Treatise*, since they count the number of ways various occurrences can combine to produce a given result.

Pascal also wrote another treatise, *Potestatum Numericarum Summa* (Sums of Numerical Powers), in which he presents his own approach to finding sums of powers formulas (he actually produces a prescription for much more general sums even than $\sum_{i=1}^{n} i^k$). We present his clearly written exposition. There we see that, armed with an ingenious idea based on the coefficients $\binom{m}{j}$ in the expansion of a binomial (i.e., $(a+b)^m = \sum_{j=0}^{m} \binom{m}{j} a^j b^{m-j}$), Pascal describes a procedure for finding sums of powers formulas. His final result is embodied in the equation

$$(k+1)\sum_{i=1}^{n} i^k = (n+1)^{k+1} - 1 - \sum_{j=0}^{k-1}\left[\binom{k+1}{j}\sum_{i=1}^{n} i^j\right].$$

Clearly one can solve here, if still tediously, for an explicit formula for the sum of kth powers, by using at each stage the already known formulas for lower exponents.

By this time in our story we will begin to discern some patterns in the sums of powers formulas for the first few values of k, which we can actually prove for general k from Pascal's equation. We can show that

$$\sum_{i=1}^{n} i^k = \frac{n^{k+1}}{k+1} + \frac{1}{2}n^k + \boxed{?}\, n^{k-1} + \cdots + \boxed{?}\, n + 0,$$

a $(k+1)$st-degree polynomial in n with zero constant term, in which we know the first two coefficients (the second term actually has a nice geometric interpretation (Exercise 1.14), which suggests the sign of the third). This leads us to hope there is a pattern to the remaining coefficients, and to wonder what they might mean in the larger picture of the relationship between discrete summation, $\sum_{i=1}^{n} i^k = \frac{n^{k+1}}{k+1} + \cdots$, and continuous summation, $\int_0^x t^k dt = \frac{x^{k+1}}{k+1}$.

Jakob Bernoulli (1654–1705) discovered the general pattern in the polynomial formulas for sums of powers. We find him explaining it in a small section of his important treatise *Ars Conjectandi* (Art of Conjecturing) on the theory of probability. Since the combination numbers, figurate numbers, and binomial coefficients are the same, it is not surprising that Bernoulli's work on sums of powers occurs in his treatise on probability theory. He discerns a general pattern in the coefficients of the polynomials, writing them in terms of the combination numbers in the arithmetical triangle and a new sequence of special numbers, which he believes occur in a predictable way throughout all the formulas for summing powers.

These new numbers soon came to be called the *Bernoulli numbers*, and ever since, they have played an important role in mathematics. They are a sequence of rational numbers, which we will denote by $B_2 = \frac{1}{6}$, $B_3 = 0$, $B_4 = -\frac{1}{30}$, ..., having a simple recursive law of formation. Bernoulli saw a pattern in the formulas in which these numbers seem to appear consistently. Specifically, he claimed, from calculating and examining the formulas explicitly up to the tenth powers, that the sums can be expressed as the following polynomials in n:

$$\sum_{i=1}^{n} i^k = \frac{n^{k+1}}{k+1} + \frac{1}{2}n^k + \frac{k}{2}B_2 n^{k-1} + \frac{k(k-1)(k-2)}{2\cdot 3\cdot 4}B_4 n^{k-3}$$
$$+ \frac{k(k-1)(k-2)(k-3)(k-4)}{2\cdot 3\cdot 4\cdot 5\cdot 6}B_6 n^{k-5}$$
$$+ \cdots + \text{ ending in a term involving } n \text{ or } n^2.$$

A critical observation here is that the Bernoulli numbers that occur are the same numbers in all the formulas, even as k varies. The pattern claimed here is clear (including that the odd Bernoulli numbers beginning with B_3 are all zero, and that the constant term in each polynomial is always zero). Observe that by setting $n = 1$ on both sides of this family of equations, we obtain

$$1 = \frac{1}{k+1} + \frac{1}{2} + \sum_{j=2}^{k} \frac{1}{k+1} \binom{k+1}{j} B_j$$

for each $k \geq 2$. The kth equation clearly allows recursive calculation of B_k from knowing the previous Bernoulli numbers[5] (Exercise 1.15).

Our chapter culminates by reading from the work of Leonhard Euler a few decades later. Euler dominated eighteenth-century mathematics, and produced seminal ideas in almost all its branches, as well as in physics. He was also perhaps the most prolific human writer of all time: his collected works are still in the process of being published, and will span close to one hundred thick volumes. Euler was particularly fascinated by the interplay between the continuous and the discrete in studying series, and the eighteenth century became a garden of discoveries about infinite series and related functions, largely thanks to Euler's genius [135, Chapter 20]. Euler's summation formula for series will bring together the sums of powers problem and the Basel problem on the infinite sum of reciprocal squares.

We have already mentioned Euler's early attraction to the famous Basel problem, to find the exact sum of the convergent series of reciprocal squares

$$\sum_{i=1}^{\infty} \frac{1}{i^2} = \frac{1}{1} + \frac{1}{4} + \frac{1}{9} + \frac{1}{16} + \frac{1}{25} + \cdots - \boxed{?} \, .$$

In a series of papers through the 1730s and beyond, apparently initially motivated largely by desire to sum this series, Euler discovered, applied, and refined his summation formula for obtaining approximations to finite and infinite sums, paradoxically by using divergent series [66, v. 14]. Since his formula was also independently discovered by the Scottish mathematician Colin Maclaurin (1698–1746), it is today called the *Euler–Maclaurin summation formula*.

Around the year 1730, the 23-year-old Euler, along with his frequent correspondents Christian Goldbach (1690–1764) and Daniel Bernoulli (1700–1782) (son of Johann, Euler's teacher), tried to find more and more accurate fractional or decimal estimates for the sum of the series of reciprocal squares. They were likely trying to guess the exact value of the sum, hoping to recognize that their approximations looked like something familiar, perhaps involving π, as had Leibniz's series. But these estimates were challenging, since the series converges very slowly. To wit, if we estimate the sum simply by calculating a partial sum $\sum_{i=1}^{n} \frac{1}{i^2}$, we may be sorely disappointed by the accuracy achieved.

---

[5] Explicit formulas for Bernoulli numbers, which do not rely on recursive knowledge about the previous numbers, are much more complicated [98].

The error is precisely the tail end of the series, which is bounded via

$$\frac{1}{n+1} = \int_{n+1}^{\infty} \frac{1}{x^2} dx < \sum_{i=n+1}^{\infty} \frac{1}{i^2} < \int_{n}^{\infty} \frac{1}{x^2} dx = \frac{1}{n},$$

by the standard method of inscribing and circumscribing rectangles of unit width along the curve $y = 1/x^2$. So if one were to add up 100 terms of the series by hand, as accurately as needed, the untallied tail end would be known only to lie between $1/100$ and $1/101$. Even taking this fully into account, the accuracy with which one would know the true sum would still only be the difference of these two numbers, i.e., about one ten-thousandth. Euler wanted far greater accuracy than this (Exercise 1.16). He first developed some clever special methods, and then in the early 1730s he hit gold with the discovery of his summation formula.

When we read Euler, we will see that his summation formula is in essence

$$\sum_{i=1}^{n} f(i) \approx C + \int^{n} f(x)dx + \frac{f(n)}{2} + B_2 \frac{f'(n)}{2!} + B_3 \frac{f''(n)}{3!} + B_4 \frac{f'''(n)}{4!} + \cdots,$$

where $\int^{n} f(x)dx$ means a fixed antiderivative without the usual constant of integration added on, but with n substituted for x, and C denotes a constant that depends on f and the antiderivative chosen, but is independent of n. The motivation we can provide at this point is twofold. First, when f is specialized to the power functions f_k, Euler's formula clearly specializes to Bernoulli's sum of powers formulas (Exercise 1.17). Second, it is obvious that the first three terms in the formula correspond to the trapezoid approximation to the integral (Exercise 1.18). It is reasonable to expect that the difference between the discrete sum on the left and the area represented by the antiderivative on the right will involve how the graph of f curves, and hence the derivatives of f; but the surprising thing is that these derivatives are all evaluated only at the single value n. We will see Euler derive his formula ingeniously from Taylor series.

One of Euler's first uses of his summation formula was to approximate the sum of the reciprocal squares. In a paper submitted to the St. Petersburg Academy of Sciences on the 13th of October, 1735, Euler applied it to approximate the sum of reciprocal squares and other series. He calculated the sum of reciprocal squares correct to twenty decimal places! Only seven and a half weeks later, Euler presented another paper, solving the famous Basel problem by demonstrating that the precise sum of the series is $\pi^2/6$. "Now, however, quite unexpectedly, I have found an elegant formula for $1 + \frac{1}{4} + \frac{1}{9} + \frac{1}{16} +$ etc., depending upon the quadrature of the circle [i.e., upon π]" [245, p. 261] (we paraphrase his proof in a footnote in the first of two sections on Euler's work). He even showed how to generalize his approach to find the exact sums of many other infinite series, such as the sum of the reciprocal fourth powers. While Euler's proof solving the Basel problem was soon criticized as lacking rigor, he was understandably convinced of the truth of his answer, partly because it so perfectly matched the highly accurate approximation from his summation formula. Later he found other, rigorously acceptable, ways of justifying his claim.

We may never know with certainty whether Euler already suspected, when he wrote his paper of October 13, that the exact sum was $\frac{\pi^2}{6}$, or whether his calculation to twenty places was actually part of guessing the answer. We do know that Daniel Bernoulli wrote to him "The theorem on the sum of the series $1 + \frac{1}{4} + \frac{1}{9} + \frac{1}{16} + \cdots = \frac{pp}{6}$ and $1 + \frac{1}{2^4} + \frac{1}{3^4} + \frac{1}{4^4} + \cdots = \frac{p^4}{90}$ is very remarkable. You must no doubt have come upon it a posteriori. I should very much like to see your solution" [10, p. 1075].

In our two sections on Euler's work we will study the summation formula in his own words from his book *Institutiones Calculi Differentialis* (Foundations of Differential Calculus), published in 1755. Here his presentation of the formula is intertwined with many of his subsequent discoveries.

In reading Euler's work, we will find that he ignores many questions we have about the rigor and validity of the mathematical steps he takes and the conclusions he draws. Not the least of these is that his summation formula usually diverges, yet still he calculates with great effectiveness using it. In this respect we should view Euler as a pioneer whose vision, brilliance, intuition, and experience about questions of convergence and divergence allowed him to excel where most mortals would stumble.

In our first selections from Euler's book we will see him derive the summation formula, analyze the Bernoulli numbers it contains, and relate these numbers to familiar power series from calculus, proving many of the most intriguing properties of the Bernoulli numbers. Finally, he applies the summation formula to give the first actual proof for Bernoulli's summation of powers formulas, thus completing the long search.

In our last section we will read how he uses his summation formula to make his remarkable approximation for the sum of reciprocal squares, before he proved that the value is $\pi^2/6$. Here as elsewhere Euler is always rechecking and verifying his results in different ways, with confirmation serving as his stabilizing rudder for confidence in further work.

In the *Institutiones* Euler also not only makes an exact determination of the sum of reciprocal squares as $\pi^2/6$, but actually finds the exact sums of all the series of reciprocal even powers, namely the series $\sum_{i=1}^{\infty} \frac{1}{i^{2k}}$ for every natural number k. Most unexpectedly, the very same Bernoulli numbers that help approximate these sums via Euler's summation formula will occur one by one in the precise formulas for the sums of each of these series. This seems a striking coincidence, but actually hints at a link between Euler's summation formula and Fourier analysis, a modern branch of mathematics that studies the representation of arbitrary functions as infinite sums of trigonometric functions of various frequencies [137, Chapter 14].

Thus wends the thread of the relationship between the continuous and the discrete through two millennia, from the ancient counting of a number of dots in comparison to the area of a triangle through Euler's approximations of sums of series in relation to integration. We see the Bernoulli numbers emerge as key to this dynamic, and arise unexpectedly in other phenomena. Their importance in many parts of mathematics has grown continually ever since Euler. Today they permeate deep results in fields ranging from number theory to differential and algebraic topology [169, Appendix B], in addition to their ongoing importance in numerical analysis via the summation formula. We discuss this further at the end of the chapter. The link the Bernoulli numbers provide between the continuous and the discrete, first unveiled by Euler, continues to be key to advances in modern mathematics.

Exercise 1.1. In the spirit of the triangular and square numbers of Figure 1.2, generalize to define pentagonal numbers, hexagonal numbers, and general polygonal numbers for any regular polygon of side n. Deduce formulas showing that sums of terms in increasing integer arithmetic progressions beginning with 1 produce the polygonal numbers, and obtain closed formulas for these.

Exercise 1.2. Write out a table of polygonal numbers and discover some more patterns from this table. Prove your conjectures [258, p. 94].

Exercise 1.3. Verify the sum of squares formula

$$1^2 + 2^2 + 3^2 + \cdots + n^2 = \frac{n(n+1)(2n+1)}{6}$$

using mathematical induction. Perhaps discover or look up some other ways to obtain the formula that do not require knowing it in advance. Your proof using mathematical induction requires this advance knowledge, which is always a drawback of induction: the result needs to be known before proof by induction is possible.

Exercise 1.4. Verify that $\frac{n^3}{3} < \sum_{i=1}^{n} i^2 < \frac{(n+1)^3}{3}$ follows from the sum of squares formula. State and prove analogous inequalities for sums of zeroth and first powers. Then make a generalizing conjecture about analogous inequalities for $\sum_{i=1}^{n} i^k$ for any positive integer k. Verify your conjecture in various situations.

Exercise 1.5. Use polar coordinates to calculate the area inside Archimedes' spiral with the fundamental theorem of calculus, and compare it with his theorem.

Exercise 1.6. Guess a formula for sums of cubes: First calculate the first six sums. Then prove by mathematical induction that your guess is correct.

Exercise 1.7. Nicomachus wrote, "When the successive odd numbers are set forth indefinitely beginning with 1, observe this: The first one makes the potential cube; the next two, added together, the second; the next three, the third; the four next following, the fourth; the succeeding five, the fifth; the next six, the sixth; and so on" [180, Book 2, Chapter 20]. State and prove his general pattern (Hint: average within the blocks), and then use it to obtain and prove the general formula for the sum of the first n cubes.

Exercise 1.8. Prove $(1 + 2 + \cdots + n)^2 = (1 + 2 + \cdots + (n-1))^2 + n^3$ by mathematical induction, and discuss how the inductive step can be interpreted with the geometry of Al-Karajī's figure.

Exercise 1.9. Prove ibn al-Haytham's equation by mathematical induction. Perhaps first try his example values of n and k.

Exercise 1.10. Prove ibn al-Haytham's equation by interchanging the order in his double summation.

Exercise 1.11. Deduce the formula for a sum of fourth powers from ibn al-Haytham's equation, by inductively substituting the known formulas for smaller values of k.

Exercise 1.12. Calculate $\int_0^a x^k dx$ by considering lower and upper sums of rectangles based on left and right endpoints of equally spaced partitions of the interval, and by using Roberval's inequalities to compute the appropriate limit.

Exercise 1.13. Prove Roberval's inequalities by interpreting the sum of powers as both an upper and lower Riemann sum for an obvious function (you may use the calculus).

Exercise 1.14. By the time we have read Pascal's work we will be able to show (Exercise 1.38) that

$$\sum_{i=1}^{n} i^k = \frac{n^{k+1}}{k+1} + \frac{1}{2}n^k + \boxed{?}\,n^{k-1} + \cdots + \boxed{?}\,n + 0.$$

There is a simple geometric interpretation of the second term. Draw a picture illustrating the difference between the region under the curve $y = x^k$ for $0 \le x \le n$ and the region of circumscribing rectangles with ends at integer values. Interpreting their areas as $\int_0^n x^k dx = \frac{n^{k+1}}{k+1}$ and $\sum_{i=1}^n i^k$, find an interpretation in the picture of how the term $\frac{1}{2}n^k$ above represents part of the region between these two, and explain what its connection is to the trapezoid rule from calculus as a numerical approximation for definite integrals. This should suggest to you the sign of the next term in the formula above. What should it be and why?

Exercise 1.15. Use Bernoulli's recursive formulas to calculate the first several Bernoulli numbers. Use them to check Bernoulli's claim against the sums of powers for which you already know formulas. Also conjecture at least one further property it appears the Bernoulli numbers may have from what you find, and then calculate a few more numbers to begin testing your conjecture.

Exercise 1.16. Put yourself in Euler's shoes and try making an educated guess for the exact sum of the reciprocal squares. First calculate a particular partial sum by hand to a certain accuracy (maybe to the tenth term for starters), bound the remainder with integrals, as in the text, and try averaging these to add to the partial sum to make a guess for the infinite sum. Then, with the sum $\frac{\pi}{4}$ of Leibniz's series as inspiration, try dividing π by your guess, to see whether you obtain approximately a whole number, or maybe a fraction with small numerator and denominator. If this does not work, try using π^2 instead. If you are using a machine to help you, discuss how you would plan your calculations if you had only your brain, a pen or pencil, and paper, like Euler. Speculate further about what Euler may have considered while doing all this, and why.

Exercise 1.17. Verify that Euler's summation formula specializes to the formulas of Bernoulli for sums of powers. Explain what the constant C is; pay special attention to the final terms of Bernoulli's formulas.

Exercise 1.18. Verify that if we use trapezoids instead of rectangles to approximate the area represented by $\int_c^n f(x)dx$, we obtain the trapezoid rule:
$$\sum_{i=c+1}^n f(i) - \left(\frac{f(n)-f(c)}{2}\right) \approx \int_c^n f(x)dx.$$

1.2 Archimedes Sums Squares to Find the Area Inside a Spiral

In 216 B.C.E., the Sicilian city of Syracuse allied itself with Carthage during the second Punic war, and thus was attacked by Rome, portending what would ultimately happen to the entire Hellenic world. During a long siege, soldiers of the Roman general Marcellus were terrified by the ingenious war machines defending the city, invented by the Syracusan Archimedes (c. 287–212 B.C.E.). These included catapults to hurl great stones, as well as ropes, pulleys, and hooks to raise and smash Marcellus's ships, and perhaps even burning mirrors setting fire to their sails. Finally though, probably through betrayal, Roman soldiers entered the city in 212 B.C.E., with orders from Marcellus to capture Archimedes alive. Plutarch relates that "as fate would have it, he was intent on working out some problem with a diagram and, having fixed his mind and his eyes alike on his investigation, he never noticed the incursion of the Romans nor the capture of the city. And when a soldier came up to him suddenly and bade him follow to Marcellus, he refused to do so until he had worked out his problem to a demonstration; whereat the soldier was so enraged that he drew his sword and slew him" [133, p. 97].

Despite the great success of Archimedes' military inventions, Plutarch says that "He would not deign to leave behind him any commentary or writing on such subjects; but, repudiating as sordid and ignoble the whole trade of engineering, and every sort of art that lends itself to mere use and profit,

Photo 1.1. Archimedes.

he placed his whole affection and ambition in those purer speculations where there can be no reference to the vulgar needs of life" [133, p. 100]. Perhaps the best indication of what Archimedes truly loved most is his request that his tombstone include a cylinder circumscribing a sphere, accompanied by the inscription of his remarkable theorem that the sphere is exactly two-thirds of the circumscribing cylinder in both surface area and volume!

Archimedes was the greatest mathematician of antiquity, and one of the top handful of all time; his achievements seem astounding even today. The son of an astronomer, he spent most of his life in Syracuse on the island of Sicily, in today's southern Italy, except for a likely period in Alexandria studying with successors of Euclid. In addition to his mathematical achievements, and contrasting with the view expressed by Plutarch, his reputation during his lifetime derived from an impressive array of mechanical inventions, from the water snail (a screw for raising irrigation water) to compound pulleys, and his fearful war instruments. Referring to his principle of the lever, Archimedes boasted, "Give me a place to stand on, and I will move the earth." When King Hieron of Syracuse heard of this and asked Archimedes to demonstrate his principle, he demonstrated the efficacy of his pulley systems by single-handedly pulling a three-masted schooner laden with passengers and freight [133]. One of his most famous, but possibly apocryphal, exploits was to determine for the king whether a goldsmith had fraudulently alloyed a supposed gold crown with cheaper metal. He is purported to have realized, while in a public bath, the principle that his floating body displaced exactly its weight in water, and realizing he could use this to solve the problem, rushed home naked through the streets shouting "Eureka! Eureka!" (I have found it).

The treatises of Archimedes contain a wide array of area, volume, and center of gravity determinations, including the equivalent of many of the best-known formulas taught in high school today. Archimedes also laid the mathematical foundation for the fields of statics and hydrodynamics and their interplay with geometry, and frequently used intricate balancing arguments. A fascinating treatise on a different topic is *The Sandreckoner*, in which he numbered the grains of sand needed to fill the universe, by developing an effective system for dealing with large numbers. Even though he calculated in the end that only 10^{63} grains would be needed, his system could actually calculate with numbers as enormous as $\left(\left(10^8 \right)^{10^8} \right)^{10^8}$. Archimedes even modeled the universe with a mechanical planetarium incorporating the motions of the sun, the moon, and the "five stars which are called the wanderers" (i.e., the known planets) [92].

We will read excerpts from Archimedes' work on the area inside a certain spiral, beginning with his preparatory study of a sum of squares. Archimedes wrote mostly in verbal style, without modern symbols for addition, equality, exponents, and parentheses. He would have used a sequence of letters like $A, B, \Gamma, \ldots, \Omega$ to represent the terms in an arithmetic progression, not the subscripted $A_1, A_2, A_3, \ldots, A_n$ we see below, which is very modern notation substituted in this English translation of Archimedes' works [7].

∞∞∞∞∞∞∞∞∞

Archimedes, from

On Spirals

Proposition 10

If $A_1, A_2, A_3, \ldots, A_n$ be n lines forming an ascending arithmetical progression in which the common difference is equal to the least term A_1, then

$$(n+1) A_n^2 + A_1 (A_1 + A_2 + A_3 + \cdots + A_n) = 3 \left(A_1^2 + A_2^2 + A_3^2 + \cdots + A_n^2 \right).$$

∞∞∞∞∞∞∞∞∞

To give a vivid sense for how much the notation has been modernized, consider the statement of the same proposition in a different modern translation, by E.J. Dijksterhuis:

If a series of any number of lines be given, which exceed one another by an equal amount, and the difference be equal to the least, and if other lines be given equal in number to these and in quantity to the greatest, the squares on the lines equal to the greatest, plus the square on the greatest and the rectangle contained by the least and the sum of all those exceeding one another by an equal amount will be the triplicate of all the squares on the lines exceeding one another by an equal amount [48, p. 122] (Exercise 1.19).

The proof Archimedes gave for this appears quite algebraic and unmotivated, but Kathe Kanim has recently found a transparent unifying geometric view that follows every step of Archimedes' proof [128], perhaps illustrating exactly what Archimedes had in mind. Figure 1.7 illustrates this view of the equality of areas claimed by Archimedes.

Notice that while Archimedes expressed this as an equality of areas, today we tend to interpret it as just a formula about numbers, ignoring the dimensionality involved. If we think of A_1 as the unit of linear measurement, replacing it by the number 1, we obtain the statement

$$(n+1)n^2 + \sum_{i=1}^{n} i = 3 \sum_{i=1}^{n} i^2$$

about numbers.

As it stands, this expresses the sum of squares in terms of the sum of first powers. But the reader may substitute the known summation formula first powers to turn this into our explicit polynomial formula for a sum of squares:

$$1^2 + 2^2 + 3^2 + \cdots + n^2 = \sum_{i=1}^{n} i^2 = \frac{n(n+1)(2n+1)}{6} .$$

What Archimedes really needed for his proof of the area inside a spiral was a pair of inequalities bounding a sum of squares (Exercise 1.20), which he states as corollaries of the proposition:

∞∞∞∞∞∞∞∞∞

Corollary. 1. *From this it is evident that*

$$n \cdot A_n^2 < 3 \left(A_1^2 + A_2^2 + \cdots + A_n^2 \right).$$

Also ... $A_n^2 > A_1 \left(A_n + A_{n-1} + \cdots + A_1 \right)$ ... .

It follows from the proposition that

$$n \cdot A_n^2 > 3 \left(A_1^2 + A_2^2 + \cdots + A_{n-1}^2 \right).$$

Corollary. 2. *All these results will hold if we substitute* similar figures *for squares on all the lines; for similar figures are in the duplicate ratio of their sides.*

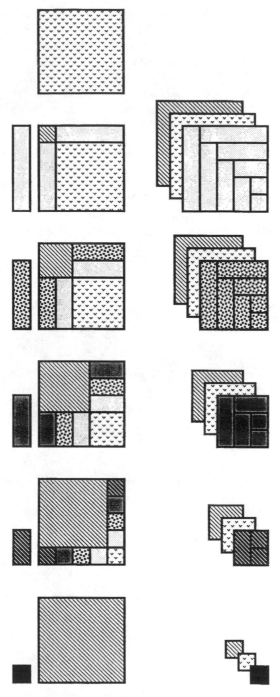

Fig. 1.7. A sum of squares.

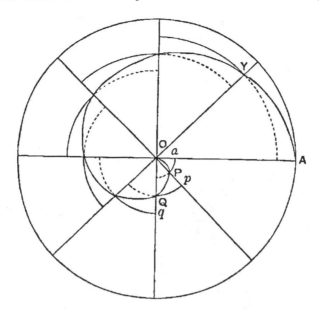

Fig. 1.8. The area of the spiral.

Now recall the spiral described by Archimedes as quoted in our introduction.

Proposition 24

The area bounded by the first turn of the spiral and the initial line is equal to one-third of the "first circle" ... .

Let O be the origin, OA the initial line, A the extremity of the first turn. Draw the "first circle," i.e., the circle with O as centre and OA as radius. Then, if C_1 be the area of the first circle, R_1 that of the first turn of the spiral bounded by OA, we have to prove that

$$R_1 = \frac{1}{3}C_1.$$

For if not, R_1 must be either greater or less than $\frac{1}{3}C_1$.

I. If possible, suppose $R_1 < \frac{1}{3}C_1$.

We can then circumscribe a figure about R_1 made up of similar sectors of circles (Figure 1.8) such that if F be the area of this figure,

$$F - R_1 < \frac{1}{3}C_1 - R_1 \quad \text{[Exercise 1.21]},$$

whence $F < \frac{1}{3}C_1$.

Let OP, OQ, ... be the radii of the circular sectors, beginning from the smallest. The radius of the largest is of course OA. The radii then form an ascending arithmetical progression in which the common difference is equal to the least term OP. If n be the number of the sectors, we have [by Proposition 10, Corollary 1]

$$n \cdot OA^2 < 3\left(OP^2 + OQ^2 + \cdots + OA^2\right);$$

and since the similar sectors are proportional to the squares on their radii, it follows that $C_1 < 3F$, or $F > \frac{1}{3}C_1$. But this is impossible, since F was less than $\frac{1}{3}C_1$. Therefore, $R_1 \not< \frac{1}{3}C_1$.

II. If possible, suppose $R_1 > \frac{1}{3}C_1$.

We can then *inscribe* a figure made up of similar sectors of circles such that if f be its area,

$$R_1 - f < R_1 - \frac{1}{3}C_1,$$

whence $f > \frac{1}{3}C_1$.

If there are $(n-1)$ sectors, their radii, as OP, OQ, ..., form an ascending arithmetical progression in which the least term is equal to the common difference, and the greatest term, as OY, is equal to $(n-1)OP$. Thus, [Proposition 10, Corollary 1]

$$n \cdot OA^2 > 3\left(OP^2 + OQ^2 + \cdots + OY^2\right),$$

whence $C_1 > 3f$, or $f < \frac{1}{3}C_1$, which is impossible, since $f > \frac{1}{3}C_1$. Therefore $R_1 \not> \frac{1}{3}C_1$.

Since then R_1 is neither greater nor less than $\frac{1}{3}C_1$,

$$R_1 = \frac{1}{3}C_1.$$

∞∞∞∞∞∞∞∞∞

Note that this artful determination of the area enclosed in the spiral by the method of exhaustion (Exercises 1.22, 1.23, 1.24) does not rely on Archimedes' precise description of a sum of squares (Exercise 1.25), but only on the inequalities from Corollary 1. The reader should verify that Archimedes' claim in Corollary 2 holds, as needed in his proof for the spiral, for similar sectors of circles (Exercise 1.26).

Exercise 1.19. Compare the translations given by Heath and Dijksterhuis. Do they say the same thing? What are the advantages and disadvantages of the verbal versus the modern symbolic approaches?

Exercise 1.20. Explain how the inequalities of Archimedes' Corollary 1 can be seen geometrically by studying Figure 1.7. Also show that they are equivalent to $\frac{n^3}{3} < \sum_{i=1}^{n} i^2 < \frac{(n+1)^3}{3}$.

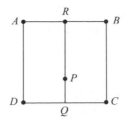

Fig. 1.9. Duplicate ratio.

Exercise 1.21. Prior to Proposition 24, Archimedes proves that one can circumscribe, about the region enclosed by the spiral, a figure made up of similar sectors of circles, as in Figure 1.8, so that the circumscribed figure exceeds the spiral by "less than any assigned area." Show that this is always possible. This is the geometric heart of the method of exhaustion.

Exercise 1.22. Select and study another of Archimedes' proofs by exhaustion [7]. Rewrite it in your own words, and explain the general plan of how the method of exhaustion works.

Exercise 1.23. Consider square $ABCD$ in Figure 1.9. Suppose that side AD begins to move toward side BC through segments RQ so that RQ remains parallel to AD. Suppose further that a point P, which begins at D, moves along RQ so that the ratio of the plane rectangle with sides of length PQ and DC to the square on DQ remains constant, so that P reaches point B when RQ coincides with BC. Prove that the quadrature (area) of the region marked off by curve DPB and sides DC, BC is one-third of the total quadrature of square $ABCD$. Your proof should be in the spirit of Archimedes' geometric constructions. Along with the geometry, use the logic of a double *reductio ad absurdum* argument to reach the final conclusion.

Exercise 1.24.
1. Why did the ancient Greeks study mathematics to such an extent that their culture produced several volumes of geometric work?
2. We have seen that Archimedes essentially used Riemann sums to find the area of the first turn of a spiral. What prevented him from developing integral calculus as practiced today?

Exercise 1.25. We do not know how Archimedes discovered his sum of squares formula, but try your hand at it like this: Compare each successive value of the sum of squares with the corresponding sum of first powers (i.e., using the same natural numbers in each case), by looking at their ratios. Capitalize on the pattern that emerges.

Exercise 1.26. Show that the sequence of sectors in Archimedes' proof are "similar," and demonstrate how this ensures that their areas satisfy the result he needs in the proof (as claimed in Corollary 2), that their areas obey the numbered inequalities of Corollary 1.

1.3 Fermat and Pascal Use Figurate Numbers, Binomials, and the Arithmetical Triangle to Calculate Sums of Powers

In the autumn of 1636, Pierre de Fermat (1601–1665) wrote to Marin Mersenne (1588–1648) in Paris [133, pp. 436, 481f], implying that he had solved "what is perhaps the most beautiful problem of all arithmetic," finding the precise sum of powers in an arithmetic progression, no matter what the power [19], [73, volume II, pp. 70, 84–85]. Fermat, who did mathematics in his spare time, was one of the truly great figures in the history of mathematics, often stating his results without proof in his letters, leaving it to others to try to work out the details and verify his claims. Some of his assertions about numbers have driven the development of mathematics for hundreds of years. More about Fermat's life can be found in our chapter on the quadratic reciprocity law between primes.

In his letter to Mersenne, Fermat wrote that he had reached his results on sums of powers by using the following theorem on "natural progressions":

∞∞∞∞∞∞∞

Pierre de Fermat, from
Letter to Mersenne. September/October, 1636, and again to Roberval, November 4, 1636

The last number multiplied by the next larger number is double the collateral triangle;

the last number multiplied by the triangle of the next larger is three times the collateral pyramid;

the last number multiplied by the pyramid of the next larger is four times the collateral triangulo-triangle;

and so on indefinitely in this same manner [19], [163, p. 230f], [73, vol. II, pp. 70, 84–85].

∞∞∞∞∞∞∞

Although Fermat characteristically does not say any more, we will be able to understand what he is claiming here, and see how it produces formulas for sums of powers. This analysis will also prepare us for reading Pascal's and Bernoulli's works on sums of powers.

First, exactly what does Fermat mean by the collateral triangle, pyramid, and triangulo-triangle, examples of what we call figurate numbers? We saw earlier that a triangular number counts the dots in a triangular figure composed of rows starting with one dot and increasing in length by one (Figure 1.2). Thus by a "collateral triangle" Fermat means the triangular number with a specified number of rows, i.e., a certain length to its side, such as the triangular number $1 + 2 + 3 = 6$ with side length three. Likewise, a pyramidal (or tetrahedral) number counts the dots in a three-dimensional pyramidal figure formed by piling shrinking triangles on top of each other as in Figure 1.5. So the collateral pyramid with side length three is the sum $1 + 3 + 6 = 10$ of the first three triangular numbers. Likewise, Fermat's four-dimensional triangulo-triangle with side length three is the sum $1 + 4 + 10 = 15$ of the first three pyramidal numbers; and so on.

Fermat's claims are not merely geometric interpretations of relationships between certain whole numbers; they are actually discrete analogues of continuous results we already know about areas and volumes. For instance, his first claim is a discrete version of the fact that for any line segment, the product of the segment with itself creates a rectangle with area twice that of the triangle with that side and height. Fermat's use of the "next larger number" for one of the discrete lengths is exactly what is necessary to ensure a precise two-to-one ratio of the discrete count of dots in a rectangular figure to the count of dots in an inscribed triangle at the corners, as we know happens for the continuous measurements of the analogous areas. This is visible geometrically in the packing of dots in the middle of Figure 1.2. Similarly, when he says "the last number multiplied by the triangle of the next larger is three times the collateral pyramid," he is expressing a delicate discrete version of the continuous fact that a three-dimensional pyramid on a triangular base has volume one-third that of the prism created with the same base and height. The reader may enjoy seeing the discrete relationship geometrically as well, by packing six discrete triangular pyramidal figures into a box figure, half of which is the discrete prism. Fermat claims that similar relationships hold in all dimensions.

Let us formally define the *figurate numbers* by their recursive *piling up* property. Here $F_{n,j}$ will denote the figurate number that is j-dimensional with n dots along each side. Call n its *side length*. For instance, $F_{3,2}$ is the planar triangular number with 3 dots per side, so by counting up its rows of dots, $F_{3,2} = 1 + 2 + 3 = 6$. The piling up property, by which we define larger figurate numbers from smaller ones, is encoded in the formula $F_{n+1,j+1} = F_{n,j+1} + F_{n+1,j}$. This formalizes the idea that to increase the side length of a $(j+1)$-dimensional figurate number from n to $n+1$, we simply add on another layer at its base, consisting of a j-dimensional figurate number of side length $n+1$ (Figure 1.5). This defining formula is called a *recursion relation*, since it defines each of the figurate numbers in terms of those of lower dimension and side length, provided we start correctly by specifying those numbers with the smallest dimension and those with the smallest side length. While we could start from dimension one, it is useful to begin with zero-dimensional figurate numbers, all of which we will define to have the value one. We thus define

$$F_{n,0} = 1 \qquad (n \geq 1),$$
$$F_{1,j} = 1 \qquad (j \geq 0),$$
$$F_{n+1,j+1} = F_{n+1,j} + F_{n,j+1} \qquad (n \geq 1, j \geq 0).$$

We must check that our starting data determine just what was intended for Fermat's figurate numbers. We have required that any figurate number with side length one will have exactly one dot, no matter what its dimension. And the recursion relation clearly produces the desired one-dimensional figurate numbers $F_{n,1} = n$ for all $n \geq 1$ from the starting data of zero-dimensional numbers.

Now Fermat's claimed formulas to Mersenne assume the form

$$nF_{n+1,1} = 2F_{n,2},$$
$$nF_{n+1,2} = 3F_{n,3},$$
$$nF_{n+1,3} = 4F_{n,4},$$

$$\cdots$$

Taking these temporarily for granted, let us see how sums of powers can be calculated, as Fermat claimed. As an example, we will derive the formula for a sum of squares. We know that any pyramidal number is obtained by piling up triangular numbers, i.e.,

$$F_{n,3} = \sum_{i=1}^{n} F_{i,2}.$$

We can also obtain closed formulas for the figurate numbers on both sides of this equality by iteratively using Fermat's formulas:

$$F_{n,3} = \frac{n}{3}F_{n+1,2} = \left(\frac{n}{3}\right)\left(\frac{n+1}{2}\right)F_{n+2,1} = \left(\frac{n}{3}\right)\left(\frac{n+1}{2}\right)\left(\frac{n+2}{1}\right),$$
$$\text{and} \quad F_{i,2} = \frac{i}{2}F_{i+1,1} = \left(\frac{i}{2}\right)\left(\frac{i+1}{1}\right).$$

So the two sides of the piling-up equation above expand to produce

$$\frac{n\,(n+1)\,(n+2)}{3\cdot 2\cdot 1} = F_{n,3} = \sum_{i=1}^{n} F_{i,2} = \sum_{i=1}^{n} \frac{i(i+1)}{2} = \frac{1}{2}\sum_{i=1}^{n} i^2 + \frac{1}{2}\sum_{i=1}^{n} i \, .$$

Now we simply substitute $n(n + 1)/2$ for $\sum_{i=1}^{n} i$, which we already know inductively, and solve for $\sum_{i=1}^{n} i^2$, clearly yielding a polynomial with leading term $n^3/3$, as expected. The reader may verify that we get the correct formula, and may continue this method to sums of higher powers (Exercise 1.27). In fact, this very method is advanced by Bernoulli in the text we will read in the next section. While it is clear that the process can be continued indefinitely, it quickly becomes impractically complicated, and it is also not clear that it yields any new general insight.

We shall now study the figurate numbers further to see why Fermat's claim about them is true, and simultaneously prepare the groundwork for reading Pascal. The reader may have noticed that the individual figurate numbers

seem familiar, from looking back to Pascal's arithmetical triangle displayed in
the introduction (Figure 1.6). The numbers shown there match the figurate
numbers, i.e., $F_{n,j}$ appears in Pascal's "parallel row" n and "perpendicular
row" $j + 1$. Indeed, in his treatise on the arithmetical triangle, Pascal defines
the numbers in the triangle by starting the process off with a 1 in the corner,
and defines the rest simply by saying that each number is the sum of the
two numbers directly above and directly to the left of it, which corresponds
precisely to the recursion relation by which we formally defined the figurate
numbers.

Perhaps the reader also already recognizes the numbers in the arithmetical triangle as binomial coefficients or combination numbers. The numbers
occurring along Pascal's ruled diagonals in Figure 1.6 appear to be the coefficients in the expansion of a binomial; for instance, the coefficients of
$(a + b)^4 = 1a^4 + 4a^3b + 6a^2b^2 + 4ab^3 + 1b^4$ occur along the diagonal Pascal labels with 5. Our modern notation for these binomial coefficients is that
$\binom{m}{j}$ denotes the coefficient of $a^{m-j}b^j$ in the expansion of $(a + b)^m$. Indeed, if
in the arithmetical triangle we index both the diagonals and their individual
entries beginning with zero, then the entry in diagonal m at column j will be
the binomial coefficient $\binom{m}{j}$. This is easy to prove (Exercise 1.28).

Since we have now identified both the figurate numbers and the binomial
coefficients as the numbers in the triangle generated by the basic recursion
relation, their precise relationship follows just by comparing their indexing:

$$F_{i+1,j} = \binom{i+j}{j} \quad \text{for } i, j \geq 0.$$

We can also calculate a closed formula in terms of factorials for the numbers
in the triangle:

$$\binom{m}{j} = \frac{m!}{j!(m-j)!} = \frac{m(m-1)\cdots(m-j+1)}{j!} \quad \text{for } 0 \leq j \leq m,$$

where the notation $i!$ (read "i factorial") is defined to mean

$$i \cdot (i-1) \cdot (i-2) \cdots 3 \cdot 2 \cdot 1,$$

and $0!$ is defined to be 1 (Exercise 1.29).

This ubiquitous triangle of numbers had already been in use for over 500
years, in places ranging from China to the Islamic world, before Pascal developed and applied its properties in his *Traité du Triangle Arithmétique* (Treatise on the Arithmetical Triangle), written by 1654 [133]. A key fact, which
Pascal called the *Twelfth Consequence*, is that neighboring numbers along a
diagonal in the triangle are always in a simple ratio:

$$(m - j)\binom{m}{j} = (j + 1)\binom{m}{j+1} \quad \text{for } j < m,$$

which is easily obtained from the factorial formula above (Exercise 1.30).

Translated into figurate numbers, Pascal's Twelfth Consequence reemerges as precisely Fermat's claim in his letter to Mersenne! For instance, letting $j = 2$ and $m = n + 2$ yields

$$n\binom{n+2}{2} = 3\binom{n+2}{3},$$

$$\text{or} \quad nF_{n+1,2} = 3F_{n,3},$$

exactly Fermat's claim that "the last number multiplied by the triangle of the next larger is three times the collateral pyramid." The reader may now easily confirm Fermat's general claim (Exercise 1.31), and we are ready to move on to Pascal's work on sums of powers.

Blaise Pascal (1623–1662) was born in Clermont-Ferrand, in central France. Even as a teenager his father introduced him to meetings of Marin Mersenne's circle of mathematical discussion in Paris. He quickly became involved in the development of projective geometry, the first in a sequence of highly creative mathematical and scientific episodes in his life, punctuated by periods of religious fervor. Around age twenty-one he spent several years developing a mechanical addition and subtraction machine, in part to help his father in

Photo 1.2. Blaise Pascal.

tax computations as a local administrator. It was the first of its kind ever to be marketed. Then for several years he was at the center of investigations of the problem of the vacuum, which led to an understanding of barometric pressure. In fact, the scientific unit of pressure is named the *pascal*. He is also known for Pascal's law on the behavior of fluid pressure.

Around 1654 Pascal conducted his studies on the arithmetical triangle and its relationship to probabilities, as we described in the introduction. His correspondence with Fermat in that year marks the beginning of probability theory. Several years later, Pascal refined his ideas on area problems via the method of indivisibles already being developed by others, and solved various problems of areas, volumes, centers of gravity, and lengths of curves.[6] After only two years of work on the calculus of indivisibles, Pascal fell gravely ill, abandoned almost all intellectual work to devote himself to prayer and charitable work, and died three years later at age thirty-nine. In addition to his work in mathematics and physics, Pascal is prominent for his *Provincial Letters* defending Christianity, which gave rise to his posthumously published *Pensées* on religious philosophy [92, 63]. Pascal was an extremely complex person, and one of the outstanding scientists of the mid-seventeenth century, but we will never know how much more he might have accomplished with more sustained efforts and a longer life.

The relation of the arithmetical triangle to counting combinations, and thus their centrality in probability theory, follows easily from the factorial formula above for the triangle's numbers. The reader may verify that $\binom{m}{j}$ represents the number of different combinations of j elements that can occur in a set of m elements (Exercise 1.32). For instance, there are $\binom{11}{5} = 462$ different teams of 5 players possible from a pool of 11 people available to play.

We have seen that the numbers in the arithmetical triangle have three interchangeable interpretations: as figurate numbers, combination numbers, and binomial coefficients. Given this multifaceted nature, it is no wonder that they arose early on, in various manners and parts of the world, and that they are ubiquitous today. The arithmetical triangle in fact overflows with fascinating patterns (Exercises 1.33, 1.34). The reader will enjoy reading Pascal's actual treatise [100, v. 30].

In about the same year as his *Treatise on the Arithmetical Triangle*, Pascal produced the text we will now study, *Potestatum Numericarum Summa* (Sums of Numerical Powers), which analyzes sums of powers in arithmetic progressions in terms of the numbers in the arithmetical triangle, interpreted as binomial coefficients. Pascal also makes the connection between these results and area problems via the method of indivisibles.

Fermat's great enthusiasm in 1636 for the problem of calculating sums of powers was not immediately embraced by others, and Pascal, although a

---

[6] Later in the seventeenth century, Leibniz, one of the two inventors of the infinitesimal calculus, which supplanted the method of indivisibles, explicitly credited Pascal's approach as stimulating his own ideas on the so-called characteristic triangle of infinitesimals in his fundamental theorem of calculus.

direct correspondent of Fermat's, was apparently unaware of Fermat's work when he made his own analysis about eighteen years later in *Sums of Numerical Powers*. Below we will see the transition from a geometric to algebraic approach almost complete, since Pascal, unlike Archimedes, Al-Karajī, al-Haytham, and even Fermat, is bent on presenting a generalized arithmetic solution for the problem, albeit still using mostly verbal descriptions of formulas, instead of modern algebraic notation. We present a translation from the French (*Sommation des puissances numériques*), published in his collected works with the Latin on facing pages [182, v. III, pp. 341–367]. We will find that Pascal obtains a compact formula directly relating sums of powers for various exponents, using binomial coefficients as the intermediary.

∞∞∞∞∞∞∞∞

Blaise Pascal, from
Sums of Numerical Powers

Remark.

Given, starting with the unit, some consecutive numbers, for example $1, 2, 3, 4$, one knows, by the methods the Ancients made known to us, how to find the sum of their squares, and also the sum of their cubes; but these methods, applicable only to the second and third degrees, do not extend to higher degrees. In this treatise, I will teach how to calculate not only the sum of squares and of cubes, but also the sum of the fourth powers and those of higher powers up to infinity: and that, not only for a sequence of consecutive numbers beginning with the unit, but for a sequence beginning with any number, such as the sequence $8, 9, 10, \ldots$. And I will not restrict myself to the natural sequence of numbers: my method will apply also to a progression having as ratio [difference] $2, 3, 4$, or any other number,—that is to say to a sequence of numbers different by two units, like $1, 3, 5, 7, \ldots$, $2, 4, 6, 8, \ldots$, or differing by three units like $1, 4, 7, 10, 13, \ldots$. And what is more, whatever the first term in the sequence may be: if the first term is 1, as in the sequence with ratio three, $1, 4, 7, 10, \ldots$: or if it is another term in the progression, as in the sequence $7, 10, 13, 16, 19$; or even if it is alien to the progression, as in the sequence with ratio three, $5, 8, 11, 14, \ldots$ beginning with 5. It is remarkable that a single general method will suffice to treat all these different cases. This method is so simple that it will be explained along several lines, and without the preparation of algebraic notations to which difficult demonstrations must have recourse. One can judge this after having read the following problem.

Definition.

Consider a binomial $A + 3$, whose first term is the letter A, and the second a number: raise this binomial to any power, the fourth for example, which gives

$$A^4 + 12.A^3 + 54.A^2 + 108.A + 81;$$

the numbers $12, 54, 108$, which are multiplying the different powers of A, and are formed by the combination of the figurate numbers with the second term, 3, of the binomial, will be called the *coefficients* of A.

Thus, *in the cited example, 12 will be the coefficient of the cube A; 54, that of the square; and 108, that of the first power.*

As for the number 81, it will be called the *pure number*.

Lemma

Suppose any number, like 14, be given, and a binomial $14 + 3$, whose first term is 14 and the second any number 3, in such a manner that the difference of the numbers 14 and $14 + 3$ will equal 3. Let us raise these numbers to a same power, the fourth for example: the fourth power of 14 is 14^4, that of the binomial, $14 + 3$, is

$$14^4 + 12.14^3 + 54.14^2 + 108.14 + 81.$$

In this expression, the powers of the first term, 14, of the binomial are obviously affected by the same coefficients as the powers of A in the expansion of $(A+3)^4$. This put down, the difference of the two fourth powers, 14^4 and

$$14^4 + 12.14^3 + 54.14^2 + 108.14 + 81,$$

is $12.14^3 + 54.14^2 + 108.14 + 81$; this difference comprises: on the one hand, the powers of 14 whose degree is less than the proposed degree 4, these powers being affected by the coefficients which the same powers of A have in the expansion of $(A + 3)^4$; on the other hand, the number 3 (*the difference of the proposed numbers*) raised to the *fourth* power [because the *absolute number* 81 is the *fourth* power of the number 3]. From this we deduce the *following Rule*:

The difference of like powers of two numbers comprises: the difference of these numbers raised to the proposed power; plus the sum of all the powers of lower degree of the smaller of the two numbers, these powers being respectively multiplied by the coefficients which the same powers of A have in the expansion of a binomial raised to the proposed power and having as first term A and as second term the difference of the given numbers.

Thus, the difference of 14^4 and 11^4 will be

$$12.11^3 + 54.11^2 + 108.11 + 81,$$

since the difference of the first powers is 3. And so forth.

A single general method for finding the sum of like powers of the terms of any progression.

Given, beginning with any term, any sequence of terms of an arbitrary progression, find the sum of like powers of these terms raised to any degree.

Suppose an arbitrary number 5 *is chosen as the first term of a progression whose ratio [difference], arbitrarily chosen, is for example three; consider, in this progression, as many of the terms as one wishes, for instance the terms* $5, 8, 11, 14$, *and raise these terms to any power, suppose to the cube. The question is to find the sum of the cubes* $5^3 + 8^3 + 11^3 + 14^3$.

These cubes are $125, 512, 1331, 2744$; **and their sum is** 4712. **Here is how one finds this sum**.

Let us consider the binomial $A + 3$ **having as first term** A **and as second term the difference of the progression**.

Raise this binomial to the fourth power, the power immediately higher than the proposed degree three; we obtain the expression

$$A^4 + 12.A^3 + 54.A^2 + 108.A + 81.$$

This admitted, we consider the number 17, *which, in the proposed progression, immediately follows the last term considered,* 14. *We take the* fourth *power of* 17, *known as* 83521, *and subtract from it:*

First: the sum 38 *of the terms considered,* $5 + 8 + 11 + 14$, *multiplied by the number* 108 *which is the coefficient of* A;

Second: the sum of the squares of the same terms $5, 8, 11, 14$, *multiplied by the number* 54, *which is the coefficient of* A^2.

And so on, in case one still has the powers of A *of lesser degree than the proposed degree* three.

With these subtractions made, one subtracts also the fourth *power of the first term proposed,* 5.

Finally one subtracts the number 3 *(ratio [difference] of the progression) itself raised to the* fourth *power and taken as many times as one considers terms in the progression, here* four *times.*

The remainder of the subtraction will be a multiple of the sum sought; it will be the product of this sum with the number 12, *which is the coefficient of* A^3, *that is to say the coefficient of the term* A *raised to the proposed power* three.

Thus, in practice, one must form the fourth power of 17, being 83521, then subtracting from it successively:

First, the sum of the terms proposed, $5 + 8 + 11 + 14$, being 38, multiplied by 108,—that is, the product 4104;

Then the sum of the squares of the same terms, $5^2 + 8^2 + 11^2 + 14^2$, or $25 + 64 + 121 + 196$, or again 406, which, multiplied by 54, gives 21924;

Then the number 5 to the *fourth* power, which is 625;

Finally the number 3 to the *fourth* power, being 81, multiplied by *four*, which gives 324. In summary one must subtract the numbers 4104, 21924, 625, 324, whose sum is 26977. Taking this sum away from 83521, there remains 56544.

The remainder thus obtained is equal to the sum sought, 4712, multiplied by 12; and, in fact, 4712 multiplied by 12 equals 56544.

The rule is, as one sees, easy to apply. Here now is how one proves it.

The number 17 raised to the *fourth* power, which one writes 17^4, is equal to

$$17^4 - 14^4 + 14^4 - 11^4 + 11^4 - 8^4 + 8^4 - 5^4 + 5^4.$$

In this expression, *only the term 17^4 appears with the single sign $+$; the other terms are in turns added and subtracted.*

But the difference of the terms 17 and 14 is 3; likewise the difference of the terms 14 and 11, and of the terms 11 and 8, and of the terms 8 and 5. Thenceforth, according to our preliminary lemma: $17^4 - 14^4$ equals $12.14^3 + 54.14^2 + 108.14 + 81$.

Likewise $14^4 - 11^4$ equals $12.11^3 + 54.11^2 + 108.11 + 81$.

Likewise $11^4 - 8^4$ equals $12.8^3 + 54.8^2 + 108.8 + 81$.

Likewise $8^4 - 5^4$ equals $12.5^3 + 54.5^2 + 108.5 + 81$.

The term 5^4 does not need to be transformed.

One then finds as the value of 17^4:

$$12.14^3 + 54.14^2 + 108.14 + 81$$
$$+ 12.11^3 + 54.11^2 + 108.11 + 81$$
$$+ 12.8^3 + 54.8^2 + 108.8 + 81$$
$$+ 12.5^3 + 54.5^2 + 108.5 + 81$$
$$+ 5^4,$$

or, *on interchanging the order of the terms:*

$$5 + 8 + 11 + 14 \text{ multiplied by } 108,$$
$$+ 5^2 + 8^2 + 11^2 + 14^2 \text{ multiplied by } 54,$$
$$+ 5^3 + 8^3 + 11^3 + 14^3 \text{ multiplied by } 12,$$
$$+ 81 + 81 + 81 + 81$$
$$+ 5^4.$$

If therefore one subtracts on both sides the sum:

$$5 + 8 + 11 + 14 \text{ multiplied by } 108,$$
$$+ 5^2 + 8^2 + 11^2 + 14^2 \text{ multiplied by } 54,$$
$$+ 81 + 81 + 81 + 81$$
$$+ 5^4;$$

There remains 17^4 diminished by the previously known quantities:

$$- 5 - 8 - 11 - 14 \text{ multiplied by } 108,$$
$$- 5^2 - 8^2 - 11^2 - 14^2 \text{ multiplied by } 54,$$
$$- 81 - 81 - 81 - 81$$
$$- 5^4;$$

which will be found equal to the sum $5^3 + 8^3 + 11^3 + 14^3$ multiplied by 12. Q.E.D.

One may thus present as follows the statement and the general solution of the proposed problem.

The sum of powers

Given, beginning with any term, any sequence of terms of an arbitrary progression, find the sum of like powers of these terms raised to any degree.

We form a binomial having A as its first term, and for its second term the difference of the given progression; we raise this binomial to the degree immediately higher than the proposed degree, and we consider the coefficients of the various powers of A in the expansion obtained.

Now we raise to the same degree the term that, in the given progression, immediately follows the last term considered. Then we subtract from the number obtained the following quantities:

First: The first term given in the progression,—that is, the smallest of the given terms,—itself raised to the same power (immediately higher than the proposed degree).

Second: The difference of the progression, raised to the same power, and taken as many times as of the terms considered in the progression.

Third: The sums of the given terms, raised to the various degrees less than the proposed degree, these sums being respectively multiplied by the coefficients of the same powers of A in the expansion of the binomial formed above.

The remainder of the subtraction thus accomplished is a multiple of the sum sought: it contains it as many times as unity is contained in the coefficient of the power of A whose degree is equal to the proposed degree.

NOTE

The reader himself will deduce practical rules that apply in each particular case. Suppose, for example, that one wishes to find the sum of a certain number of terms in the natural sequence [i.e., of natural numbers] beginning with an arbitrary number: here is the rule that one deduces from our general method:

In a natural progression beginning with any number, the square of the number immediately above the last term, diminished by the square of the first term and the number of terms given, is equal to double the sum of the stated terms.

Suppose given a sequence of any consecutive numbers whose first term is arbitrary, for example the *four* numbers $5, 6, 7, 8$: I say that $9^2 - 5^2 - 4$ equals the double of $5 + 6 + 7 + 8$.

One will easily obtain analogous rules giving the sums of powers of higher degrees and which apply to all progressions.

Conclusion.

Any who are a little acquainted with the doctrine of *indivisibles* will not fail to see what profit one may make from the preceding results for the determination of curvilinear areas. These results permit the immediate squaring of all types of parabolas and an infinity of other curves.

If then we extend to continuous quantities the results found for numbers, by the method expounded above, we will be able to state the following rules:

Rules relating to the natural progression beginning with unity.

The sum of a certain number of lines is to the square of the largest as 1 is to 2.

The sum of the squares of the same lines is to the cube of the largest as 1 is to 3.

The sum of their cubes is to the fourth power of the largest as 1 is to 4.

General rule relating to the natural progression beginning with unity.

The sum of like powers of a certain number of lines is to the immediately greater power of the largest among them as unity is to the exponent of this same power.

I will not pause here for the other cases, because this is not the place to study them. It will be enough for me to have cursorily stated the preceding rules. One can discover the others without difficulty by relying on the principle that *one does not increase a continuous magnitude when one adds to it, in any number one wishes, magnitudes of a lower[7] order of infinitude*. Thus points add nothing to lines, lines to surfaces, surfaces to solids; or—to speak in numbers as is proper in an arithmetical treatise,—roots do not count in regard to squares, squares in regard to cubes, and cubes in regard to square-squares. In such a way one must disregard, as nil, quantities of smaller order.

I have insisted on adding these few remarks, familiar to those who practise indivisibles, in order to bring out the always wonderful connection that nature, in love with unity, establishes between objects distant in appearance. It appears in this example, where we see the calculation of the *dimensions of continuous magnitudes* joined with the *summation of numerical powers*.

∞∞∞∞∞∞∞∞∞

Pascal's approach to sums of powers is rich with detail, and ends with his view on how this topic displays the connection between the continuous and the discrete. His idea of using a sum of equations in which one side "telescopes" via cancellations is masterful, and is a tool widely used in mathematics today. Like

---

[7] The French version mistakenly says "higher" here.

al-Karajī for exponent three, and al-Haytham for exponents four and higher, Pascal presents a rule obtained by generalizable example, but expanded in scope on two fronts: to arithmetic progressions with arbitrary differences and to those beginning with any number. The reader is urged to consider whether his example convinces one of the general rule, and then try applying it to obtain the sum of fourth powers in his progression $5, 8, 11, 14$ (Exercise 1.35), and finally use it to obtain a formula for the sum of the first n fifth powers (Exercise 1.36). Doing this displays a clear advantage over al-Haytham's equation. Pascal's only requires us to substitute directly our known formulas for sums with previous exponents, and then solve immediately, without having first to expand and rearrange al-Haytham's double summation and then apply previous formulas a second time before solving. In this sense we can say that Pascal's prescription represents the first explicit recipe for sums of powers.

Although it requires knowledge of the formulas for all previous exponents, Pascal's is an attractive formulation. Let us transcribe his verbal prescription into modern notation for how to find the sum of the first n kth powers:

$$(k+1)\sum_{i=1}^{n} i^k = (n+1)^{k+1} - 1^{k+1} - n \cdot 1^{k+1} - \sum_{j=1}^{k-1}\binom{k+1}{j}\sum_{i=1}^{n} i^j \, .$$

We call this *Pascal's equation*. The reader may verify that his method generalizes to produce what he claims in his verbal prescription for more general progressions (Exercise 1.37).

We can use Pascal's equation to confirm patterns that have slowly been emerging throughout the chapter, namely that sums-of-powers polynomials have a particular degree and predictable leading and trailing coefficients:

$$\sum_{i=1}^{n} i^k = \frac{n^{k+1}}{k+1} + \frac{1}{2}n^k + \boxed{?}\,n^{k-1} + \cdots + \boxed{?}\,n + 0 \text{ for } k \geq 1.$$

We leave it to the reader to confirm these features and even to push one step further to discover and confirm a simple pattern for the coefficients of n^{k-1} in the polynomial formulas (Exercise 1.38). We can also use Pascal's equation to prove that sums of powers satisfy Roberval's inequalities discussed in the introduction (Exercise 1.39).

The patterns in the polynomial coefficients begin to reveal more of the connection between the continuous and the discrete. The term $\frac{n^{k+1}}{k+1}$ is the area $\int_0^n x^k dx$ under the curve $y = x^k$ between 0 and n. The left side of the above equation is the area of a right-endpoint approximating sum of rectangles for this area, and thus the rest constitutes "correcting terms" interpolating between the area under the curve and the sum of rectangles. The term $\frac{1}{2}n^k$ amounts to improving the right-endpoint approximation to a trapezoidal approximation, and the next term also has interpretation as a further correction (Exercise 1.14). We can speculate that the other coefficients in these polynomials continue to follow an interesting pattern. We pursue this in the next episode of our story, emerging at the turn of the eighteenth century in the work of Jakob Bernoulli (1654–1705).

We end this section by remarking on Pascal's *Conclusion* about indivisibles and the squaring (area) of higher parabolas. Clearly for him the connection between "dimensions of continuous magnitudes" and "summation of numerical powers" is striking and subtle, and was probably a prime motivation for his investigations on sums of powers in an era when many were vying to square higher parabolas and other curves. His view is that for continuous quantities, terms of "lower order of infinitude" (i.e., lesser dimension) add nothing, and one must "disregard [them] as nil," so that the sums of powers formulas above become

$$\sum_{i=1}^{n} i^k \approx \frac{n^{k+1}}{k+1},$$

which is his statement about summing continuous quantities. Today we recognize this as analogous to our integration formula

$$\int_0^n t^k dt = \frac{n^{k+1}}{k+1}$$

for the area under a higher parabola. Turning these analogies into a tight logical connection between discrete summation formulas and continuous area results was part of the long struggle to define and rigorize calculus, which began with the classical Greek mathematics exemplified by Archimedes, and lasted until well into the nineteenth century (Exercises 1.40, 1.41, 1.42).

Exercise 1.27. Finish deducing and checking the sum of squares formula derived from Fermat's claims about figurate numbers, and carry the procedure forward to obtain the formula for a sum of cubes from Fermat's claims. Discuss what would be involved in carrying this to higher powers.

Exercise 1.28. Show that the coefficients in the expansion of a binomial satisfy the starting data and the recursion relation of the arithmetical triangle. In other words, if for all $m \geq 0$ we write $(a+b)^m = \sum_{j=0}^{m} \binom{m}{j} a^{m-j} b^j$, show that these coefficients satisfy the starting data $\binom{m}{0} = \binom{m}{m} = 1$, and Pascal's recursion relation $\binom{m+1}{j} = \binom{m}{j} + \binom{m}{j-1}$. (Hint: write $(a+b)^{m+1} = (a+b)(a+b)^m$.)

Exercise 1.29. Show that

$$\binom{m}{j} = \frac{m!}{j!(m-j)!} \quad \text{for } 0 \leq j \leq m.$$

(Hint: Show that this factorial formula satisfies the starting data and the recursion relation of the arithmetical triangle.)

Exercise 1.30. Prove Pascal's Twelfth Consequence from the factorial formula for binomial coefficients.

Exercise 1.31. State Fermat's general claim about figurate numbers, and prove it from Pascal's Twelfth Consequence.

Exercise 1.32. Prove that the number of distinct five-card hands possible from a standard deck of fifty-two playing cards is $\binom{52}{5}$. Then prove that $\binom{m}{j}$ is the number of different combinations of j elements that can occur in a set of m elements. Hint: First see why

$$m\,(m-1)\cdots(m-j+1)$$

is the number of different ways of selecting a sequence of j elements from a sequence with m elements (where different orderings of the same elements count as different sequences).

Exercise 1.33. Prove that the sum of the numbers in each row of the arithmetical triangle is a power of two. Hint: binomial theorem.

Exercise 1.34. Find a pattern of your own in the arithmetical triangle, and then prove that it holds.

Exercise 1.35. Apply Pascal's method to obtain the sum of the fourth powers in the progression $5, 8, 11, 14$, and then check your answer by direct calculation.

Exercise 1.36. Apply Pascal's method to obtain the polynomial formula for the sum of the fifth powers in a natural progression beginning with one, i.e., $\sum_{i=1}^{n} i^5$.

Exercise 1.37. Write out Pascal's general result, in modern notation, and provide a proof (based on the method of his example) to justify his general prescription for a sum of powers of any arithmetic progression, i.e., with arbitrary difference and beginning with any number. Include a modern formulation of his algorithm, and apply it to compute some examples.

Exercise 1.38. In the text and exercises we have obtained explicit polynomial formulas for sums of powers up to exponent five. From these we conjecture

$$\sum_{i=1}^{n} i^k = \frac{n^{k+1}}{k+1} + \frac{1}{2}n^k + \boxed{?}\,n^{k-1} + \cdots + \boxed{?}\,n + 0.$$

Use Pascal's equation to prove these observed patterns for all k. (Hint: mathematical induction. You will need a strong form of induction, in which you assume the truth of all preceding statements, not just the one prior to the one you are trying to verify. Why is this stronger form of mathematical induction a valid method of proof?) Then push one step further to conjecture and prove a pattern for the coefficient of n^{k-1} in the formulas.

Exercise 1.39. Prove the inequality $\sum_{i=1}^{n} i^k < (n+1)^{k+1}/(k+1)$ of Roberval using Pascal's equation. Can you also use it to prove his second inequality $n^{k+1}/(k+1) < \sum_{i=1}^{n} i^k$, which is harder to show?

Exercise 1.40. Compare and contrast the work of Archimedes and Pascal toward finding the area of a bounded curvilinear region in the plane. Cite specific contributions of each author toward the solution of this problem. Explain what specific mathematical constructions from the era of Fermat and Pascal facilitate a discussion of "area under a curve."

Exercise 1.41. Use

$$\sum_{i=1}^{n} i^k = \frac{n^{k+1}}{k+1} + \frac{1}{2}n^k + \cdots$$

as obtained in Exercise 1.38 to prove that

$$\lim_{n \to \infty} \frac{\sum_{i=1}^{n} i^k}{n^{k+1}} = \frac{1}{k+1},$$

a result known as Wallis's theorem (discussed further in the next section). Utilize Wallis's theorem and the modern definition of the integral as a limit of approximating sums to calculate

$$\int_0^x t^k \, dt = \frac{x^{k+1}}{k+1} \text{ for any real } x > 0.$$

Discuss how this supports what Pascal is arguing in his *Conclusion*.

Exercise 1.42. Read about the work of Fermat and Pascal, and discuss how close each came to the modern idea of integration and the fundamental theorem of calculus? What prevented them from developing calculus?

1.4 Jakob Bernoulli Finds a Pattern

Jakob (Jacques, James) Bernoulli (1654–1705) was one of two spectacular mathematical brothers in a large family of mathematicians spanning several generations. The Bernoulli family had settled in Basel, Switzerland, when fleeing the persecution of Protestants by Catholics in the Netherlands in the sixteenth century.

Jakob at first studied mathematics against the will of his father, who wanted him to become a minister, and then traveled widely to learn from prominent mathematicians and scientists in France, the Netherlands, and England. He was appointed professor of mathematics in Basel, and he and his younger brother Johann (Jean, 1667–1748) were among the first to fully absorb Gottfried Leibniz's (1646–1716) newly invented methods of calculus, and to apply them to solve many fascinating mathematical questions. For instance, in 1697 Jakob used a differential equation to solve the brachistochrone problem, i.e., to find the curve down which a frictionless bead will slide from

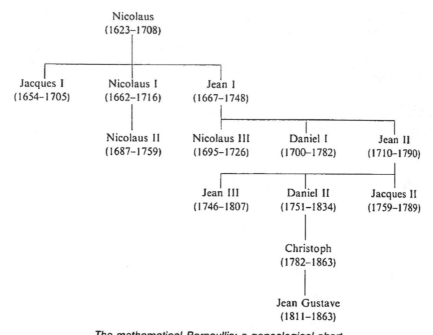

The mathematical Bernoullis: a genealogical chart.

Fig. 1.10. The mathematical Bernoulli family [18, p. 416].

one point to another in the least time.[8] His method began a new mathematical field, the calculus of variations, in which one seeks among all curves the one that maximizes or minimizes some property [133, pp. 547–549]. Bernoulli also used the calculus to discover numerous wonderful properties of the logarithmic spiral, leading him to request that this "spira mirabilis" be engraved on his tombstone [18, p. 417f], [258, pp. 148–153]. And he worked and published much on infinite series, including his unsuccessful attempts to find the infinite sum of reciprocal squares, discussed in the introduction.

Jakob wrote the earliest substantial book on probability theory, *Ars Conjectandi* (The Art of Conjecturing). Its posthumous publication in 1713 contained much original work, including the pattern we have been seeking in the

---

[8] The brachistrochrone problem was posed as a challenge in 1696 by Johann Bernoulli, and solved independently by Newton, Leibniz, both Johann and Jakob Bernoulli, and L'Hospital. They were amazed that the solution curve turned out to be already familiar in another context. Johann wrote, "With justice we admire Huygens because he first discovered that a heavy particle traverses a cycloid in the same time, no matter what the starting point may be. But you will be struck with astonishment when I say that this very same cycloid, the tautochrone of Huygens, is the brachistochrone we are seeking" [133, pp. 547–549], [135, p. 575], [211]. See our curvature chapter regarding the cycloid and Huygens's solution of the tautochrone problem in his work on creating the perfect pendulum clock.

Photo 1.3. Jakob Bernoulli.

formulas for sums of powers. It should not surprise us that Jakob connects probability theory and sums of powers, for we have learned that the figurate numbers, binomial coefficients, and combination numbers are simply different interpretations of the numbers in the arithmetical triangle of Pascal.

While Pascal's equation displayed a compact connection between sums of powers formulas for various exponents, its recursive nature still prevents quick and easy calculation of the polynomial formulas representing $\sum_{i=1}^{n} i^k$ for various values of k. Nor did it reveal any general pattern in all the coefficients of these polynomials, even though we suspect there is one. Bernoulli addresses both issues in his short addendum[9] on sums of powers (*Summae Potestatum*) in a chapter of *Ars Conjectandi* on permutations and combinations [15, v. 3, pp. 164–167], [16], [232, pp. 85–90]. Here the *Bernoulli numbers* first appear,

---

[9] We are indebted to Daniel E. Otero for this translation from Latin.

JACOBI BERNOULLI,

Profeff. Bafil. & utriufque Societ. Reg. Scientiar.
Gall. & Pruff. Sodal.

MATHEMATICI CELEBERRIMI,

ARS CONJECTANDI,

OPUS POSTHUMUM.

Accedit

TRACTATUS
DE SERIEBUS INFINITIS,

Et EPISTOLA Gallicè fcripta

DE LUDO PILÆ
RETICULARIS.

FESTINA LENTE.

BASILEÆ,
Impenfis THURNISIORUM, Fratrum.

cɮↃ Iↄcc XIII.

Photo 1.4. *Ars Conjectandi.*

and our knowledge takes a tremendous leap forward. We will comment in detail after our source, but we note as an aid beforehand that Bernoulli uses the integral sign to represent finite summations!

∞∞∞∞∞∞∞∞

Jakob Bernoulli, from
The Art of Conjecturing
Part Two

A THEORY OF PERMUTATIONS AND COMBINATIONS
On combinations of particular numbers of things; which leads to figurate numbers and their properties

[...] *Scholium.* We note in passing that many (among others, Faulhaber and Remmelin from Ulm, Wallis, Mercator in his *Logarithmotechnia,* Prestet) have engaged themselves in the study of figurate numbers. But I have found no one who has given a universal and scientific demonstration of this property. Wallis put forward his fundamental methods in the *Arithm. Infinitorum,* where he investigates inductively[10] the ratios of the series of Squares, Cubes, or other powers of the natural numbers to the series, having as many terms, of the largest of these powers. From this he moves [...] to the study of Triangular, Pyramidal, and the remaining figurate numbers. But it would have been more convenient and appropriate in the nature of things had he instead first prepared a treatise on the figurate numbers, with universal and accurate demonstrations, and then later continued the investigation of sums of powers. For after all, the method of demonstration by induction is not particularly scientific, and besides, each series requires its own special methods. Those series which should be considered first, by general estimation, and whose natures are most fundamental and simple, are seen to be the figurate numbers, which are generated by addition, while the powers are generated by multiplication. Moreover, the series of figurates, beginning with their respective zeros, have exact fractional ratios with the series having the same number of constant terms equal to the largest of these,[11] which is not necessarily

---

[10] That is, by inductive, as opposed to deductive, reasoning. Bernoulli is not referring here to the method of proof called mathematical induction, which is entirely different.

[11] The reader may wish to decipher this claim, and see that it is actually equivalent to Fermat's claims in his letter to Mersenne. Hint: Bernoulli means that to obtain an unchanging fractional ratio $1/r$ using the sum of any series of r-dimensional figurate numbers in the numerator, one should prefix the sequence with r zeros for determining the number of constant terms in the denominator of the ratio. He then contrasts this with sums of powers, and compares with the infinite case; can you see what he is getting at?

so for the powers (at least not in a finite number of terms, regardless how many zeros, by excess or defect, are prefixed to it). Furthermore, from the knowledge of the sum of figurates, it is no more difficult to determine the sums of powers, and so the author has concluded from these first ideas, as I will now do most briefly.

Let there be given the series of natural numbers from unity: $1, 2, 3, 4, 5$, etc., up to n, and suppose that we ask for the sums of these, or of their squares, their cubes, etc.: In the Table of Combinations[12] the indefinite term in the [...][13] third column is found to be

$$\frac{n-1 \cdot n-2}{1 \cdot 2} = \frac{nn - 3n + 2}{2},$$

and the sum of all the terms (that is, all $\frac{nn-3n+2}{2}$) is

$$\frac{n \cdot n - 1 \cdot n - 2}{1 \cdot 2 \cdot 3} = \frac{n^3 - 3nn + 2n}{6};$$

this gives $\int \overline{\frac{nn - 3n + 2}{2}}$ or $\int \frac{1}{2}nn - \int \frac{3}{2}n + \int 1 = \frac{n^3 - 3nn + 2n}{6}$.

So $\int \frac{1}{2}nn = \frac{n^3 - 3nn + 2n}{6} + \int \frac{3}{2}n - \int 1.$

But $\int \frac{3}{2}n = \frac{3}{2} \int n =$ (by what was shown above) $\frac{3}{4}nn + \frac{3}{4}n,$

and $\int 1 = n$; substituting these above gives

$$\int \frac{1}{2}nn = \frac{n^3 - 3nn + 2n}{6} + \frac{3nn + 3n}{4} - n = \frac{1}{6}n^3 + \frac{1}{4}nn + \frac{1}{12}n,$$

and by doubling, $\int nn$ (the sum of the squares of all n)

$$= \frac{1}{3}n^3 + \frac{1}{2}nn + \frac{1}{6}n.$$

[...][14] And by proceeding to higher powers in turn, we easily build up the following formulas:

---

[12] That is, the arithmetical triangle.

[13] We omit Bernoulli's derivation of the sum of first powers, moving directly to a sum of squares.

[14] Bernoulli continues on to a sum of cubes.

Sums of Powers[15]

$$\int n = \frac{1}{2}nn + \frac{1}{2}n.$$

$$\int nn = \frac{1}{3}n^3 + \frac{1}{2}nn + \frac{1}{6}n.$$

$$\int n^3 = \frac{1}{4}n^4 + \frac{1}{2}n^3 + \frac{1}{4}nn.$$

$$\int n^4 = \frac{1}{5}n^5 + \frac{1}{2}n^4 + \frac{1}{3}n^3 * - \frac{1}{30}n.$$

$$\int n^5 = \frac{1}{6}n^6 + \frac{1}{2}n^5 + \frac{5}{12}n^4 * - \frac{1}{12}nn.$$

$$\int n^6 = \frac{1}{7}n^7 + \frac{1}{2}n^6 + \frac{1}{2}n^5 * - \frac{1}{6}n^3 * + \frac{1}{42}n.$$

$$\int n^7 = \frac{1}{8}n^8 + \frac{1}{2}n^7 + \frac{7}{12}n^6 * - \frac{7}{24}n^4 * + \frac{1}{12}nn.$$

$$\int n^8 = \frac{1}{9}n^9 + \frac{1}{2}n^8 + \frac{2}{3}n^7 * - \frac{7}{15}n^5 * + \frac{2}{9}n^3 * - \frac{1}{30}n.$$

$$\int n^9 = \frac{1}{10}n^{10} + \frac{1}{2}n^9 + \frac{3}{4}n^8 * - \frac{7}{10}n^6 * + \frac{1}{2}n^4 * - \frac{3}{20}nn.$$

$$\int n^{10} = \frac{1}{11}n^{11} + \frac{1}{2}n^{10} + \frac{5}{6}n^9 * - 1n^7 * + 1n^5 * - \frac{1}{2}n^3 * + \frac{5}{66}n.$$

Indeed, a pattern can be seen in the progressions herein [Exercise 1.43], which can be continued by means of this rule. Suppose that c is the value of any power; then the sum of all n^c or

$$\int n^c = \frac{1}{c+1}n^{c+1} + \frac{1}{2}n^c + \frac{c}{2}An^{c-1} + \frac{c \cdot c - 1 \cdot c - 2}{2 \cdot 3 \cdot 4}Bn^{c-3}$$

$$+ \frac{c \cdot c - 1 \cdot c - 2 \cdot c - 3 \cdot c - 4}{2 \cdot 3 \cdot 4 \cdot 5 \cdot 6}Cn^{c-5}$$

$$+ \frac{c \cdot c - 1 \cdot c - 2 \cdot c - 3 \cdot c - 4 \cdot c - 5 \cdot c - 6}{2 \cdot 3 \cdot 4 \cdot 5 \cdot 6 \cdot 7 \cdot 8}Dn^{c-7} \dots \text{ \& so on,}$$

where the value of the power n continues to decrease by two until it reaches n or nn. The uppercase letters A, B, C, D, etc., in order, denote the coefficients of the final term of $\int nn$, $\int n^4$, $\int n^6$, $\int n^8$, etc., namely

$$A = \frac{1}{6}, \quad B = -\frac{1}{30}, \quad C = \frac{1}{42}, \quad D = -\frac{1}{30}.$$

---

[15] There is an error in the original published Latin table of sums of powers formulas. The last coefficient in the formula for $\int n^9$ should be $-\frac{3}{20}$, not $-\frac{1}{12}$; we have corrected this here.

These coefficients are such that, when arranged with the other coefficients of the same order, they add up to unity: so, for D, which we said signified $-\frac{1}{30}$, we have

$$\frac{1}{9} + \frac{1}{2} + \frac{2}{3} - \frac{7}{15} + \frac{2}{9}(+D) - \frac{1}{30} = 1.$$

By means of these formulas, I discovered in under a quarter hour's work that the tenth (or quadrato-sursolid) powers of the first thousand numbers from unity, when collected into a sum, yield

$$91409924241424243424241924242500.$$

Clearly this renders obsolete the work of Ismael Bulliald, who wrote so as to thicken the volumes of his *Arithmeticae Infinitorum* with demonstrations involving immense labor, unexcelled by anyone else, of the sums of up to the first six powers (which is only a part of what we have superseded in a single page).

◯◯◯◯◯◯◯◯◯◯

Interestingly, while Bernoulli indicates his familiarity with the work of Johann Faulhaber (mentioned in our introduction), John Wallis (1616–1703), and Nicolaus Mercator (1620–1687) on sums of powers, he mentions neither Fermat nor Pascal. Wallis had studied sums of powers in his *Arithmetica Infinitorum* of 1655, with the same motivation as Fermat and Pascal, to find the areas under higher parabolas. Bernoulli contrasts Wallis's work with his own, including comparing sums of figurate numbers with sums of powers. While finite sums of powers do not behave as nicely as sums of figurate numbers, Bernoulli's subsequent formulas shed light on the nature of the difference between them by providing a precise expression for $\sum_{i=1}^{n} i^k$ as a polynomial in n.

Notice Bernoulli's summation notation as he proceeds to analyze sums of powers. The expression after the integral indicates both the general term and the ending index, i.e., he writes $\int n^7$ for $\sum_{i=1}^{n} i^7$ (Exercise 1.45). He also uses an asterisk to indicate "missing" terms, i.e., monomials with zero coefficient.

Bernoulli first shows how to derive sum formulas for the first few exponents, using his knowledge of the arithmetical triangle, by exactly the same method we presented when considering Fermat's claim to have solved the problem. He presents the results of calculation in a table of polynomials for sums up to the tenth powers. And now suddenly he claims:

> Indeed, a pattern can be seen in the progressions herein which can be continued by means of this rule:

Perhaps readers will delight in discovering this pattern for themselves (Exercise 1.43) before studying Bernoulli's description of it.

The reader should check that in modern notation, Bernoulli is claiming

$$\sum_{i=1}^{n} i^k = \frac{n^{k+1}}{k+1} + \frac{n^k}{2} + \sum_{j=2}^{k} \frac{1}{k+1} \binom{k+1}{j} B_j n^{k+1-j} \text{ for } k \geq 1,$$

where we have represented the special sequence of numbers that Bernoulli calls $A, B, C, D, \ldots$ by B_{2m} for $l \geq 1$, and $B_{2m+1} = 0$. These are the numbers anticipated since the beginning of the chapter.[16] We shall soon see Leonhard Euler (1707–1783) accomplish amazing feats with these numbers. It was he who would christen them *Bernoulli numbers*. We will see Euler prove Bernoulli's claimed patterns in the next section.

Bernoulli also claims that he can compute his special sequence of numbers $A, B, C, D, \ldots$. First he notes that they

in order, denote the coefficients of the final term of $\int nn$, $\int n^4$, $\int n^6$, $\int n^8$.

Indeed, we notice that the coefficient of n in the general formula he gives is always the first occurrence of a new Bernoulli number in the process. And he says:

> These coefficients are such that, when arranged with the other coefficients of the same order, they add up to unity.

Here he is simply evaluating both sides of his general formula at $n = 1$. Since the left side is then 1, the kth formula simplifies to

$$1 = \frac{1}{k+1} + \frac{1}{2} + \sum_{j=2}^{k} \frac{1}{k+1} \binom{k+1}{j} B_j.$$

Since the last term in the sum is the newest Bernoulli number B_k, one can solve for it in terms of the previous ones. Thus the Bernoulli numbers are recursively defined by these formulas. He gives as an example the computation of $D = B_8 = -\frac{1}{30}$ from the formula for $k = 8$ and the previous numbers. While this still leaves a step-by-step aspect to the determination of sums of powers formulas, the process is now greatly simplified. Moreover, we see a general pattern in the relationship between the coefficients for different values of k, since the Bernoulli numbers are the same in the formulas for all k.

How might we attempt to verify the general validity of the pattern Bernoulli guessed? Since Pascal gave us an equation relating the sums of kth powers to those of lower powers, we should be able to proceed by strong mathematical induction on k, by simply substituting all the formulas of Bernoulli's into Pascal's equation to verify the inductive claim at each stage. All but one of Bernoulli's formulas substituted in Pascal's equation are assumed true inductively, and the kth is thus shown true by verifying the equality itself (Exercise 1.46).

---

[16] The evidence suggests that around the same time, Takakazu Seki (1642?–1708) in Japan also discovered the same numbers [210, 257].

Before we explore the world opened by Euler using Bernoulli numbers, we encourage the reader to look for another pattern in these numbers, one not even mentioned by Bernoulli (Exercise 1.47).

Exercise 1.43. Guess, as did Bernoulli, the complete pattern of coefficients for sums of powers formulas just from the examples in Bernoulli's table. Clearly the pattern is to be sought down each column of Bernoulli's table. The key is to multiply each column of numbers by a common denominator, and then compare with the arithmetical triangle (computing the sequence of successive differences in a column, and the successive differences in that sequence, etc., may also help). Can you also express the general rule for calculating the special numbers $A, B, C, D, \ldots$, which Bernoulli introduces? Hint: What happens when $n = 1$?

Exercise 1.44. See whether you can duplicate Bernoulli's claim that he calculated (by hand, of course) the sum of the tenth powers of the first thousand numbers in less than a quarter of an hour.

Exercise 1.45. Why do we use the notation $\sum_{i=1}^{n} i^7$ today, instead of Bernoulli's $\int n^7$, for sums? What are the advantages and disadvantages of the two notations?

Exercise 1.46. Prove Bernoulli's claimed formulas by strong mathematical induction, in the manner suggested in the text, using Pascal's equation, Bernoulli's claims, and the Bernoulli numbers as defined recursively. At some point in your calculations you may need to prove and use the identity $\binom{a}{b}\binom{b}{c} = \binom{a}{c}\binom{a-c}{b-c}$. Hint: When substituting Bernoulli's claims into Pascal's equation, verify equality by calculating and comparing the coefficients for an arbitrary power of n on each side of the equation.

Exercise 1.47. What do you conjecture about the signs of the Bernoulli numbers? Compute several more Bernoulli numbers to see whether your conjecture has promise. This conjecture will be addressed by the work of Euler in the next section.

1.5 Euler's Summation Formula and the Solution for Sums of Powers

> Euler calculated without any apparent effort, just as men breathe, as eagles sustain themselves in the air. Arago [258, p. 354].

Leonhard Euler (1707–1783) towered over eighteenth-century mathematics and was one of the greatest mathematicians of all time. His overall life and work are discussed in our chapter on prime numbers and the quadratic reciprocity law.

Euler spent the first part of his mathematical career at the newly organized St. Petersburg Academy of Sciences, where he arrived from his hometown of Basel in May of 1727. The famous Basel problem, to find the exact sum $\sum_{i=1}^{\infty} \frac{1}{i^2}$ of the reciprocal squares, was very much in the air among mathematicians, and it was not long before Euler found better and better approximations to the true sum. Since the series converges rather slowly, calculating individual partial sums is not very fruitful, and finding good approximations to the actual sum required ingenuity. As early as 1731 Euler found an application of calculus showing that the sum was 1.644934..., considerably better than any previous results. By 1732 he was developing his *summation formula*, and applying it to find closed expressions for certain finite summations, such as a sum of squares $\sum_{i=1}^{n} i^2$, but not yet to approximate sums of infinite series like $\sum_{i=1}^{\infty} \frac{1}{i^2}$.

Then in a paper[17] presented to the St. Petersburg Academy on October 13, 1735 [66, v. 16, part 2, p. XX], [66, v. 14, pp. 108–123], Euler shows how to use just a few terms of his diverging summation formula to find incredibly accurate approximations for sums of infinite series. For the sum of reciprocal squares he obtains $\sum_{i=1}^{\infty} \frac{1}{i^2} \approx 1.644\,934\,066\,848\,226\,436\,47$, which is indeed accurate in all twenty places![18] We are left in awe that just a few terms of a diverging formula can so closely approximate this sum. Paradoxically, Euler's summation formula, even though it usually diverges, provides breathtaking acceleration of approximations for partial and infinite sums of many slowly converging or diverging series.

Evidence from his earlier papers suggests that he had not yet guessed the exact sum from his approximations, but that by the time he presented his twenty-place approximation he felt sure that the true sum was $\pi^2/6$, and was probably searching for a way to prove it. This was a completely different challenge from finding close approximations, yet less than eight weeks later he presented his first proof solving the Basel problem, in the paper *De summis serierum reciprocarum*[19] [66, v. 16, part 2, p. XXII], [66, v. 14, pp. 73–86].

---

[17] *Inventio Summae Cuiusque Seriei Ex Dato Termino Generali.*

[18] The twenty-place accuracy he gave can be achieved with his summation formula, although he did not show all the detailed calculations to support it beyond the fourteenth place.

[19] Euler reasoned like this. First write $\sin x = x \left(1 - \frac{x^2}{3!} + \frac{x^4}{5!} - \cdots \right)$, and treat the power series as an infinite polynomial with leading term 1. Factor into linear factors, each of the form $\left(1 - \frac{x}{r}\right)$, corresponding to the roots $r = \pm\pi, \pm 2\pi, \pm 3\pi, \ldots$ of $\frac{\sin x}{x}$. Thus

$$\left(1 - \frac{x^2}{3!} + \frac{x^4}{5!} - \cdots \right)$$
$$= \left(1 - \frac{x}{\pi}\right)\left(1 + \frac{x}{\pi}\right)\left(1 - \frac{x}{2\pi}\right)\left(1 + \frac{x}{2\pi}\right)\left(1 - \frac{x}{3\pi}\right)\left(1 + \frac{x}{3\pi}\right)\cdots$$
$$= \left(1 - \frac{x^2}{\pi^2}\right)\left(1 - \frac{x^2}{4\pi^2}\right)\left(1 - \frac{x^2}{9\pi^2}\right)\cdots .$$

In his October 13 paper, Euler also demonstrates how to derive Bernoulli's conjectured formulas for sums of powers from his summation formula, although he appears at that time to have been unaware of Bernoulli's prior empirical discovery of the patterns in the sums of powers formulas. Thus Euler simultaneously uses his summation formula to lead him both to the sum of the most sought-after infinite series of the day, and to prove the general sums of powers formulas.

Euler's papers on his summation formula, related infinite series, and the Bernoulli numbers are discussed individually in a 1935 essay by Georg Faber in Euler's collected works [66, v. 16, part 2, pp. VII–XXXIX]. Euler's work is also discussed from diverse points of view in many modern books [54], [97, pp. 119–136], [104, II.10], [106, Chapter XIII], [116, p. 197f], [137, Chapter XIV], [245, p. 184, 257–285], [258, p. 338f].

We will read about the summation formula from Euler's book on the differential calculus, *Institutiones Calculi Differentialis* (Foundations of Differential Calculus), published in 1755 [66, v. 10], [67], during the second part of his career, at the Berlin Academy of Sciences. *Institutiones* presents his mature view of the summation formula, its applications and relation to Bernoulli's numbers, and also the unexpected additional connection he discovered between the Bernoulli numbers and the exact sums of the infinite series of reciprocal even powers.

Euler was a wizard with infinite series. Much of his book is actually devoted to the relationship between differential calculus and infinite series, in contrast to the emphasis in today's calculus books. In the second part of the book, Euler presents his way of finding sums of series, first finite and then infinite, via his summation formula. There is a published English translation of the first part [68], but not the second part, of the *Institutiones*. We have translated [69], [136, Eneström 212] much more of his work on the summation formula from part two than we present here, and we encourage the reader to explore there many additional aspects that we briefly mention in what follows. In our first section on Euler's work, we will see Euler derive his summation formula, analyze the nature of its Bernoulli numbers in connection with trigonometric functions, and prove Bernoulli's sums of powers formulas.

When Euler presented this amazing factorization, it was criticized for lacking rigor. Some said there might be another unknown factor in the product, one with no real zeros. This criticism was justified, but Euler's expansion is correct, and later he found acceptable means to justify it. Now imagine multiplying out the infinite product, and isolating contributions to the coefficient of x^2, which together are $-\frac{1}{\pi^2}\sum_{i=1}^{\infty}\frac{1}{i^2}$. Matching this with the coefficient $-\frac{1}{3!}$ in the power series on the other side, Euler deduces

$$\frac{1}{1^2} + \frac{1}{2^2} + \frac{1}{3^2} + \cdots + \frac{1}{i^2} + \cdots = \frac{\pi^2}{6}.$$

In his paper, Euler uses this matching idea to find the sums of quite a variety of infinite series (Exercises 1.48, 1.49).

INSTITUTIONES
CALCULI
DIFFERENTIALIS

CUM EIUS VSU

IN ANALYSI FINITORUM

AC

DOCTRINA SERIERUM

AUCTORE

LEONHARDO EULERO

ACAD. REG. SCIENT. ET ELEG. LITT. BORUSS. DIRECTORE
PROF. HONOR. ACAD. IMP. SCIENT. PETROP. ET ACADEMIARUM
REGIARUM PARISINAE ET LONDINENSIS
SOCIO.

IMPENSIS
ACADEMIAE IMPERIALIS SCIENTIARUM
PETROPOLITANAE
1755.

Photo 1.5. *Foundations of Differential Calculus.*

Euler's derivation of the summation formula rests on two ideas. First, use Taylor series from calculus, and adeptness with summations, to relate the sum of the values of a function at finitely many successive integers to similar sums involving the derivatives of the function. It may seem that this just makes things more complicated, but his second idea will remedy this.

∞∞∞∞∞∞∞∞

Leonhard Euler, from
Foundations of Differential Calculus

Part Two, Chapter 5
On Finding Sums of Series from the General Term

103. Suppose y is the general term of a series, belonging to the index x, and thus y is any function of x. Further, suppose Sy is the summative term of this series, expressing the aggregate of all terms from the first or another fixed term up to y, inclusive. The sums of the series are calculated from the first term, so that if $x = 1$, y is the first term, and likewise Sy yields this first term; alternatively, if $x = 0$, the summative term Sy vanishes, because no terms are being summed. With these stipulations, the summative term Sy is a function of x that vanishes if one sets $x = 0$.[20] ...

105. Consider a series whose general term, belonging to the index x, is y, and whose preceding term, with index $x - 1$, is v; because v arises from y, when x is replaced by $x - 1$, one has[21]

$$v = y - \frac{dy}{dx} + \frac{ddy}{2dx^2} - \frac{d^3y}{6dx^3} + \frac{d^4y}{24dx^4} - \frac{d^5y}{120dx^5} + \text{etc.}$$

If y is the general term of the series

$$
\begin{array}{ccccccc}
1 & 2 & 3 & 4 & \cdots & x-1 & x \\
a + & b + & c + & d + & \cdots + & v & + y
\end{array}
$$

---

[20] Today we might think that so far Euler only has in mind that y and Sy are sequences indexed by natural numbers x, except that he does refer to y as *any function* of x. In a moment we will see that for him the word *function* definitely means much more than just a sequence. The reader should reflect on what Euler has in mind.

[21] Euler expresses the value v of his function at $x - 1$ in terms of its value y at x and the values of all its derivatives, also implicitly evaluated at x. This uses Taylor series for the function, with increment -1, so he is tacitly assuming that this all makes sense, i.e., that his function equals its Taylor series. Note also that the symbols x and y are being used, respectively, to indicate the final value of an integer index and the final value of the function evaluated there, and also more generally as a variable and function of that variable. Today we would find this much too confusing to dare write this way.

and if the term belonging to the index 0 is A, then v, as a function of x, is the general term of the series

$$
\begin{array}{cccccccc}
1 & 2 & 3 & 4 & 5 & \cdots & & x \\
A + a & + b & + c & + d & + & \cdots & + v \, ,
\end{array}
$$

so if Sv denotes the sum of this series, then $Sv = Sy - y + A$. If one sets $x = 0$, then $Sy = 0$ and $y = A$, so Sv vanishes.

106. Because

$$
v = y - \frac{dy}{dx} + \frac{ddy}{2dx^2} - \frac{d^3y}{6dx^3} + \text{etc.,}
$$

one has, from the preceding,

$$
Sv = Sy - S\frac{dy}{dx} + S\frac{ddy}{2dx^2} - S\frac{d^3y}{6dx^3} + S\frac{d^4y}{24dx^4} - \text{etc.,}
$$

and, because $Sv = Sy - y + A$,

$$
y - A = S\frac{dy}{dx} - S\frac{ddy}{2dx^2} + S\frac{d^3y}{6dx^3} - S\frac{d^4y}{24dx^4} + \text{etc.,}
$$

or equivalently

$$
S\frac{dy}{dx} = y - A + S\frac{ddy}{2dx^2} - S\frac{d^3y}{6dx^3} + S\frac{d^4y}{24dx^4} - \text{etc.}
$$

Thus if one knows the sums of the series whose general terms are $\frac{ddy}{dx^2}$, $\frac{d^3y}{dx^3}$, $\frac{d^4y}{dx^4}$, etc., one can obtain the summative term of the series whose general term is $\frac{dy}{dx}$. The constant A must then be such that the summative term $S\frac{dy}{dx}$ disappears when $x = 0$, and this condition makes it easier to determine than saying that it is the term belonging to the index 0 in the series whose general term is y.

∞∞∞∞∞∞∞∞∞

To summarize, Euler's final equation here relates the sum of the values of the derivative of y at the integers from 1 to x to the sums of all the (infinitely many) higher derivatives of y at these same numbers, and also involves the value of y itself at x. In addition, there is an unknown constant A. This may seem rather overwhelming, but Euler immediately illustrates practical application for this equation in §§107–108 by selecting the power function $y = x^{n+1}/(n+1)$. This has the advantage that the sums are all just multiples of sums of powers, and they vanish after some point in the equation. He is left with a finite expression for Sx^n (i.e., for $\sum_{i=1}^{x} i^n$), with many similarities to Pascal's equation earlier for a sum of powers in terms of sums of lower powers (Exercises 1.50, 1.51). He applies this inductively from $n = 0$ upwards to calculate the closed formulas for sums of powers of the natural numbers explicitly up through the sum of fourth powers (Exercise 1.52).

Of course, this was a very special choice of function, in which the summations on the right were considered "easier", and there was only a finite number of them. In general, the sums on the right will be no easier to determine than the one on the left, and there will be infinitely many of them. But Euler has something up his sleeve.

His second idea, brilliant in scope, is to eliminate all the summations on the right side of this equation, successively substituting for them by using the very equation itself on each of them, applied one at a time to higher and higher derivatives, and keeping track of the terms thus created. This will produce a formula on the right whose terms involve only the function and its derivatives at the value x. No summations from 1 to x remain on the right, leaving only the single summation from 1 to x on the left side. This process yields a formula, a function of x, for this summation. The result will be a first version of the *Euler–Maclaurin summation formula*.

Euler's idea is somewhat similar in spirit to what we have already practiced in inductive calculations with the equations of Pascal and Bernoulli. Applied to a power function, it could conceivably even yield Bernoulli's polynomial formula for a sum of powers!

Euler begins with a little shift in perspective in his basic equation above, to view the left side as the sum of values of the primary function of interest, rather than of its derivative.

∞∞∞∞∞∞∞∞∞

109. Since from the above one has

$$S\frac{dy}{dx} = y\,[-A] + \frac{1}{2}S\frac{ddy}{dx^2} - \frac{1}{6}S\frac{d^3y}{dx^3} + \frac{1}{24}S\frac{d^4y}{dx^4} - \frac{1}{120}S\frac{d^5y}{dx^5} + \text{etc.},$$

if one sets $\frac{dy}{dx} = z$, then $\frac{ddy}{dx^2} = \frac{dz}{dx}$, $\frac{d^3y}{dx^3} = \frac{ddz}{dx^2}$, etc. And because $dy = z\,dx$, y will be a quantity whose differential is $z\,dx$, and this one writes as $y = \int z\,dx$. Now the determination of the quantity y from z according to this formula assumes the integral calculus; but we can nevertheless make use of this expression $\int z\,dx$, if for z we use no function other than that whose differential is $z\,dx$ from above. Thus substituting these values yields

$$Sz = \int z\,dx + \frac{1}{2}S\frac{dz}{dx} - \frac{1}{6}S\frac{ddz}{dx^2} + \frac{1}{24}S\frac{d^3z}{dx^3} - \text{etc.},$$

adding to it a constant value such that when $x = 0$, the sum Sz also vanishes.

∞∞∞∞∞∞∞∞∞

Next Euler prepares the substitutions he plans to make.

∞∞∞∞∞∞∞∞∞

110. But if in the expressions above one substitutes the letter z in place of y, or if one differentiates the preceding equation, which yields the same, one obtains

$$S\frac{dz}{dx} = z + \frac{1}{2}S\frac{ddz}{dx^2} - \frac{1}{6}S\frac{d^3z}{dx^3} + \frac{1}{24}S\frac{d^4z}{dx^4} - \text{etc.};$$

but using $\frac{dz}{dx}$ in place of y one obtains

$$S\frac{ddz}{dx^2} = \frac{dz}{dx} + \frac{1}{2}S\frac{d^3z}{dx^3} - \frac{1}{6}S\frac{d^4z}{dx^4} + \frac{1}{24}S\frac{d^5z}{dx^5} - \text{etc.}$$

Similarly, replacing y successively by the values $\frac{ddz}{dx^2}$, $\frac{d^3z}{dx^3}$ etc., produces

$$S\frac{d^3z}{dx^3} = \frac{ddz}{dx^2} + \frac{1}{2}S\frac{d^4z}{dx^4} - \frac{1}{6}S\frac{d^5z}{dx^5} + \frac{1}{24}S\frac{d^6z}{dx^6} - \text{etc.},$$

$$S\frac{d^4z}{dx^4} = \frac{d^3z}{dx^3} + \frac{1}{2}S\frac{d^5z}{dx^5} - \frac{1}{6}S\frac{d^6z}{dx^6} + \frac{1}{24}S\frac{d^7z}{dx^7} - \text{etc.},$$

and so forth indefinitely.

∞∞∞∞∞∞∞∞

Now Euler is ready to make all the substitutions. To find the resulting formula, he first notes that his substitutions must produce a certain form, with only some unknown coefficients to be determined. Then he sets up and solves a linear system for these coefficients.

∞∞∞∞∞∞∞∞

111. Now when these values for $S\frac{dz}{dx}$, $S\frac{ddz}{dx^2}$, $S\frac{d^3z}{dx^3}$ are successively substituted in the expression

$$Sz - \int zdx + \frac{1}{2}S\frac{dz}{dx} - \frac{1}{6}S\frac{ddz}{dx^2} + \frac{1}{24}S\frac{d^3z}{dx^3} - \text{etc.},$$

one obtains an expression for Sz, composed of the terms $\int zdx$, z, $\frac{dz}{dx}$, $\frac{ddz}{dx^2}$, $\frac{d^3z}{dx^3}$ etc., whose coefficients are easily obtained as follows. One sets

$$Sz = \int zdx + \alpha z + \frac{\beta dz}{dx} + \frac{\gamma ddz}{dx^2} + \frac{\delta d^3z}{dx^3} + \frac{\varepsilon d^4z}{dx^4} + \text{etc.},$$

and substitutes for these terms the values they have from the previous series, yielding[22]

$$\int zdx = Sz - \tfrac{1}{2}S\frac{dz}{dx} + \tfrac{1}{6}S\frac{ddz}{dx^2} - \tfrac{1}{24}S\frac{d^3z}{dx^3} + \tfrac{1}{120}S\frac{d^4z}{dx^4} - \text{etc.}$$

$$\alpha z = \quad + \alpha S\frac{dz}{dx} - \tfrac{\alpha}{2}S\frac{ddz}{dx^2} + \tfrac{\alpha}{6}S\frac{d^3z}{dx^3} - \tfrac{\alpha}{24}S\frac{d^4z}{dz^4} + \text{etc.}$$

$$\tfrac{\beta dz}{dx} = \qquad\qquad \beta S\frac{ddz}{dx^2} - \tfrac{\beta}{2}S\frac{d^3z}{dx^3} + \tfrac{\beta}{6}S\frac{d^4z}{dx^4} - \text{etc.}$$

$$\tfrac{\gamma ddz}{dx^2} = \qquad\qquad\qquad \gamma S\frac{d^3z}{dx^3} - \tfrac{\gamma}{2}S\frac{d^4z}{dx^4} + \text{etc.}$$

$$\tfrac{\delta d^3z}{dx^3} = \qquad\qquad\qquad\qquad \delta\, S\frac{d^4z}{dx^4} - \text{etc.}$$

etc.

---

[22] These lines are obtained by rearranging the equations in §§109–110, and multiplying them by $\alpha, \beta, \gamma, \ldots$.

Since these values, added together, must produce Sz, the coefficients α, β, γ, δ etc. are defined by the sequence of equations

$$\alpha - \frac{1}{2} = 0, \quad \beta - \frac{\alpha}{2} + \frac{1}{6} = 0, \quad \gamma - \frac{\beta}{2} + \frac{\alpha}{6} - \frac{1}{24} = 0,$$

$$\delta - \frac{\gamma}{2} + \frac{\beta}{6} - \frac{\alpha}{24} + \frac{1}{120} = 0, \quad \varepsilon - \frac{\delta}{2} + \frac{\gamma}{6} - \frac{\beta}{24} + \frac{\alpha}{120} - \frac{1}{720} = 0,$$

$$\zeta - \frac{\varepsilon}{2} + \frac{\delta}{6} - \frac{\gamma}{24} + \frac{\beta}{120} - \frac{\alpha}{720} + \frac{1}{5040} = 0 \quad \text{etc.}$$

Euler believes that the pattern determining these coefficients is clear to us. It is typical of his work to arrange the results of calculations, and to write out enough of them, to make a general pattern clear and convincing. Today we would tend to write general formulas using arbitrary indices attached to unknowns to describe Euler's pattern. The pattern by which the coefficients above are inductively determined is reminiscent of the way Bernoulli described an inductive pattern determining the Bernoulli numbers. We shall see that the similarity is more than coincidental.

112. So from these equations the successive values of all the letters α, β, γ, δ etc. are defined; they are

$$\alpha = \frac{1}{2}, \quad \beta = \frac{\alpha}{2} - \frac{1}{6} = \frac{1}{12}, \quad \gamma = \frac{\beta}{2} - \frac{\alpha}{6} + \frac{1}{24} = 0,$$

$$\delta = \frac{\gamma}{2} - \frac{\beta}{6} + \frac{\alpha}{24} - \frac{1}{120} = -\frac{1}{720}, \quad \varepsilon = \frac{\delta}{2} - \frac{\gamma}{6} + \frac{\beta}{24} - \frac{\alpha}{120} + \frac{1}{720} = 0, \text{ etc.,}$$

and if one continues in this fashion one finds that alternating terms vanish. The third, fifth, seventh letters, and so on, in fact all odd terms except the first, are zero, so that this series appears to contradict the law of continuity by which the terms proceed. A rigorous proof is especially needed that all odd terms except the first vanish.

Before Euler shows how to apply his summation formula to derive his results for various choices of the function z and of the initial and final indices in a summation, he will study closely the coefficients $\alpha, \beta, \gamma, \ldots$ in the formula itself, and discover their properties and intimate connections with other mathematics.

Euler states confidently that every odd term in this sequence of numbers vanishes, except for the first. He sets out to prove this and other features of these numbers. Euler uses what is today called a *generating function*, namely he creates a formal power series whose coefficients are chosen to be the sequence of numbers in question. He proceeds to show that the nature of the sequence of numbers determines certain properties that this "function" must obey. This is a powerful technique much used in modern mathematics. Entire books exist on the theory and applications of generating functions [138, 246]. The question of convergence of the power series for a generating function is not always important, since it is only the combinatorial properties of its coefficients that are relevant, i.e., it is being manipulated formally algebraically, as a polynomial of infinite degree.

$\infty\infty\infty\infty\infty\infty$

113. Because the letters are determined from the preceding by a constant law, they form a recurrent series. In order to develop this, consider the series

$$1 + \alpha u + \beta u^2 + \gamma u^3 + \delta u^4 + \varepsilon u^5 + \zeta u^6 + \text{etc.},$$

and set its value $= V$, so it is clear[23] that this recurrent series arises from the development of the fraction

$$V = \frac{1}{1 - \frac{1}{2}u + \frac{1}{6}u^2 - \frac{1}{24}u^3 + \frac{1}{120}u^4 - \text{etc.}} .$$

And when this fraction is resolved in a different way in an infinite series according to the powers of u, then necessarily the same series

$$V = 1 + \alpha u + \beta u^2 + \gamma u^3 + \delta u^4 + \varepsilon u^5 + \text{etc.}$$

will always result. In this fashion a different rule for determining the letters α, β, γ, δ etc. results.

$\infty\infty\infty\infty\infty\infty$

Now begins the fun. Euler can recognize the denominator above in terms of a transcendental function he knows well, allowing him to bring to bear the beautiful and powerful relationships between familiar transcendental functions, their power series, and calculus.

$\infty\infty\infty\infty\infty\infty$

114. Because one has

$$e^{-u} = 1 - u + \frac{1}{2}u^2 - \frac{1}{6}u^3 + \frac{1}{24}u^4 - \frac{1}{120}u^5 + \text{etc.},$$

where e denotes the number whose hyperbolic logarithm[24] is one, then

$$\frac{1 - e^{-u}}{u} = 1 - \frac{1}{2}u + \frac{1}{6}u^2 - \frac{1}{24}u^3 + \frac{1}{120}u^4 - \text{etc.},$$

---

[23] To see that what Euler says is clear, multiply both sides by the denominator and formally carry out the multiplication of the two infinite polynomials.

[24] Think about why Euler calls this the "hyperbolic" logarithm, by considering the area under the hyperbola $y = 1/x$.

and thus

$$V = \frac{u}{1 - e^{-u}}.$$

Now one removes from this series the second term $\alpha u = \frac{1}{2}u$, so that

$$V - \frac{1}{2}u = 1 + \beta u^2 + \gamma u^3 + \delta u^4 + \varepsilon u^5 + \zeta u^6 + \text{etc.},$$

whence

$$V - \frac{1}{2}u = \frac{\frac{1}{2}u\left(1 + e^{-u}\right)}{1 - e^{-u}}.$$

Multiplying numerator and denominator by $e^{\frac{1}{2}u}$ yields[25]

$$V - \frac{1}{2}u = \frac{u\left(e^{\frac{1}{2}u} + e^{-\frac{1}{2}u}\right)}{2\left(e^{\frac{1}{2}u} - e^{-\frac{1}{2}u}\right)},$$

and converting the quantities $e^{\frac{1}{2}u}$ and $e^{-\frac{1}{2}u}$ into series gives

$$V - \frac{1}{2}u = \frac{1 + \frac{u^2}{2\cdot4} + \frac{u^4}{2\cdot4\cdot6\cdot8} + \frac{u^6}{2\cdot4\cdot6\cdot8\cdot10\cdot12} + \text{etc.}}{2\left(\frac{1}{2} + \frac{u^2}{2\cdot4\cdot6} + \frac{u^4}{2\cdot4\cdot6\cdot8\cdot10} + \text{etc.}\right)},$$

or

$$V - \frac{1}{2}u = \frac{1 + \frac{u^2}{2\cdot4} + \frac{u^4}{2\cdot4\cdot6\cdot8} + \frac{u^6}{2\cdot4\cdots12} + \frac{u^8}{2\cdot4\cdots16} + \text{etc.}}{1 + \frac{u^2}{4\cdot6} + \frac{u^4}{4\cdot6\cdot8\cdot10} + \frac{u^6}{4\cdot6\cdots14} + \frac{u^8}{4\cdot6\cdots18} + \text{etc.}}.$$

115. Since no odd powers occur in this fraction, likewise none can occur in its expansion; because $V - \frac{1}{2}u$ equals the series

$$1 + \beta u^2 + \gamma u^3 + \delta u^4 + \varepsilon u^5 + \zeta u^6 + \text{etc.},$$

the coefficients of the odd powers γ, ε, η, ι etc. all vanish. And so it is clear why the even-ordered terms after the second all equal zero in the series $1 + \alpha u + \beta u^2 + \gamma u^3 + \delta u^4 +$ etc., for otherwise the law of continuity would be violated. Thus

$$V = 1 + \frac{1}{2}u + \beta u^2 + \delta u^4 + \zeta u^6 + \theta u^8 + \varkappa u^{10} + \text{etc.},$$

---

[25] Here arise early occurrences of hyperbolic trigonometric functions, which provide a very helpful way to view and work with functions like this. They were first studied comprehensively in 1768 by Johann Lambert [133, p. 570], [135, p. 404], after initial discovery by Euler and others. The hyperbolic cosine is $\cosh x = \sum_{k=0}^{\infty} x^{2k} / (2k)!$, by analogy with the Taylor series for $\cos x$. Note that $\cosh x = \left(e^x + e^{-x}\right)/2$. Similarly, $\sinh x = \sum_{k=0}^{\infty} x^{2k+1} / (2k+1)! = \left(e^x - e^{-x}\right)/2$. Now note that Euler's $V - (1/2)u = (u/2)\cosh(u/2) / \sinh(u/2) = (u/2)\coth(u/2)$, and keep this in mind for the next two footnotes. Much of what Euler does here could be phrased in terms of properties of hyperbolic trigonometric functions.

and if the letters β, δ, ζ, θ, $\varkappa$ have been determined by the development of the above fraction, one obtains the summative term Sz of the series, whose general term $= z$ corresponds to the index x, expressed as

$$Sz = \int z\,dx + \frac{1}{2}z + \frac{\beta\,dz}{dx} + \frac{\delta\,d^3 z}{dx^3} + \frac{\zeta\,d^5 z}{dx^5} + \frac{\theta\,d^7 z}{dx^7} + \text{etc.}$$

∞∞∞∞∞∞∞∞

Having shown that only the odd derivatives appear in the summation formula, Euler now claims that the remaining constants β, δ, ζ, θ, ... alternate in sign. In order to show this, and then to relate these coefficients to the Bernoulli numbers, Euler introduces two closely related sequences. First he sets

$$A = \beta; \quad B = -\delta; \quad C = \zeta; \quad D = -\theta; \quad \text{etc.,}$$

with the expectation that this new sequence is entirely positive. When he makes these substitutions in the final expression for $V - \frac{1}{2}u$ obtainable from §115, and simultaneously replaces u^2 with $-u^2$ both there and in the quotient of power series[26] for $V - \frac{1}{2}u$ that culminates §114, he can equate the two resulting expressions for $V - \frac{1}{2}u$. Continuing in his words:

∞∞∞∞∞∞∞∞

118. ... [I]n order to determine the letters A, B, C, D etc., we consider the series

$$1 \qquad Au^2 - Bu^4 - Cu^6 - Du^8 - Eu^{10} - \text{etc.,}$$

which arises from the development of the fraction

$$\frac{1 - \frac{u^2}{2\cdot4} + \frac{u^4}{2\cdot4\cdot6\cdot8} - \frac{u^6}{2\cdot4\cdots12} + \frac{u^8}{2\cdot4\cdots16} - \text{etc.}}{1 - \frac{u^2}{4\cdot6} + \frac{u^4}{4\cdot6\cdot8\cdot10} - \frac{u^6}{4\cdot6\cdots14} + \frac{u^8}{4\cdot6\cdots18} - \text{etc.}},$$

or consider the series[27]

$$\frac{1}{u} - Au - Bu^3 - Cu^5 - Du^7 - Eu^9 - \text{etc.} = s,$$

which arises from the development of the fraction

$$s = \frac{1 - \frac{u^2}{2\cdot4} + \frac{u^4}{2\cdot4\cdot6\cdot8} - \frac{u^6}{2\cdot4\cdots12} + \text{etc.}}{u - \frac{u^3}{4\cdot6} + \frac{u^5}{4\cdot6\cdot8\cdot10} - \frac{u^7}{4\cdot6\cdots14} + \text{etc.}}.$$

---

[26] That is, replace u by iu in the hyperbolic trigonometric expression and its expression as a quotient of power series, yielding a new quotient of power series for $(iu/2)\coth(iu/2)$, to follow next.

[27] Since we know that this division by u produces $(i/2)\coth(iu/2)$, see if you can reduce this to something more familiar using the definitions of cosh and sinh in terms of the exponential function, combined with Euler's identity $e^{i\theta} = \cos\theta + i\sin\theta$. In doing so you will anticipate Euler's next step, which he does by examining the quotient of power series directly.

But since

$$\cos\frac{1}{2}u = 1 - \frac{u^2}{2\cdot 4} + \frac{u^4}{2\cdot 4\cdot 6\cdot 8} - \frac{u^6}{2\cdot 4\cdots 12} + \text{etc.},$$

$$\sin\frac{1}{2}u = \frac{u}{2} - \frac{u^3}{2\cdot 4\cdot 6} + \frac{u^5}{2\cdot 4\cdot 6\cdot 8\cdot 10} - \frac{u^7}{2\cdot 4\cdots 14} + \text{etc.},$$

we have

$$s = \frac{\cos\frac{1}{2}u}{2\sin\frac{1}{2}u} = \frac{1}{2}\cot\frac{1}{2}u.$$

Thus if one converts the cotangent of the arc $\frac{1}{2}u$ into a series, according to the powers of u, the values of the letters A, B, C, D, E, etc. are revealed.

∞◌∞◌∞◌∞◌∞◌∞

In §119 Euler confirms that all these numbers are positive, using a power series approach to the nonlinear differential equation that the cotangent function satisfies by dint of its derivative formula. The differential equation also produces a set of quadratic recursive formulas for these numbers quite different in nature from those he obtained earlier for $\alpha, \beta, \gamma, \ldots$. The new formulas suggest that the fractions obtained for A, B, C, D, E, $\ldots$ have fast-growing denominators, and he writes:

∞◌∞◌∞◌∞◌∞◌∞

120. But because the denominators of these fractions become very large, and substantially impede calculation, we want instead of the letters A, B, C, D, etc. to introduce new ones:[28]

$$A = \frac{\alpha}{1\cdot 2\cdot 3}, \quad B = \frac{\beta}{1\cdot 2\cdot 3\cdot 4\cdot 5}, \quad C = \frac{\gamma}{1\cdot 2\cdot 3\cdots 7},$$

$$D = \frac{\delta}{1\cdot 2\cdot 3\cdots 9}, \quad E = \frac{\varepsilon}{1\cdot 2\cdot 3\cdots 11}, \quad \text{etc.}$$

∞◌∞◌∞◌∞◌∞◌∞

Euler's choice to alter the denominators of $A, B, C, D, E, \ldots$ by the selected factorials is a delicate one, since the effects propagate down the list under the recursive quadratic formulas of §119, which he adapts to make calculations of his new $\alpha, \beta, \gamma, \delta, \varepsilon, \ldots$. However, he asserts that calculating with these new formulas is eminently manageable (in fact, he gives the result for seven more steps than we display below). Moreover, he is about to explain that these are almost Bernoulli's numbers!

---

[28] **Caution:** These new symbols α, β, $\ldots$ are **completely different** from the α, β, $\ldots$ used earlier. Today we would expect a mathematical writer to avoid confusion by not using the same symbol to mean two different things in such close proximity. Euler wasn't easily confused.

∞∞∞∞∞∞∞∞∞

121. ... If one finds the values of the letters α, β, γ, δ, etc. according to this rule, which entails little difficulty in calculation, then one can express the summative term of any series, whose general term $= z$ corresponding to the index x, in the following fashion:

$$Sz = \int z\,dx + \frac{1}{2}z + \frac{\alpha\,dz}{1\cdot 2\cdot 3\,dx} - \frac{\beta\,d^3 z}{1\cdot 2\cdot 3\cdot 4\cdot 5\,dx^3} + \frac{\gamma\,d^5 z}{1\cdot 2\cdots 7\,dx^5}$$
$$-\frac{\delta\,d^7 z}{1\cdot 2\cdots 9\,dx^7} + \frac{\varepsilon\,d^9 z}{1\cdot 2\cdots 11\,dx^9} - \frac{\zeta\,d^{11} z}{1\cdot 2\cdots 13\,dx^{11}} + \text{etc.}$$

As far as the letters α, β, γ, δ, etc. are concerned, one obtains the following values:

$$
\begin{array}{lll}
\alpha = \frac{1}{2} & \text{or} & 1\cdot 2\alpha = 1 \\
\beta = \frac{1}{6} & & 1\cdot 2\cdot 3\beta = 1 \\
\gamma = \frac{1}{6} & & 1\cdot 2\cdot 3\cdot 4\gamma = 4 \\
\delta - \frac{3}{10} & & 1\cdot 2\cdot 3\cdots 5\delta = 36 \\
\varepsilon = \frac{5}{6} & & 1\cdot 2\cdot 3\cdots 6\varepsilon = 600 \\
\zeta = \frac{691}{210} & & 1\cdot 2\cdot 3\cdots 7\zeta = 24\cdot 691 \\
\eta = \frac{35}{2} & & 1\cdot 2\cdot 3\cdots 8\eta = 20160\cdot 35 \\
\theta = \frac{3617}{30} & & 1\cdot 2\cdot 3\cdots 9\theta = 12096\cdot 3617
\end{array}
$$

$\cdots$ $\cdots$

∞∞∞∞∞∞∞∞∞

Clearly Euler is interested in discovering yet further features of the coefficients. In fact, he is working to determine which multiples of them will be whole numbers. After calculating by hand the first fifteen new fractions $\alpha, \beta, \gamma, \ldots$, he illustrates on the right side of the table that if he multiplies them by a simple pattern of factorials, the results are always whole numbers. In other words, each denominator is a divisor of the corresponding factorial. He does not actually say anything about this, but we can imagine that he might know it to be a general pattern. It is not hard to prove.

In the next section Euler claims to relate these numbers to the special numbers in Bernoulli's formulas for finite sums of powers. Recall that we saw in our Bernoulli source that each new Bernoulli number arises first as the coefficient of the first power of the variable in a sum of even powers.

∞∞∞∞∞∞∞∞∞

122. These numbers have great use throughout the entire theory of series. First, one can obtain from them the final terms in the sums of even powers, for which we noted above (in §63 of part one) that one cannot obtain them, as one can the other terms, from the sums of earlier powers. For the even powers, the

last terms of the sums are products of x and certain numbers, namely for the 2nd, 4th, 6th, 8th, etc., $\frac{1}{6}$, $\frac{1}{30}$, $\frac{1}{42}$, $\frac{1}{30}$ etc. with alternating signs. But these numbers arise from the values of the letters α, β, γ, δ, etc., which we found earlier, when one divides them by the odd numbers 3, 5, 7, 9, etc. These numbers are called the Bernoulli numbers after their discoverer Jakob Bernoulli, and they are

$$\frac{\alpha}{3} = \frac{1}{6} = \mathfrak{A} \qquad \frac{\iota}{19} = \frac{43867}{798} = \mathfrak{J}$$

$$\frac{\beta}{5} = \frac{1}{30} = \mathfrak{B} \qquad \frac{\varkappa}{21} = \frac{174611}{330} = \mathfrak{K} = \frac{283\cdot617}{330}$$

$$\frac{\gamma}{7} = \frac{1}{42} = \mathfrak{C} \qquad \frac{\lambda}{23} = \frac{854513}{138} = \mathfrak{L} = \frac{11\cdot131\cdot593}{2\cdot3\cdot23}$$

$$\frac{\delta}{9} = \frac{1}{30} = \mathfrak{D} \qquad \frac{\mu}{25} = \frac{236364091}{2730} = \mathfrak{M}$$

$$\frac{\varepsilon}{11} = \frac{5}{66} = \mathfrak{E} \qquad \frac{\nu}{27} = \frac{8553103}{6} = \mathfrak{N} = \frac{13\cdot657931}{6}$$

$$\frac{\zeta}{13} = \frac{691}{2730} = \mathfrak{F} \qquad \frac{\xi}{29} = \frac{23749461029}{870} = \mathfrak{O}$$

$$\frac{\eta}{15} = \frac{7}{6} = \mathfrak{G} \qquad \frac{\pi}{31} = \frac{8615841276005}{14322} = \mathfrak{P}$$

$$\frac{\theta}{17} = \frac{3617}{510} = \mathfrak{H} \qquad \text{etc.}$$

When we compare Euler's numerical values with Bernoulli's formulas for sums of powers, we see that indeed the numbers $\mathfrak{A}, \mathfrak{B}, \mathfrak{C}, \ldots$, which Euler defines by respectively dividing his $\alpha, \beta, \gamma, \ldots$ by $3, 5, 7, \ldots$, do appear to agree (modulo alternating signs) with the numbers appearing in Bernoulli's formulas. In other words, recalling the notation we introduced for Bernoulli's numbers in the previous section, $\mathfrak{A} = B_2$, $\mathfrak{B} = -B_4$, $\mathfrak{C} = B_6$, $\ldots$. But is this really a valid general pattern? Euler derives quadratic recursive formulas for $\mathfrak{A}, \mathfrak{B}, \mathfrak{C}, \ldots$ in §§119–123, but these are nothing like the linear recursive formulas that Bernoulli gave for his numbers. Neither has Euler yet related $\mathfrak{A}, \mathfrak{B}, \mathfrak{C}, \ldots$ to formulas for sums of powers. So we are not yet convinced that Euler's numbers fully agree with Bernoulli's. Fortunately, Euler will confirm their equality in §§130–132 below, as part of his first major application of the summation formula, to sums of numerical powers[29] (Exercise 1.53).

---

[29] We saw earlier, in §121, Euler's interest in the integrality properties of his new numbers, in particular the nature of their denominators. These properties are fascinating and useful. For instance, in 1840 it was proven by Clausen and von Staudt that, when reduced to lowest terms, the denominator of B_{2n} is precisely the product of all primes p for which $p-1$ divides $2n$. The observation implicit in Euler's table in §121 is a weaker version of this (Exercise 1.54). The numerators of Bernoulli numbers are more elusive, but just as important. In 1850 Ernst Kummer (1810–1893) proved Fermat's last theorem (that $x^p + y^p = z^p$ has no solutions in natural numbers for $p > 2$) for all prime exponents p that do not divide the numerator of any $B_{2n}/2n$ for $2n < p-1$. In the latter half of the twentieth century, both numerators and denominators of Bernoulli numbers have provided answers to important questions in the study of global shapes in high-dimensional surfaces, in the fields of differential and algebraic topology [169, Appendix B]. The reader is invited to prove a number-theoretic integrality property of Bernoulli numbers in Exercise 1.55.

But first Euler uses $\mathfrak{A}, \mathfrak{B}, \mathfrak{C}, \ldots$ to solve a problem dear to his heart, determining the precise sums of all the infinite series

$$1 + \frac{1}{2^{2n}} + \frac{1}{3^{2n}} + \frac{1}{4^{2n}} + \frac{1}{5^{2n}} + \text{etc.,}$$

of reciprocal even powers. He does this in §§124–125 by relating these series to the cotangent function, and thus to $\mathfrak{A}, \mathfrak{B}, \mathfrak{C}, \ldots$, since we have already seen him relate $A, B, C, \ldots$, and thus $\mathfrak{A}, \mathfrak{B}, \mathfrak{C}, \ldots$, to the cotangent. His impressive result is that

$$\sum_{i=1}^{\infty} \frac{1}{i^{2n}} = \frac{(-1)^{n+1} B_{2n} 2^{2n-1}}{(2n)!} \pi^{2n} \quad \text{for all } n \geq 1.$$

Then in §129 Euler uses this wonderful relationship to establish asymptotically how the Bernoulli numbers change in size as one moves outward in the sequence, which will be important in applying the summation formula later. He simply observes that when n is large, the series sums are increasingly close to one. So from the formula above, the ratio of consecutive Bernoulli numbers is asymptotically approximately[30]

$$\frac{B_{2n+2}}{B_{2n}} \approx -\frac{(2n+2)(2n+1)}{4\pi^2} \approx -\frac{n^2}{\pi^2}.$$

And he comments that the Bernoulli numbers thus "form a highly diverging sequence, which grows more strongly than any geometric sequence of growing terms," i.e., faster than r^n for any fixed r (Exercise 1.56). This completes Euler's analysis of the properties of the Bernoulli numbers themselves, and he is now ready to turn his summation formula toward applications.

Euler begins by clearly displaying for the first time the actual prominence and values of the Bernoulli numbers in the formula.

∞∞∞∞∞∞∞∞∞∞

130. Thus if one has found the numbers α, β, γ, δ etc., or $\mathfrak{A}$, $\mathfrak{B}$, $\mathfrak{C}$, $\mathfrak{D}$ etc., then given a series whose general term z is a function of its index x, the summative term Sz can be expressed as follows:

$$Sz = \int z\,dx + \frac{1}{2}z + \frac{1}{6} \cdot \frac{dz}{1 \cdot 2dx} - \frac{1}{30} \cdot \frac{d^3 z}{1 \cdot 2 \cdot 3 \cdot 4dx^3}$$
$$+ \frac{1}{42} \cdot \frac{d^5 z}{1 \cdot 2 \cdot 3 \cdots 6dx^5} - \frac{1}{30} \cdot \frac{d^7 z}{1 \cdot 2 \cdot 3 \cdots 8dx^7}$$
$$+ \frac{5}{66} \cdot \frac{d^9 z}{1 \cdot 2 \cdot 3 \cdots 10dx^9} - \frac{691}{2730} \cdot \frac{d^{11} z}{1 \cdot 2 \cdot 3 \cdots 12dx^{11}}$$
$$+ \frac{7}{6} \cdot \frac{d^{13} z}{1 \cdot 2 \cdot 3 \cdots 14dx^{13}} - \frac{3617}{510} \cdot \frac{d^{15} z}{1 \cdot 2 \cdot 3 \cdots 16dx^{15}}$$
$$\cdots \text{ etc.}$$

---

[30] By asymptotically approximate equality $a \approx b$ we mean here that the ratio a/b approaches 1 as n approaches infinity.

Thus if one knows the integral $\int z dx$, or the quantity whose differential is $= z dx$, one finds the summative term by means of continuing differentiation. One must not neglect that a constant value must always be added to this expression, of a nature that the sum will $= 0$ when x becomes 0.

131. If now z is an integral rational function[31] of x, so that the derivatives eventually vanish, then the summative term is represented by a finite expression. We illustrate this by some examples.

First example.
Find the summative term of the following series.

$$1 \quad 2 \quad 3 \quad 4 \quad 5 \qquad\qquad x$$

$$1 + 9 + 25 + 49 + 81 + \cdots + (2x - 1)^2$$

Since here $z = (2x - 1)^2 = 4xx - 4x + 1$, one has

$$\int z dx = \frac{4}{3}x^3 - 2x^2 + x,$$

because from this, differentiation produces $4xx dx - 4x dx + dx = z dx$. Further differentiation yields

$$\frac{dz}{dx} = 8x - 4, \quad \frac{ddz}{dx^2} = 8, \quad \frac{d^3 z}{dx^3} = 0, \quad \text{etc.}$$

So the summative term sought equals

$$\frac{4}{3}x^3 - 2x^2 + x + 2xx - 2x + \frac{1}{2} + \frac{2}{3}x - \frac{1}{3} \pm \text{Const.},$$

in which the constant must remove the terms $\frac{1}{2} - \frac{1}{3}$, so

$$S(2x - 1)^2 = \frac{4}{3}x^3 - \frac{1}{3}x = \frac{x}{3}(2x - 1)(2x + 1).$$

So if one sets $x = 4$, the sum of the first four terms is given by

$$1 + 9 + 25 + 49 = \frac{4}{3} \cdot 7 \cdot 9 = 84.$$

∽∽∽∽∽∽∽∽∽

The reader may carry out a similar example in Exercise 1.57.

We arrive at Euler's first application of the summation formula, in which he proves Bernoulli's sums of powers formulas, simultaneously convincing us that the numbers occurring in his summation formula, $\mathfrak{A}$, $\mathfrak{B}$, $\mathfrak{C}$, $\mathfrak{D}$, etc., really are the same as those that arose via a different recursion relationship in Bernoulli's sums of powers formulas. He will show us that Bernoulli's sums of powers formulas are simply special instances of the summation formula itself.

---

[31] By this Euler means a polynomial.

ⓍⒹⓍⒹⓍⒹⓍⒹⓍⒹⓍⒹⓍⒹ

132. From this general expression for the summative term, the sum for powers of natural numbers, that we communicated in the first part (§§29 and 61), but which we could not prove at that time, follows very easily. Let us set $z = x^n$, so that $\int z\,dx = \frac{1}{n+1}x^{n+1}$, and differentiating,

$$\frac{dz}{dx} = nx^{n-1}, \quad \frac{ddz}{dx^2} = n(n-1)x^{n-2}, \quad \frac{d^3z}{dx^3} = n(n-1)(n-2)x^{n-3},$$

$$\frac{d^5z}{dx^5} = n(n-1)(n-2)(n-3)(n-4)x^{n-5},$$

$$\frac{d^7z}{dx^7} = n(n-1)\cdots(n-6)x^{n-7}, \text{etc.}$$

From this we deduce the following summative term corresponding to the general term x^n:

$$Sx^n = \frac{1}{n+1}x^{n+1} + \frac{1}{2}x^n + \frac{1}{6}\cdot\frac{n}{2}x^{n-1} - \frac{1}{30}\cdot\frac{n(n-1)(n-2)}{2\cdot3\cdot4}x^{n-3}$$

$$+\frac{1}{42}\cdot\frac{n(n-1)(n-2)(n-3)(n-4)}{2\cdot3\cdot4\cdot5\cdot6}x^{n-5}$$

$$-\frac{1}{30}\cdot\frac{n(n-1)\cdots\cdots\cdots(n-6)}{2\cdot3\cdots8}x^{n-7}$$

$$+\frac{5}{66}\cdot\frac{n(n-1)\cdots\cdots\cdots(n-8)}{2\cdot3\cdots10}x^{n-9}$$

$$-\frac{691}{2730}\cdot\frac{n(n-1)\cdots\cdots\cdots(n-10)}{2\cdot3\cdots12}x^{n-11}$$

$$+\frac{7}{6}\cdot\frac{n(n-1)\cdots\cdots\cdots(n-12)}{2\cdot3\cdots14}x^{n-13}$$

$$-\frac{3617}{510}\cdot\frac{n(n-1)\cdots\cdots\cdots(n-14)}{2\cdot3\cdots16}x^{n-15}$$

$$\cdots$$

etc.

This expression differs from the former [i.e., that in §§29 and 61] only in that here we have introduced the Bernoulli numbers $\mathfrak{A}$, $\mathfrak{B}$, $\mathfrak{C}$, $\mathfrak{D}$ etc., whereas above we used the numbers α, β, γ, δ etc.; the agreement is clear. Thus here we have been able to give the summative terms for all powers up to the thirtieth, inclusive; if we wanted to perform this investigation via other means, lengthy and tedious calculations would be necessary.

ⓍⒹⓍⒹⓍⒹⓍⒹⓍⒹⓍⒹⓍⒹ

Of course for any fixed n, the derivatives beyond the nth will all vanish, so this sum is finite. Moreover, as Euler discussed above in §§131 and 132, the formula must be adjusted by a constant to yield 0 when $x = 0$. In other words, the constant term coming from the nth derivative on the right side must be removed. Thus Euler has proved precisely Bernoulli's claimed polynomial formula for the sum $\sum_{i=1}^{x} i^n$. Are Euler's numerical coefficients $\frac{1}{2}, \frac{1}{6}, -\frac{1}{30}$, etc., necessarily all the same as the numbers Bernoulli claimed in his recursive formulas? Yes, because by setting $x = 1$ here, just as Bernoulli did, we see that Euler's coefficients are governed by the same recursive formulas as Bernoulli's.[32]

Thus Euler's summation formula has completed the solution of the ancient problem of finding formulas for sums of powers, reducing it to understanding the Bernoulli numbers. But, as we shall see in the last section of our chapter, Euler also had in mind many other applications for his summation formula, in particular for finding sums of infinite series, like the reciprocal squares of the Basel problem.

Exercise 1.48. In his paper *De summis serierum reciprocarum* of the early 1730s [66, v. 14, pp. 73–86], Euler uses infinite product expansions to find sums of many series by ingenious methods. We have seen how he matches the coefficients of x^2 on both sides of the equation

$$\left(1 - \frac{x^2}{3!} + \frac{x^4}{5!} - \cdots\right) = \frac{\sin x}{x}$$

$$= \left(1 - \frac{x^2}{\pi^2}\right)\left(1 - \frac{x^2}{4\pi^2}\right) \cdots \left(1 - \frac{x^2}{i^2\pi^2}\right) \cdots \left(1 - \frac{x^2}{j^2\pi^2}\right) \cdots$$

to obtain $\sum_{i=1}^{\infty} i^2 = \frac{\pi^2}{6}$. Let us go further, as did Euler, to match the coefficients of x^4. Multiplying out the right side and matching with the left produces $\frac{1}{5!} = \frac{1}{\pi^4} \sum_{j>i\geq1} \frac{1}{i^2 j^2}$. This allowed him to find the sum of $\sum_{i\geq1} \frac{1}{i^4}$ by doing some "infinite algebra" to relate these two sums via the binomial theorem:

$$\left(\sum_{i\geq1} \frac{1}{i^2}\right)^2 = \left(\frac{1}{1^2} + \frac{1}{2^2} + \cdots + \frac{1}{i^2} + \cdots + \frac{1}{j^2} + \cdots\right)^2$$

$$= \sum_{i\geq1} \frac{1}{i^4} + 2 \sum_{j>i\geq1} \frac{1}{i^2 j^2} \ .$$

Use this equation to find the exact value of $\sum_{i\geq1} \frac{1}{i^4}$.

This coefficient matching was the beginning of an inductive process aiming for sums of higher and higher reciprocal even powers.

Exercise 1.49. Modify Euler's approach by using the cosine function instead of the sine, and factoring it according to its roots. Carry out an analysis similar

---

[32] Formally one can actually reverse one's entire view, to derive Euler's general summation formula from Bernoulli's formulas for sums of powers, by expanding the function z of x in its Taylor series and applying Bernoulli's formulas to the sums of the various powers of x in the expansion.

to what was done above for the sine, and find the precise sums of the following series:

$$(a) \sum_{n=0}^{\infty} (2n+1)^{-2} \qquad (b) \sum_{n=0}^{\infty} (2n+1)^{-4}.$$

Exercise 1.50. The similarity between the last equation Euler derives in our excerpt from §106 and what we called *Pascal's equation* near the end of the previous section is more than one might expect. Apply Euler's equation to $y = x^{k+1}$, write the terms using binomial coefficients, and compare it with Pascal's equation. Notice that adding the two together causes half the sums of powers on the right to cancel, and thus leads to an equation in which half the powers do not appear, enabling quicker calculations. For instance, use the resulting equation to find the formula for a sum of cubes without ever knowing a sum of squares.

Exercise 1.51. Prove that the polynomial formula for $\sum_{i=1}^{n} i^k$ always has $n(n+1)(2n+1)$ as a factor for k even and at least 2, and always has $n^2(n+1)^2$ as a factor for k odd and at least 3. Hint: Prove each by induction on k using the equation derived in Exercise 1.50. Analyze the roots of the function consisting of the combined terms in the equation that are not sums of powers greater than one. Show that it has the desired linear factors. Use calculus to confirm repeated roots.

Exercise 1.52. Following Euler, use his equation derived in §106 to obtain the formula for a sum of fourth powers from the known formulas for lower powers. Pay attention to the constant A.

Exercise 1.53. Combine the excerpts from sections 114 to 122 to obtain what is called the generating function for the Bernoulli numbers:

$$\frac{x}{e^x - 1} = 1 - \frac{x}{2} + \sum_{n=2}^{\infty} \frac{B_n}{n!} x^n.$$

(Hint: Beware of Greek letters bearing two different meanings.) Then calculate the Bernoulli numbers B_2, B_3, B_4 by expanding the left side in a power series, and comparing with the right. (Hint: Expand e^x and perform long division algebraically.)

Exercise 1.54. Show that the observation implicit in Euler's table in §121 about the denominators of Bernoulli numbers is a consequence of the theorem of Clausen and von Staudt that, when reduced to lowest terms, the denominator of B_{2n} is the product of all primes p for which $p - 1$ divides $2n$.

Exercise 1.55. Prove that for each n, $2^{2n}(2^{2n} - 1)B_{2n}/2n$ is an integer. Hint: From Euler's two expressions for s in §118, and the relationship to Bernoulli numbers he gives in succeeding sections, derive a power series expression for $\cot z$ where the coefficients are in terms of Bernoulli numbers. Now derive the trigonometric identity $\tan z = \cot z - 2 \cot 2z$, and use it to obtain a power series for $\tan z$. Finally, consider the successive derivatives of $\tan z$ at 0. (More generally, in this result 2^{2n} can be replaced by k^{2n} for any integer k [169, Appendix B].)

Exercise 1.56. Use Euler's determination of the sums of reciprocal even powers to obtain the estimate

$$B_{2n} \approx \frac{(-1)^{n+1}(2n)!}{2^{2n-1}\pi^{2n}}$$

for the Bernoulli numbers, and explore the accuracy of this estimate against some actual Bernoulli numbers.

Exercise 1.57. Euler's second example in §131 is to find a closed formula for a sum of odd cubes

$$1 + 27 + 125 + 343 + \cdots + (2x - 1)^3.$$

Carry out the details, and check your answer against the first four terms added by hand, as he did.

Then work out how Pascal's technique would compute this latter sum.

1.6 Euler Solves the Basel Problem

In the next chapter of the book *Institutiones Calculi Differentialis*, Euler embarks on new territory, applying the summation formula to obtain extraordinarily accurate approximations for sums of a variety of infinite series and their finite partial sums, as well as for other values of interest, such as π. We present here his inspiring approximation for the Basel problem on the sum of reciprocal squares, and mention some of his other applications that the reader may explore. Finally we discuss the profound influence on the development of modern mathematics from Euler's study of sums of reciprocal powers and Bernoulli numbers.

∞∞∞∞∞∞∞∞

Leonhard Euler, from
Foundations of Differential Calculus

Part Two, Chapter 6
On the summing of progressions via infinite series

140. The general expression that we found in the previous chapter for the summative term of a series, whose general term corresponding to the index x is z, namely

$$Sz = \int z\,dx + \frac{1}{2}z + \frac{\mathfrak{A}dz}{1 \cdot 2dx} - \frac{\mathfrak{B}d^3z}{1 \cdot 2 \cdot 3 \cdot 4dx^3} + \frac{\mathfrak{C}d^5z}{1 \cdot 2 \cdots 6dx^5} - \text{etc.},$$

actually serves to determine the sums of series whose general terms are integral rational functions of the index x, because in these cases one eventually arrives at vanishing differentials. On the other hand, if z is not such a function of x, then the differentials continue without end, and there results an infinite series that expresses the sum of the given series up to and including the term whose index $= x$. The sum of the series, continuing without end, is thus given by taking $x = \infty$, and one finds in this way another infinite series equal to the original.

141. If one sets $x = 0$, the expression represented by the series must vanish, as we already noted; and if this does not occur, one must add to or take away from the sum a constant amount, so that this requirement is satisfied. If this is the case, then when $x = 1$ one obtains the first term of the series, when $x = 2$ the sum of the first and second, when $x = 3$ the sum of the first three terms of the series, etc. Because in these cases the sum of the first, first two, first three, etc. terms is known, this is also the value of the infinite series expressing the sum; and thus one is placed in a position to sum innumerably many series.

142. Since when a constant value is added to the sum, so that it vanishes when $x = 0$, the true sum is then found when x is any other number, then it is clear that the true sum must likewise be given whenever a constant value is added that produces the true sum in any particular case. Thus suppose it is not obvious, when one sets $x = 0$, what value the sum assumes and thus what constant must be used; one can substitute other values for x, and through addition of a constant value obtain a complete expression for the sum. Much will become clear from the following.

⁐⁐⁐⁐⁐⁐⁐⁐⁐

The first application Euler makes of his summation formula to an infinite series, in §§142a–144, is to the diverging harmonic series $\sum_{i=1}^{\infty} \frac{1}{i}$. This leads to the constant bearing his name, today denoted by γ, arguably the most important special constant in all of mathematics after π and e. The Euler constant is the limiting difference between $\sum_{i=1}^{x} \frac{1}{i}$ and $\ln x$ as x approaches infinity. Euler extracts from the summation formula (which also diverges) an approximation of γ accurate to 15 places, and then easily obtains the sum of the first thousand terms of the diverging harmonic series to 13 places. In fact it is clear from what he writes that one could use his approach to find the constant γ to whatever accuracy desired, and then apply the summation formula to find the value of arbitrarily large finite harmonic sums to that same accuracy.[33]

Let us now see how Euler applies the summation formula to that old puzzle, the Basel problem.

⁐⁐⁐⁐⁐⁐⁐⁐⁐

148. After considering the harmonic series we wish to turn to examining the series of reciprocals of the squares, letting

$$s = 1 + \frac{1}{4} + \frac{1}{9} + \frac{1}{16} + \cdots + \frac{1}{xx}.$$

---

[33] Approximations of this nature are relevant even today, since we still do not know whether Euler's constant γ is rational or irrational!

Since the general term of this series is $z = \frac{1}{xx}$, then $\int z\,dx = -\frac{1}{x}$, the differentials of z are

$$\frac{dz}{2dx} = -\frac{1}{x^3}, \quad \frac{ddz}{2 \cdot 3dx^2} = \frac{1}{x^4}, \quad \frac{d^3 z}{2 \cdot 3 \cdot 4dx^3} = -\frac{1}{x^5} \quad \text{etc.,}$$

and the sum is

$$s = C - \frac{1}{x} + \frac{1}{2xx} - \frac{\mathfrak{A}}{x^3} + \frac{\mathfrak{B}}{x^5} - \frac{\mathfrak{C}}{x^7} + \frac{\mathfrak{D}}{x^9} - \frac{\mathfrak{E}}{x^{11}} + \text{etc.,}$$

where the added constant C is determined from one case in which the sum is known. We therefore wish to set $x = 1$. Since then $s = 1$, one has

$$C = 1 + 1 - \frac{1}{2} + \mathfrak{A} - \mathfrak{B} + \mathfrak{C} - \mathfrak{D} + \mathfrak{E} - \text{etc.,}$$

but this series alone does not give the value of C, since it diverges strongly.

◯◯◯◯◯◯◯◯◯◯◯

On the face of it, these formulas seem absurd. The expression Euler obtains for the "constant" is clearly a divergent series, because of the growth of the Bernoulli numbers, which we saw Euler analyze earlier. In fact, the summation formula above diverges for every value of x because of the supergeometric swiftness of growth that he established (Exercise 1.58). Euler, however, is not fazed by this seemingly dismal situation. He has a plan for obtaining from such divergent series highly accurate approximations for both very large finite sums and infinite sums.

Euler's idea is to add up the terms in the summation formula only "until they begin to diverge" [245, p. 261], but not beyond. For those unfamiliar with the theory of divergent series, this may seem preposterous, but in fact it has sound theoretical underpinnings, discovered later. Euler's entire approach was ultimately fully vindicated [104, 106, 116, 135, 137]. Euler himself was probably sure of his work, despite its apparently shaky foundations in divergent series, because he was continually checking and rechecking his results by a variety of theoretical and computational methods, boosting his confidence in their correctness from many different angles. This is a hallmark of an excellent mathematician. When doing mathematics one should continually be checking one's results against the store of existing knowledge. Let us see Euler do this for the sum of reciprocal squares.

First he notes that for this particular function, he already knows the value of C by other means.

◯◯◯◯◯◯◯◯◯◯◯

Above [§125] we demonstrated that the sum of the series to infinity is $= \frac{\pi\pi}{6}$, and therefore setting $x = \infty$, and $s = \frac{\pi\pi}{6}$, we have $C = \frac{\pi\pi}{6}$, because then all other terms vanish. Thus it follows that

$$1 + 1 - \frac{1}{2} + \mathfrak{A} - \mathfrak{B} + \mathfrak{C} - \mathfrak{D} + \mathfrak{E} - \text{etc.} \; = \frac{\pi\pi}{6}.$$

∞∞∞∞∞∞∞∞

We need not fret about exactly what Euler means when he claims that the sum of an obviously diverging series has a specific value. He assigns values to divergent series based on the expressions from which he creates them, as in this example. This too was a forerunner of things to come, as later mathematicians developed various meaningful ways of assigning values to divergent series [106, 137].

Next Euler pretends he doesn't already know the sum of the infinite series of reciprocal squares, and approximates it using his summation formula, thereby performing a cross-check on both methods.

∞∞∞∞∞∞∞∞

149. If the sum of this series were not known, then one would need to determine the value of the constant C from another case, in which the sum were actually found. To this aim we set $x = 10$ and actually add up ten terms, obtaining[34]

$$s = 1,549767731166540690.$$

$$\text{Further, add } \frac{1}{x} = 0,1$$

$$\text{subtr.} \quad \frac{1}{2xx} = 0,005$$
$$\overline{1,644767731166540690}$$

$$\text{add} \quad \frac{\mathfrak{A}}{x^3} = 0,000166666666666666$$
$$\overline{1,644934397833207356}$$

$$\text{subtr.} \quad \frac{\mathfrak{B}}{x^5} = 0,000000333333333333$$
$$\overline{1,644934064499874023}$$

$$\text{add} \quad \frac{\mathfrak{C}}{x^7} = 0,000000002380952381$$
$$\overline{1,644934066880826404}$$

$$\text{subtr.} \quad \frac{\mathfrak{D}}{x^9} = 0,000000000033333333$$
$$\overline{1,644934066847493071}$$

$$\text{add} \quad \frac{\mathfrak{E}}{x^{11}} = 0,000000000000757575$$
$$\overline{1,644934066848250646}$$

$$\text{subtr.} \quad \frac{\mathfrak{F}}{x^{13}} = 0,000000000000025311$$
$$\overline{1,644934066848225335}$$

$$\text{add} \quad \frac{\mathfrak{G}}{x^{15}} = 0,000000000000001166$$

$$\text{subtr.} \quad \frac{\mathfrak{H}}{x^{17}} = \frac{71}{1,644934066848226430} = C.$$

This number is likewise the value of the expression $\frac{\pi\pi}{6}$, as one can find by calculation from the known value of π. From this it is clear that, although the series $\mathfrak{A}$, $\mathfrak{B}$, $\mathfrak{C}$, etc., diverges, it nevertheless produces a true sum.

---

[34] Note that Euler uses commas (as still done in Europe today) rather than points, for separating the integral and fractional parts of a decimal.

ᏳᏳᏳᏳᏳᏳᏳᏳᏳᏳ

Recall that this series diverges due to the rapid growth of Bernoulli numbers. But note that the terms he actually calculates appear to decrease rapidly, giving the initial appearance, albeit illusory, that the series converges. A closer examination of the terms shows that their decrease is slowing in a geometric sense, which hints at the fact that the series actually diverges. Recall that Euler intends to sum terms only "until they begin to diverge." How does he decide when this occurs? Notice that the series alternates in sign, and thus the partial sums bounce back and forth, apparently at first narrowing in, then broadening out as the terms themselves eventually increase due to rapid growth of the Bernoulli numbers. Euler knows to stop before the smallest bounce, with the expectation that the true sum he seeks always lies between any partial sum and the next one, and is thus bracketed most accurately if he stops just before the smallest term included. It is striking that the summation formula behaves exactly this way for many functions, including all the ones Euler is interested in. In fact, these are examples of what we today call *asymptotic series*, which are divergent, but diverge more and more slowly for larger and larger values of x and can be used for valid approximations [106, 116, 137], [135, Chapter 47]. Today asymptotic series are important in much of pure and applied mathematics, in particular the application of differential equations to applied problems.

Nineteenth-century mathematicians wrestled with the validity, theory, and usefulness of divergent series. Two (divergent) views reflected this struggle:

> The divergent series are the invention of the devil, and it is a shame to base on them any demonstration whatsoever. By using them, one may draw any conclusion he pleases and that is why these series have produced so many fallacies and so many paradoxes. ... I have become prodigiously attentive to all this, for with the exception of the geometrical series, there does not exist in all of mathematics a single infinite series the sum of which has been determined rigorously. In other words, the things which are most important in mathematics are also those which have the least foundation. ... That most of these things are correct in spite of that is extraordinarily surprising. I am trying to find a reason for this; it is an exceedingly interesting question. Niels Abel (1802–1829) 1826 [135, p. 973f].

> The series is divergent; therefore we may be able to do something with it. Oliver Heaviside (1850–1925) [135, p. 1096].

The terms in Euler's calculation above actually continue to decrease for several further steps after those he shows, which allows quite a number of additional places of accuracy in determining C if one wishes. Notice that the choice of $x = 10$ heavily influences how much accuracy can be obtained for C. A smaller choice for x would cause the summation formula to begin to diverge much sooner, and with a larger final bounce, yielding less known accuracy, while a larger x would ensure much slower divergence and great bounding accuracy for the answer, at the expense of having to compute a longer partial sum to get the calculation off the ground. In Euler's 1735 paper, in which he approximates the sum of reciprocal squares to 20 places, he also uses $x = 10$. It is profitable to explore how one can approximate the sum to any desired degree of accuracy by appropriate choices for x and for the number of terms used from the summation formula (Exercise 1.59).

The trade-offs involved in these choices are typical when one is using the summation formula, both for determining C and for applying the formula to find partial sums as well. When Euler applied the summation formula to analyze $C = \gamma$ for the harmonic series, he also used it to calculate the partial sum of the first million terms to 13 places. The reader may easily do even better for partial sums of reciprocal squares (Exercise 1.60). Delightfully, for calculating partial sums, the larger x is, the better and quicker the result from the summation formula.

As we have discussed above, Euler was able to determine the exact sums of all the infinite series of reciprocal even powers of the natural numbers in terms of the Bernoulli numbers and π. He also would have loved to find similar formulas for the reciprocal odd powers, and in §§150–153 he explores these using his summation formula. He produces highly accurate decimal approximations for the sums of reciprocal odd powers all the way through the fifteenth, hoping to see a pattern analogous to that for even powers, namely simple fractions times the relevant power of π. The first such converging series is the sum of reciprocal cubes $\sum_{i=1}^{\infty} \frac{1}{i^3}$. Euler computes it accurately to seventeen decimal places. He is disappointed, however, to find that it does not appear to be an obvious rational multiple of π^3, nor does he have better luck with the other odd powers (Exercise 1.61).

The reader may wish to explore still other applications Euler makes of his summation formula. In §§154–156 Euler uses the inverse tangent and cotangent functions to approximate π to seventeen decimal places with his summation formula, and remarks that it is amazing that one can approximate π so accurately with such an easy calculation. This is an enticing topic for further investigation (Exercise 1.62).

Finally, in §§157–162 Euler uses the summation formula to approximate both sums of logarithms and then (by exponentiating) large factorials in the forms known as *Stirling's approximation* and *Stirling's series*. These in turn lead to approximations for large binomial coefficients. For instance, if one tosses 100 coins, the probability that exactly equal numbers of coins will land heads and tails is the ratio $\binom{100}{50}/2^{100}$, but computing this number accurately is clearly a challenge. Euler explicitly computes this probability to be 1 in $12.56451\ldots$, which could be useful for a bet with friends, or in Las Vegas today.

We have now looked in some detail at how Euler derived his summation formula, studied the Bernoulli numbers, proved Bernoulli's sums of powers formulas, and applied the summation formula to approximate sums of series of reciprocal powers. We end by describing the impact today generated by Euler's study of sums of reciprocal powers and Bernoulli numbers.

Euler succeeded in finding the precise sums of all infinite series of reciprocal even powers in terms of π and Bernoulli numbers. He tried, but failed, to find the sum of the reciprocal cubes and other odd powers. Even today we know little about these sums of odd powers, although not for lack of trying. All these series are instances of the single formula

$$\zeta(z) = \sum_{i=1}^{\infty} \frac{1}{i^z}.$$

Euler recognized that this function, ζ ("zeta") of z, is extremely important, not just for natural number values for z, like those we have been considering, but for arbitrary real values of z, which he realized could actually make sense.

The zeta function is in fact key to many of the secrets of numbers, in particular the distribution of prime numbers. The fundamental connection to primes noticed by Euler is embodied in his product formula

$$\zeta(z) = \prod_{p \text{ prime}} (1 - p^{-z})^{-1},$$

where the infinite product it taken over all prime numbers. Euler's product formula is not hard to demonstrate, at least formally (Exercise 1.63). The distribution of prime numbers is a most elusive subject. Euler himself wrote:

> Mathematicians have tried in vain to this day to discover some order in the sequence of prime numbers, and we have reason to believe that it is a mystery into which the human mind will never penetrate. To convince ourselves, we have only to cast a glance at tables of primes, which some have taken the trouble to compute beyond a hundred thousand, and we should perceive at once that there reigns neither order nor rule [258, p. 301].

Euler studied both $\zeta(z)$ and related series for values of z ranging over both positive and negative integers, and even for certain fractional values of z. He compared the values of these series with each other for various combinations of z. Even when such series diverge, as do the series above for $\zeta(z)$ when $z \leq 1$, Euler sought to give them meaning, by interpreting them as limiting values of convergent power series. Euler came up with an amazing claim, which provided a systematic comparison between different values of the zeta function.

Before stating Euler's claim, we must mention that he had already found a way to generalize the factorial function $n!$ to a function which we shall call $\Pi(n)$. This new function is valid for all real numbers n except the negative integers, and satisfies the same property $\Pi(n) = n\Pi(n-1)$ as the factorial for all real numbers n (except the nonpositive integers).[35] Euler then discovered such spectacular interpolation results as $\Pi(-1/2) = \sqrt{\pi}$ [56, pp. 7–8],

---

[35] Today we often use a slightly reparametrized version of this function, called the *gamma function*, defined by $\Gamma(n+1) = \Pi(n)$, even though the function $\Pi(n)$ seems more natural.

[104, pp. 259f, 360], [135, pp. 422–424]. Making use of this generalization Π of the factorial, Euler's claim about ζ essentially has the form

$$\zeta\left(1-z\right) = \pi^{-z}2^{1-z}\Pi(z-1)\cos\left(\frac{\pi z}{2}\right)\zeta(z).$$

Today this is called the *functional equation* of the zeta function, and it can be written in many slightly different forms [10], [56, pp. 12–15]. The equation relates certain values of zeta to certain other values, as did the functional equation $\Pi\left(n\right) = n\Pi\left(n-1\right)$ for the generalized factorial. Specifically, notice that comparing the value at $1-z$ to that at z involves reflection around $z = \frac{1}{2}$ (Exercise 1.64). Indeed some of the alternative forms of the functional equation emphasize this by expressing precisely an invariance under this symmetry [56, pp. 12–15]. Since from our Euler selections we know the values Euler calculated for $\zeta\left(2n\right)$ in terms of Bernoulli numbers (see after section 122), the functional equation now provides us the values of ζ at corresponding reflected negative integers (Exercise 1.65).

Euler essentially verified the functional equation for all integer values of z, and also for $z = \frac{1}{2}$ and $\frac{3}{2}$, by computing both sides independently. His computations for integers z used all the techniques and tricks he could muster from calculus to interpret certain diverging series as meaningful limits of converging power series representations of functions. For $z = \frac{1}{2}$ he used his previous determination that $\Pi\left(-1/2\right) = \sqrt{\pi}$, and for $z = \frac{3}{2}$ he even boldly approximated the relevant diverging series via the Euler–Maclaurin summation formula! All these results provide support, but not proof, for his claimed functional equation. A lovely description of Euler's work on the zeta function and how he was led to his claim is given in [10], and further engaging exposition about the function can be found in [106, pp. 23–26], [245, pp. 272–276], and of course in Euler's own writings. The early history of the development of the zeta function is described in general in [204].

For a hundred years Euler's functional equation for the zeta function was forgotten. Then in an 1859 memoir, Bernhard Riemann (1826–1866) provided the first proof of the functional equation. Today we call ζ the *Riemann zeta function*, due to his pathbreaking advances. Riemann first showed that the function $\zeta(z)$, which as defined by Euler for real z via $\sum_{i=1}^{\infty} 1/i^z$ converges only for $z > 1$, actually has a unique meaningful extension to all the complex numbers except $z = 1$. Of course Riemann's extension must in general be expressed in other ways than Euler's [56, pp. 9–11]. This is done by a modern method known as *analytic continuation*, and the function is expressed by integration using the complex numbers (see Exercise 1.66). The extended complex function ζ is of a type that we today call *complex analytic*, which means that is is amenable to all the tools of calculus, and also that it is expressible locally by power series. (This is the case, too, for $\Pi(n)$ wherever it is defined.) We warmly recommend [56] as a historically oriented introduction to Riemann's work and the fascinating world it leads to.

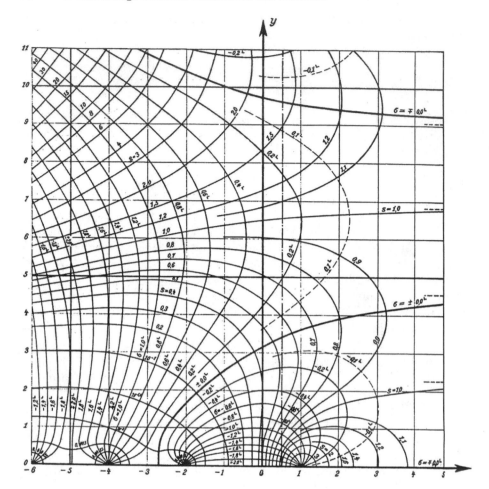

Fig. 1.11. Contours of the Riemann zeta function.

Although the discrete pattern of primes is still elusive today, as Euler predicted 250 years ago, some of their mystery has been unveiled by the complex-analytic study of the zeta function. In particular, Riemann showed that the location of its zeros provides information about the distribution of primes. He proved that, in the complex plane, all the zeros of the zeta function, except for certain obvious ones, called the *trivial zeros*, on the negative real axis (Exercise 1.67), lie in the vertical strip of numbers with real part between zero and one, inclusive. He then used this fact to estimate the number of primes in any interval of numbers.

Figures 1.11, 1.12, 1.13 [124] display information about zeta. As a complex-valued function of a complex variable, zeta is hard to display in a single picture. The displayed contour graphs sketch on the domain plane the level

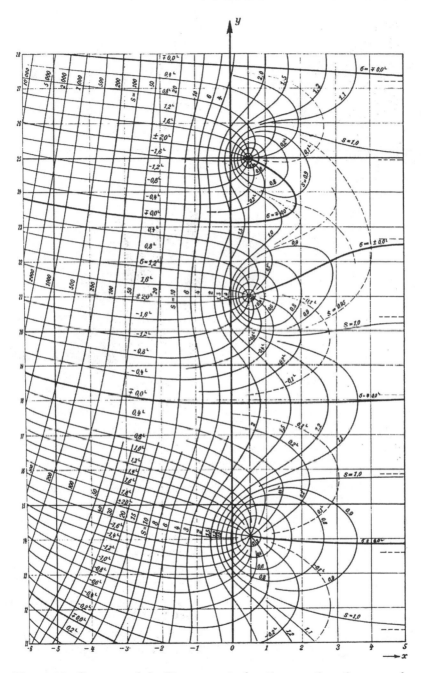

Fig. 1.12. Contours of the Riemann zeta function, continuation upwards.

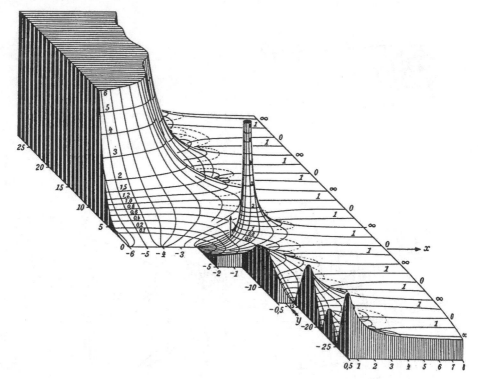

Fig. 1.13. Relief of the modulus of the Riemann zeta function. The lines of steepest gradient denoted by 0, on the right-hand side of the figure, come from a zero; those denoted by ∞ come from an infinity.

curves of the absolute value s (modulus) and of the argument σ (polar angle) of zeta. The argument level curves are labeled in right-angle units measured from the positive real axis. The relief graph is of the modulus of zeta. These contain a wealth of features, beginning with the trivial zeros along the negative real axis, the pole at 1 where ζ is undefined, and further zeros appearing along the vertical line with real part equal to one-half, called the *critical line*. The reader is invited to study these and also many wonderful graphical and interactive Internet pages about the Riemann zeta function (Exercise 1.68).

By the mid-nineteenth century, Riemann and others were focusing on a conjecture, first formulated by Carl Gauss (1777–1855) and Adrien-Marie Legendre (1752–1833) around the turn of the century. One form of this conjecture stated that if one lets $\pi(x)$ denote the number of primes not exceeding x, then $\lim_{x\to\infty} \frac{\pi(x)}{x/\ln x} = 1$. That is, the proportion $\frac{\pi(x)}{x}$ of prime numbers not exceeding x is approximately $1/\ln x$ [56, 226, 258].

Near the end of the century, in 1896, Jacques Hadamard (1865–1963) and Charles-Jean de la Vallée Poussin (1866–1962) each succeeded in proving this conjecture, now known as the *prime number theorem*, one of the most fundamental properties we know about primes. In fact, if we could obtain even more precise information about the location of the zeros of the zeta function, we would know much more about prime numbers. Riemann made his own conjecture, that in the plane of complex numbers all the nonreal zeros of zeta actually lie on the critical line. This conjecture, known today as the *Riemann hypothesis*, is perhaps the most famous unsolved problem in all of mathematics, and has driven the development of much of the modern branch of mathematics known as analytic number theory [6, 56, 226, 258]. Thus Euler's seminal work on sums of reciprocal powers, the Bernoulli numbers, and the zeta function formed the nucleus leading to some of the most vital research going on in mathematics today.

Exercise 1.58. Prove that the supergeometric rate of growth of the Bernoulli numbers, obtained by Euler in §129, forces his infinite series in §148 for a sum of reciprocal squares to diverge for all x.

Exercise 1.59. Using $x = 10$ in his summation formula, Euler gave the sum of reciprocal squares to 20 places. Do this by taking enough terms in the summation formula, and check your results against the exact sum, which we know is $\pi^2/6$. Then explore how to increase the accuracy in two different ways, either by using more terms of the summation formula, providing it does not start to diverge, or by increasing x, the number of terms of the sum of reciprocal squares calculated on the left side. How do these two approaches interact? Explain why this method can provide any arbitrary accuracy desired.

Exercise 1.60. Use the summation formula to find the sum of the first million terms in the sum of reciprocal squares to 17 decimal places. Hint: With C already determined to this accuracy, set $x = 1,000,000$ and add up the summation formula, as long as it does not begin to diverge. Compare this with how long it would take to find this sum by instead actually adding up the individual one million terms. Finally, can you think of how to adapt the summation formula to obtain any desired degree of accuracy for any partial sum?

Exercise 1.61. Follow in Euler's footsteps by using his summation formula to approximate $\sum_{i=1}^{\infty} \frac{1}{i^3}$ to seventeen decimal places. Euler obtained 1.202 056 903 159 594 28. Study whether it appears to be a simple rational multiple of π^3, as Euler had hoped it might be.

Exercise 1.62. Analyze Euler's approximation of π in §§154–156 [66, v. 10], [67], and extend it to compute a few more decimal places. Discuss its efficiency.

Exercise 1.63. Derive the Euler product formula

$$\zeta(z) = \prod_{p \text{ prime}} (1 - p^{-z})^{-1},$$

at least formally, by expanding each factor in the product as a geometric series, recalling that $\frac{1}{1-x} = \sum_{i=0}^{\infty} x^i$.

Exercise 1.64. Since the two evaluations $\zeta(1 - z)$ and $\zeta(z)$ in the functional equation for the zeta function must agree at the point of reflection $z = \frac{1}{2}$, check that all the remaining factors in the equation cancel out when $z = \frac{1}{2}$.

Exercise 1.65. Use the functional equation to determine the values of the zeta function at the negative odd integers.

Exercise 1.66. For the reader who has studied complex analysis: First read the description in [56, pp. 9–10] of the contour integral expressing the zeta function, valid on the entire complex plane, with a simple pole at $z = 1$. Then read and write a justification for every step of the explicit calculation there of the values of the zeta function at the nonpositive integers in terms of the Bernoulli numbers [56, pp. 11–12].

Exercise 1.67. Use the functional equation to determine the values of the zeta function at the negative even integers. Why can't you use the equation in the other direction to find the values of zeta at the positive odd integers? Can you use the functional equation to find the value of zeta at zero?

Exercise 1.68. Copy Figures 1.11, 1.12, 1.13 and color them to see some of their remarkable features. For each z in the complex plane, $\zeta(z)$, itself another point in the complex plane, has an angle, or argument, whose level curves are labeled with σ in Figures 1.11, 1.12. Interpret this angle as a color on the color wheel, and color the points z on the domain with the color corresponding to $\zeta(z)$. What do you notice is different between the zeros and the pole of ζ? Describe some other mathematical features you notice from your colorings.

2

Solving Equations Numerically: Finding Our Roots

2.1 Introduction

The formula

$$x_{n+1} = x_n - \frac{f(x_n)}{f'(x_n)} \tag{2.1}$$

is one of the most widely used algorithms in computers today, from the guidance systems for rockets to the calculation of orbits of heavenly bodies. From an initial guess x_0, one proceeds recursively to find $x_1, x_2, x_3, \dots$. This algorithm is often called *Newton's method*, but in this book we call it *Simpson's fluxional method*, since it first appears in a textbook of Thomas Simpson in 1740 using fluxions [214].

By applying the algorithm to the polynomial $f(x) = x^3 - 1$, whose well-known roots make the root-finding transparent, we obtain the following values when it is started at $x_0 = 2$.

n	x_n	$f(x_n)$	$f'(x_n)$	$f(x_n)/f'(x_n)$	x_{n+1}
0	2	7.0	12.0	0.58333 3333	1.41666 6667
1	1.41666 6667	1.84317 1296	6.02083 3333	0.30613 2257	1.11053 4410
2	1.11053 4410	0.36960 7290	3.69986 0026	0.09989 7641	1.01063 6768
3	1.01063 6768	0.03322 5093	3.06416 0032	0.01052 5211	1.00011 1557
4	1.00011 1557	0.00033 4709	3.00066 9381	0.00011 1544	1.00000 0012
⋮	⋮	⋮	⋮	⋮	⋮

The last column appears to be converging to 1, a root of f, with the number of significant digits doubling at each step. In general, these iterates should approach a root r of any reasonable function f, that is, $\lim_{n\to\infty} x_n = r$ with $f(r) = 0$.

An intuitive explanation of this algorithm is illustrated in Figure 2.1, where we have graphed the function f, together with several iterations. The hypotenuse of each triangle has a slope: $f'(x_n) = f(x_n)/(x_n - x_{n+1})$, from

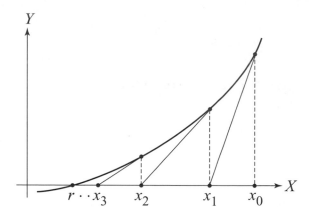

Fig. 2.1. Simpson's method (not to scale).

which equation the earlier formula (2.1) is immediately derived by solving for x_{n+1}.

How fast is this algorithm? Remarkably fast! This is part of its strong appeal. The technical term is *quadratic convergence*; this means that once the algorithm is underway, the number of new significant digits generally doubles at each iteration, as already observed in the example.

But how good is Simpson's method? Does it always converge to a root? And is it the root we want if there is more than one? Despite the method's apparent success in practice, one can easily construct all sorts of nasty counterexamples (Exercise 2.1). In our example of $f(x) = x^3 - 1$, the derivative $f'(x)$ is 0 at $x = 0$; so the tangent line will never meet the X-axis, and thus, with $x_0 = 0$, the iteration scheme (2.1) stops abruptly (Figure 2.2). Worse yet, there are infinitely many such deadbeats, i.e., values of x_0 that fail ever to reach a root. For instance, to the left of 0 where $x_0 = -1/\sqrt[3]{2}$, the tangent line in Figure 2.1 meets the X-axis at $x_1 = 0$, taking us back to the previous dead end! Now it is easy to keep moving leftward, finding more and more deadbeats. We simply work backwards by setting x_{n+1} to a previous undesirable and solving for x_n in (2.1). But what about the two well-known complex roots of $x^3 - 1$, and points in the complex plane that converge toward them, or fail to?

The swirling haunting pattern of "Fractal with Basins" on the color insert[1] displays the fate of initial points in the complex plane, and comes from a detailed examination of how Simpson's formula (2.1) bounces points around before aiming them at a root, if it ever does. The function is again $f(z) = z^3 - 1$, written now with a complex variable z. It sets its three roots, $1, -\frac{1}{2} + \frac{\sqrt{3}}{2}i$, and $-\frac{1}{2} - \frac{\sqrt{3}}{2}i$, symmetrically about the origin; their placement generates

---

[1] The fractal images were created with the freeware program Fractint, available on the internet.

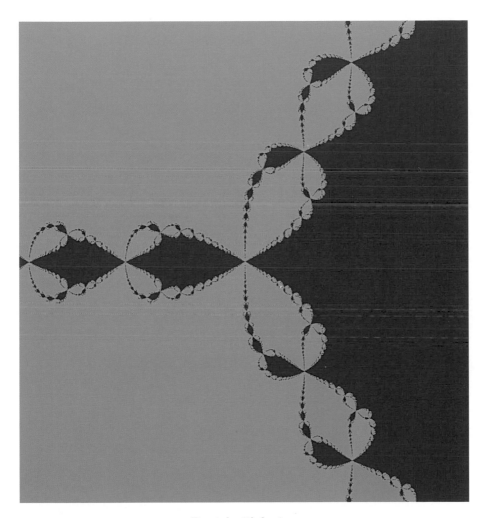

Fractal with basins.

Fractal with iterations.

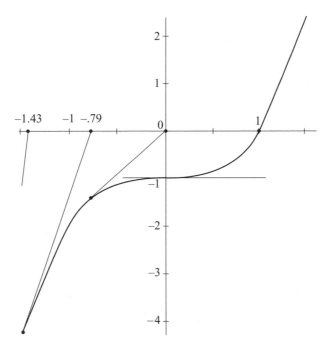

Fig. 2.2. Simpson's method fails.

rotational symmetry of the whole figure about the origin, which is at the center of the figure. The blue, green and aqua regions are the "attractor basins" for each of the three roots. Each attractor basin consists of all the initial points z_0 for which Simpson's method converges to a particular root. For instance, the basin of attraction for the root 1 is shown in blue, i.e., it consists of those initial points z_0 in the complex plane whose iteration will eventually lead to 1. Note the self-similarity of the flea-like blobs that repeat over and over in different sizes. The common boundary of the basins, consisting of those points for which the algorithm doesn't converge, is essentially invisible, but forms all the geometric interest in the picture.

Only in 1981 did Stephen Smale (1932–) prove that, in a probabilistic sense, the algorithm almost always converges [216]. Even better, he showed how to modify the algorithm to decrease the chances of a wrong answer or nonconvergence, by as much as one desires, but not totally eliminate the chances of bad behavior. The fractal pattern illustrates how diverse points may converge unexpectedly to roots far away, and how points arbitrarily close together may converge to roots far apart.

This chapter tells the story from antiquity to the present of how the problem of finding roots of functions has been tackled by diverse cultures, with their many differences in style, notation, method, and proof. Eight episodes highlight certain aspects of such numerical analysis, exemplified by the changing nature of algorithms and what each culture is willing to accept as numbers. These are snapshots of the art of coaxing numbers out of apparently intractable problems, at least intractable for the time they were posed.

Our episodes split naturally into two groups. The first four are exact methods and meant only for polynomials of limited degree: the Rhind papyrus of the Egyptians [35] written in hieroglyphics, Babylonian tablets [240] using the base 60, the work of the Persian 'Umar Khayyam (1044–1123/1124) [191] invoking conic sections, and the book of the Italian Girolamo Cardano (1501–1561) [31] employing a spatial version of the binomial theorem. Their approaches vary considerably. Roughly, the first is arithmetical; the second algebraic; 'Umar is mainly geometrical; and Cardano is geometrical algebra. But these classifications are slippery, and the beauty of going back to historical sources is to appreciate how fluid these modern distinctions are in ancient writing. Modern notation will help to explain them.

The last group starts with three algorithms that have much in common: Qin Chiu-Shao's (1202–1261) completion of powers [157] in Section 2, and Isaac Newton's (1642–1727) proportional method [178] in Section 3, culminating in Thomas Simpson's (1710–1761) fluxional method [214] in Section 4; it ends in Section 5 with Stephen Smale's justification [216] of Simpson's method, together with a flourish of chaos theory, of which fractals, like that generated by $x^3 - 1$, are a special manifestation. Their techniques are numerical, proceeding by incremental adjustments towards a solution. They work for polynomials of all degrees; in fact, the last two work for a wide range of arbitrary functions. All three algorithms have a similar modus operandi: linearize to find the next increment. But the linearization is done quite differently in each, with the result that the rate of convergence is linear for Qin, quadratic for Simpson, and somewhere in between for Newton. We present their unique approaches in their own words.

As noted before, recorded attempts and successes at solving equations stretch back almost four millennia. This is a story of false credit, dead ends and surprise, such as the unsolvability by radicals of polynomial equations of degree greater than four ([150, Chapter 5]). It is a long story, with too many strands to relate them all; neglected are how to locate and isolate roots within broad intervals before iteration begins, so as to ensure convergence to the right root. It is a braided stream, twisting and turning unexpectedly, with some rivulets drying up historically, such as the method of Qin, and others gathering strength, eroding the banks of accepted notions of numbers, and overflowing into new concepts of negative, imaginary and transcendental numbers, such as Cardano's confrontation with complex numbers.

It would be nice to claim that these episodes progress in an unbroken chain of advancing mathematics. But so much of the historical record is missing that we can not support such a claim. Although historical connections between the algorithms are tenuous, the reader should observe the generally increasing sophistication of the arguments. Let us read these for the variety of methods and the different styles of the various cultures. Numerical methods, geometrical proofs, algebraic formulas and even the calculus all make their entrances. Not all of the methods are purely one style; 'Umar's and Cardano's are mixed.

As the reader already knows, solving an equation may be tricky. What do we mean by a solution, and what is acceptable as a number: a whole number, a fraction, a decimal expansion (finite or infinite in extent), a closed formula, or even a geometrical entity? This leads to the question, when do algorithms force one into accepting new numbers? The gradual uncovering of anomalous solutions, such as negative, irrational and complex, drove in turn the evolution of the notion of number. Which leads to another question: how does the nature of the algorithm and the existence of a solution depend on what numbers are available? Conversely, what is a particular culture willing to accept as a solution? The notion of number changes over time and the idea of what we are doing also changes from solving verbal equations verbally through displaying the roots as algebraic formulas to iterating increments calculated from derivatives. There is much that we take for granted today that was murky in the past. For this reason, the road to finding roots of functions by Simpson's fluxional method is often cloudy and muddy (Exercise 2.2).

The mathematical techniques viable at a given time often determine the nature of algorithms. Simpson could conceive of (2.1) only after the invention of the calculus and the emergence of the notion of a function. The endpoints of this story are a study in contrasts. Compare the photograph in Figure 2.3 of the Egyptian hieroglyphics for solving only linear equations to the first color fractal image illustrating the basins of attraction of Simpson's fluxional method, applicable to all polynomials and most functions. One could not draw this picture without the concept of complex numbers. In fact, one could not create it without computers.

LINEAR EQUATIONS IN ANCIENT EGYPT. Let's start with the simplest equation: a linear equation, $ax = b$. It is easy to solve: just perform a division. But this is easy only if one has the concept of a fraction, which the ancient Egyptians did not. They found solving even linear equations to be a trial [35]. They had ratios of two numbers but not fractions that represented single numbers. What they did have were only *unit fractions*: fractions with 1 in the numerator, such as $\frac{1}{2}$, $\frac{1}{5}$, or $\frac{1}{17}$, when expressed in modern notation. This makes elementary arithmetic operations difficult to perform. The result of dividing one number by another had to be expressed as an integer plus a sum of unit fractions.

Unit fractions would appear to be a natural step on the way from whole numbers to numbers in between them. If one divides and it does not come out even, one could add a small division of the unit interval, and then a smaller one as necessary, and so on. For example, dividing 15 by 4 would give $3 + \frac{1}{2} + \frac{1}{4}$.

We know about unit fractions from the Rhind papyrus (Figure 2.3), written by a scribe about 1650 B.C.E. during the Hyksos domination of Egypt after

the Middle Kingdom, and transcribed and translated into English by Chace,
Bull and Manning [35]. The scribe's first words are these.

∞⑆∞⑆∞⑆∞⑆∞⑆∞

Accurate rendering. The entrance into the knowledge of all existing things and
all obscure secrets. This book was copied in the year 33, in the fourth month of
the inundation season, under the majesty of the king of Upper and Lower Egypt,
'A-user-Rê', endowed with life, in likeness to writings of old made in the time of
the king of Upper and Lower Egypt, Ne-ma'et-Rê'. It is the scribe A'h-mosè who
copies this writing.

∞⑆∞⑆∞⑆∞⑆∞⑆∞

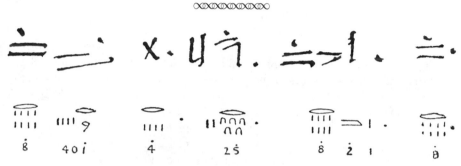

Fig. 2.3. From the Rhind mathematical papyrus; part of the division of 2 by 13 —
hieratic text, followed by its hieroglyphic transcription, and then its transliteration,
all reading from right to left.

We explain how some of the arithmetic operations were performed in ancient Egypt. To multiply two whole numbers, they would double the multiplicand repeatedly and then see which of the corresponding powers of 2 would
add up to the multiplier; they would add the corresponding multiples of the
multiplicand to get the product. For example, to calculate 11 times 13, start
doubling 13:

$$\begin{aligned}
\rightarrow\ & 1 \text{ times } 13 \text{ is } \quad 13 \\
\rightarrow\ & 2 \text{ times } 13 \text{ is } \quad 26 \\
& 4 \text{ times } 13 \text{ is } \quad 52 \\
\rightarrow\ & 8 \text{ times } 13 \text{ is } 104
\end{aligned}$$

We observe that $1 + 2 + 8 = 11$ (those rows marked $\rightarrow$), and so $13 + 26 + 104$
must be 11×13, that is, 143. So one multiplies by repeated addition, without
needing to learn the multiplication table. Today, with the logic of modern
computers in view, we see that, in effect, the multiplier is being expressed
in binary bits 1011 and the corresponding multiples of the multiplicand are
added.

How did a scribe divide and figure out the quotient as a sum of unit
fractions and integers? It was patterned after their method of multiplication
but considerably trickier, with an element of luck or skill, depending on how

this algorithm will impress you. Let us try to divide 2 by 13. Instead of successively multiplying by 2 as in the last example, one divides both sides of the ratio 13 : 1 by successive powers of 2, obtaining these ratios:

$$
\begin{array}{ll}
13 & : 1 \\
6 + \frac{1}{2} & : \frac{1}{2} \\
3 + \frac{1}{4} & : \frac{1}{4} \\
\rightarrow 1 + \frac{1}{2} + \frac{1}{8} : \frac{1}{8}
\end{array}
$$

The first line 13 : 1 means thirteen parts to the whole; or, in a contemporary view, the right number is one-thirteenth of the left. So, if this ratio is preserved by certain operations to follow, and as a result 2 could be found on the left, then on the right would be our answer, merely two parts of the thirteen parts to the whole, i.e., $2 \div 13$. But the process is not straightforward. The general idea of the algorithm is to find fractional parts of 13 on the left that will add up to 2; then the corresponding unit fractions on the right will sum to $2 \div 13$. The left side $(1 + \frac{1}{2} + \frac{1}{8} = 1\frac{5}{8})$ has in the last line become less than 2, so all we need to do is find in some other way further unit fractions that will top it out at 2. From experience the scribe would multiply the ratio 13 : 1 by 4 and 8 and invert, obtaining two more ratios that are still proportional to 13 : 1 :

$$
\begin{array}{ll}
& 13 : 1 \\
\rightarrow \frac{1}{4} & : \frac{1}{52} \\
\rightarrow \frac{1}{8} & : \frac{1}{104}
\end{array}
$$

Appropriate rows (those marked $\rightarrow$) now sum on the left sides of the ratios to 2:

$$
\left(1 + \frac{1}{2} + \frac{1}{8}\right) + \frac{1}{4} + \frac{1}{8} = 2;
$$

and so their corresponding unit fractions on the right must sum[2] to $2 \div 13$:

$$
2 \div 13 = \frac{1}{8} + \frac{1}{52} + \frac{1}{104}.
$$

Thus the Egyptians felt they knew what $2 \div 13$ *is* only when they had identified it as $\frac{1}{8} + \frac{1}{52} + \frac{1}{104}$, whereas for us it is the reverse! Calculating other quotients requires more insight, and probably some trial and error.

By playing with them, the reader will become fascinated with these archaic unit fractions. In this spirit, at the end of this section, we pose exercises about unit fractions, some of which A'h-mosè probably never thought about (Exercises 2.3, 2.4 and 2.5). There are even open problems; see [103, pp. 158–166]. For those whose appetite has been whetted and who wish to learn more, there is Peet's [183] short but scholarly work on the Rhind papyrus, and more

---

[2] In other words, by proportion of ratios, 13 is to 1 as 2 is to $\frac{1}{104} + \frac{1}{52} + \frac{1}{8}$.

generally there are Neugebauer's [176] and van der Waerden's [240] books on ancient science. At roughly the same time and 1400 kilometers to the east, more elaborate problems were being solved.

QUADRATIC EQUATIONS IN BABYLON. In the fertile valley of the Tigris and Euphrates rivers, about 1700 B.C.E., during the reign of Hammurabi in the first Babylonian dynasty, a teacher, or perhaps a student—we don't know— impresses his stylus into wet clay, recording his solution to a geometrical problem. But his solution is totally numerical: no diagram is ever drawn. Typically such problems were couched in everyday language, such as dividing money and finding people's ages, suggesting that school boys wrote the clay tablets as exercises. His method is revealed in a specific problem, using words and numbers to solve it. Coefficients are expressed as numbers rather than letters. There was no symbolism for equations with literals as coefficients: no a, b, c, x, y, $+$, $=$, etc.; this had to wait until François Viète (1540–1603) in 1591 [241].

With hindsight, we glimpse fragments of the quadratic formula peeking through the solution of a system of two equations in two unknowns. As the Babylonians had a workable system of notation with the base 60, their algorithms were effective, provided the coefficients and the root were positive. We will show how their "completing the square" was accomplished differently than we would today, although the numerical calculations amount to the same.

Let us use some modern notation to understand the algorithm in our own terms, even though this will leave open what the writer really had in mind. Let l and w denote the length and width of a field. The problem is to solve the system[3]

$$lw + l - w = 183 \text{ and } l + w = 27. \qquad (2.2)$$

Without knowing their language, we may nevertheless recognize these numbers as they appear on the clay tablet in Figure 2.4, copied by O. Neugebauer [176], and appearing also in B. L. van der Waerden [240]. At the beginning of line 6 the cuneiform for the number 183 occurs; it is in two groups, each of three vertical stylus impressions, meaning $3 \cdot 60 + 3$, and illustrating their use of the base 60. But it would be troublesome to have to express 27 as twenty-seven stylus marks, so each group of 10 marks is represented by a horizontal mark. Thus, at the beginning of line 8 is the cuneiform for the number $27 = 2 \cdot 10 + 7$. The remaining numbers developed by their algorithm may also be recognized.

The author of the tablet tells us to add the equations in (2.2), and then to add 2 to the second equation:

$$lw + 2l = 210 \text{ and } l + w + 2 = 29.$$

One might solve for w in the second equation and substitute the result into the first, and then solve the resulting quadratic equation (Exercise 2.6). But the

---

[3] These equations appear rather artificial since they violate dimensionality.

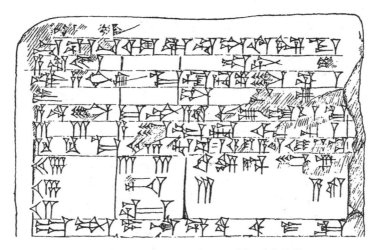

Fig. 2.4. Top of cuneiform tablet AO8862.

Babylonians had a smoother way of proceeding, and did not need to remember the quadratic formula. Noting that $lw + 2l = l(w + 2)$, it seems reasonable to introduce a new variable $w' = w + 2$ so that the system would simplify to

$$lw' = 210 \text{ and } l + w' = 29. \tag{2.3}$$

Such a system, with the product and sum being given, was standard for the times, with a standard method of solution. To solve, still in modern language, introduce the average α of the two variables, $\alpha = (l + w')/2 = 14\frac{1}{2}$, and also the increment δ needed to get the unknowns back later on:

$$l = \alpha + \delta \text{ and } w' = \alpha - \delta.$$

So $(\alpha + \delta)(\alpha - \delta) = lw' = 210$, and thus $\alpha^2 - \delta^2 = 210$. Knowing α, one easily gets δ and the solution. The writer appears to do just this by computing in succession, with words and numbers but no symbols,

$$\alpha^2 = \left(14\frac{1}{2}\right)^2 = 210\frac{1}{4}, \quad \delta^2 \equiv \alpha^2 - 210 = \frac{1}{4}, \quad \text{and so} \quad \delta = \frac{1}{2}.$$

The writer then adds and subtracts to obtain the solution:

$$l = \alpha + \delta = 15, \ w' = \alpha - \delta = 14, \text{ and } w = w' - 2 = 12. \tag{2.4}$$

More solutions are possible (Exercise 2.7).

Next we jump in distance 1200 kilometers further to the east to Persia (now Iran) and in time about 2700 years later to solve all cubic equations geometrically.

'UMAR KHAYYĀM SOLVES CUBIC EQUATIONS BY INTERSECTING CONIC SECTIONS. Solving cubic equations is harder than solving quadratic equations, whether geometrically, graphically, by a formula, or iteratively by successive approximations. The first exact method we present, due to 'Umar al Khayyām (1044–1123/24), finds a positive root geometrically at the intersection of two conics [191]. Although the Hellenic Greeks solved special cases of cubic equations in this way, 'Umar's book is the first systematic study of all cubic equations. His nonnumerical method doesn't lie in a direct path to numerical solutions, a solution is a line segment; but it is an important counterpoint to the other methods, showing the variety of what it means to "solve" an equation. It is also a significant step in the development of algebra, and a significant contribution from Islamic civilization. This selection also illustrates how algebra evolved in Persia, leading to its further development in Europe.

Abu'l Fatḥ 'Umar ibn Ibrahīm al-Khayyāmī was born in Nishapur, Persia. In his name, the phrase 'ibn Ibrahīm' means son of Abraham, and 'al Khayyāmī' means the tent maker, not that 'Umar was a tent maker but that some of his predecessors probably were. His reputation as an astronomer and mathematician and his notoriety as a philosopher and poet spread throughout both the Islamic world and Europe. Today in the West, he is known simply as Omar Khayyam, and remembered mainly for his Rubā'iyāt, a book of epigrammatic verse quatrains. To learn more about 'Umar and his times, read [190] and [191]. As a source for other contributions of Islamic civilization to mathematics, the first book is a good antidote to a Eurocentric view.

We read an excerpt from the work Risāla fi-l-barāhīn 'ala masā'il al-jabr w'al muqābala (Treatise on Demonstrations of Problems of al-Jabr and al Muqābala), written originally in Arabic. 'Umar proves geometrically that a solution of a cubic equation can be found at the intersection of two conic sections. In the spirit of his times he finds only positive solutions. To avoid negative coefficients he classifies cubic equations into 16 species, one of which we look at in detail. See [125] for more background. This translation comes from [247]. 'Umar starts with comments about the philosophy of algebra and geometry.

∝∝∝∝∝∝∝∝∝∝

One of the mathematical processes required in that branch of philosophy known as mathematical is the art of al-jabr and al-muqābala, designed for the extraction of numerical and areal unknowns, and there are kinds of it in which you require very hard kinds of introductions, (and which are) impossible to solve by most people who consider them.

∝∝∝∝∝∝∝∝∝∝

"Al-Jabr", Arabic meaning literally "restoration", and referring to the transferring of a term from one side of an equation to the other, found its way into Spanish, where it also means setting a broken bone. Into the nineteenth century in Spain, signs could still be found advertising "Algebrista y Sangrador": bone setter and blood letter. "Al-Muqābal" is the reduction of both sides of an equation by the subtraction of equal quantities (i.e.,

مصادرُ ودراساتٌ في تاريخ الرياضيات العربية   ٣

رسائل الخيام الجبرية

حققها وترجمها وقدم لها

رشدي راشد   احمد جبار

جامعة حلب

معهد التراث العلمي العربي

١٩٨١

Photo 2.1. *L'Oeuvre Algébrique d'al-Khayyām.* [191]

cancellation). "Numerical and areal unknowns" mean unknowns that represent one-dimensional numbers in linear equations or two-dimensional quantities in quadratic equations.

In the next paragraph, 'Umar tells us what the reader should know beforehand. For the *Elements* see [111]; for the *Data* see [112, pp. 421–425]; for the *Conics* see [110]. But he does not explicitly cite the propositions and theorems taken from these sources. So some imagination will be needed to follow his proof.

∞∞∞∞∞∞∞∞

It must be realized that this treatise cannot be understood except by one who has mastered Euclid's Elements and his book called the Data and two books of Apollonius on the Conics, and anyone who does not know (any) one of these three cannot understand (the present treatise).

⋮

And it is the single dimension, i.e., the root, which is in quantities; when the side is taken with its square, we have the two dimensions which are the surface; and the square in quantities is the square surface; then there are the three dimensions which make the solid; and the solid cube in quantities is the solid which is surrounded by six square faces; and as there is no other dimension, there does not occur in them, i.e., quantities, the square of the square, nor anything which comes above that.

∞∞∞∞∞∞∞∞

Four-dimensional space seemed impossible to 'Umar to visualize, and hence he dismisses equations of degree higher than three. Today mathematics and physics routinely accept higher-dimensional spaces. See our curvature chapter for Riemann's resolution of this enigma.

He next tells how he intends to combine the four possible terms: "the number, the thing, the square and cube," meaning respectively the constant term, the unknown, the square of the unknown, and its cube; in our modern language, d, cx, bx^2 and x^3. Note how the coefficients are not explicitly mentioned; but implicitly they multiply the powers. The words "edge," "thing," and "root" for the unknown are used somewhat interchangeably.

∞∞∞∞∞∞∞∞

The books of the algebraists contain, out of these four geometrical relations, i.e., pure numbers, edges, squares, and cubes, three equations between the numbers, the edges, and squares. For our part, we shall mention the ways by which it is possible to find the unknown by equations involving the four powers; i.e., the number, the thing, the square, and the cube: and beyond which, we said, none (no power) could occur in quantities.

⋮

The remaining six of the 12 kinds are: (a) cube plus root equal to a number, (b) cube plus a number equal to a root, (c) a number plus root equal to cube, (d) cube plus square equal to number, (e) cube plus a number equal to square, (f) a number plus a square equal to a cube. There is nothing in their books about these six kinds except a discourse on one of them, which is incomplete, and I shall explain that one and prove it geometrically, but not algebraically. It is impossible to prove these six except by the properties of conic sections.

∞∞∞∞∞∞∞∞

This paragraph of 'Umar is one of several classifying all possible kinds of polynomial equations in one unknown of degree no more than three. It is necessary since negative numbers were not known. Each case must be treated separately. On which side of the equation each term goes also determines which conics are to be used. This tedium clearly blocks much further progress in solving more general equations of higher degree or with two or more unknowns. We will look at how 'Umar solves case (a), which is $x^3 + cx = d$ in modern algebraic language.

'Umar thinks geometrically: the coefficient c is thought of as an area, namely a square with side AB in Figure 2.5. Similarly d is thought of as a volume that is realized as a rectangular solid with sides of lengths AB, AB and BG. For 'Umar all terms of an equation must have the same dimension; in this case, they must be three-dimensional solids.

His argument is couched entirely in words, except for pairs of letters representing segments, such as AB. To help ourselves understand him, we introduce letters for the coefficients and the unknown and rephrase his words in an intermediate language, whose use will become clearer as we proceed: • for geometric multiplication such as an edge times an area yielding a solid, and & for the addition of solids. Thus, in this language, "numbers, roots, squares

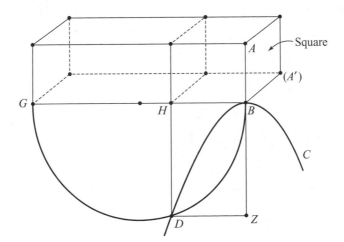

Fig. 2.5. Intersection of two conics.

and cube" mean respectively d, $c \bullet x$, $b \bullet x \bullet x$ and $x \bullet x \bullet x$, and the equation to be solved becomes $x \bullet x \bullet x$ & $c \bullet x = d$.

We now turn to his verbal argument, paraphrased, reordered, and condensed a bit. The curve BDG is a semicircle with diameter BG. The curve CBD is a parabola given by the square of DZ being equal to the product of BZ by AB. The parameter for the parabola is put on the opposite side of the equation from what we would expect today: $DZ \bullet DZ = AB \bullet BZ$; in modern notation: $x^2 = \sqrt{c}y$ since $c = AB \bullet AB$. This may be rewritten as a proportion, $AB : DZ :: DZ : HD$. In any circle we have the well-known proportion, $BH : HD :: HD : HG$. (For a review of proportions see Exercises 2.8 and 2.9.)

These two proportions combine, together with the equality of DZ with BH, to give the catenated proportion

$$AB : BH :: BH : HD :: HD : HG.$$

Next, does 'Umar have in mind a proposition telling him that proportions are transitive and may be multiplied edgewise (the first ratio by itself and then the second ratio by the third)? If so, his next step is justified:

$$AB \bullet AB : BH \bullet BH :: BH \bullet HD : HD \bullet HG.$$

If a further proposition allows him to cancel HD in the last ratio, then equivalently he may write

$$AB \bullet AB \bullet HG = BH \bullet BH \bullet BH.$$

Interchanging sides and adding the solid $AB \bullet AB \bullet BH$ to each yields the desired equation:

$$BH \bullet BH \bullet BH \ \& \ AB \bullet AB \bullet BH = AB \bullet AB \bullet HG \ \& \ AB \bullet AB \bullet BH$$
$$= AB \bullet AB \bullet BG.$$

Since $AB \bullet AB \bullet BG = d$, this shows that BH is the solution x in $x^3 + cx = d$.

It is significant that the solution is a line segment and not a number. Today we might reply, so what? Why not just measure the segment? We easily identify the line segment, a geometrical entity, with its length, a number. But this was not always so, certainly not with the classical Greeks. We are left with the question, what is acceptable as a solution (Exercises 2.10 and 2.11)?

Another Persian mathematician went further than 'Umar in studying cubic equations. Sharaf al-Din al Ṭūsī (d. 1213) found simple algebraic conditions that predicted in each case the number of positive solutions [133, pp. 262–263].

CARDANO SOLVES CUBIC EQUATIONS BY EXTRACTING ROOTS. We break chronological order by placing the selection of Cardano before that of Qin. Cardano's solving of cubic equations is the last of the exact methods that we study for solving polynomial equations; and Qin is the first of a sequence of approximate methods.

Girolamo Cardano—born in 1501 in Pavia, south of Milan, and died in 1576 in Rome—wrote his ground-breaking work, *The Great Art, or the Rules of Algebra* [31], in 1545, within two years of the great Renaissance works of Nicolaus Copernicus and Andreas Vesalius. Cardano was a typical polymath of the time: he also wrote books about medicine, astronomy and philosophy as well as additional volumes on mathematics. He did not wholly originate the methods often attributed to him for solving cubic equations; key ideas were due to Scipione del Ferro and Niccolò Tartaglia; and Lodovico Ferrari discovered how to solve biquadratic equations. A priority dispute over solving cubic equations led to a mathematical duel [133, p. 361]. We leave it to the reader to delve into the acrimony surrounding this controversial episode in the history of mathematics.

Like 'Umar of the previous section, Cardano split his solution of cubic equations into cases, solving them all, whereas their predecessors solved only special cases. At this time, cases were necessary, since only positive numbers,

Photo 2.2. Cardano.

representing lengths of line segments, were true numbers; hence, coefficients
had to be juggled from side to side so that no negatives occur. Also, both their
approaches were geometrical. Cardano reduces the problem to determining
two line segments whose difference in length is the desired root. This method
relies on a geometrical version of the binomial theorem. Unlike 'Umar, who
solved cubic equations by intersecting conics, Cardano thinks in terms of
rectilinear solids. He presents the first part of his proof geometrically for a
specific cubic equation with integral coefficients.

Cardano then translates his geometrical construction into an algebraic algorithm, expressed in words, the essential features of which persist to this
day. This tells how to find these two lengths. It is the basis of our modern
formula for solving cubic equations exactly as the difference of two cube roots
of expressions in square roots and the coefficients of the polynomial. However,
Cardano did not express his algorithm in modern symbols, which he did not
have. His algorithm forced him to confront negative, algebraic and imaginary
numbers, thus pushing the evolution of new number systems. In case there
are three real roots, his algorithm produces complex numbers inside the cube
roots, although the final answers will be real [133, pp. 366–367]! In summary,
Cardano's contribution was to solve all cases by providing needed reductions
from unknown cases to known cases, to write up and systematize the presentation, and to puzzle over the nature of the solutions he obtained. We
present one of his cases here; another is to be found in [150, pp. 224–232].
This translation is taken from [232, pp. 63–67].

Cardano entitles his Chapter XI *Concerning a Cube and "Things" Equal
to a Number*, and poses this case as a problem.

∞∞∞∞∞∞∞

For example, let the cube of AB and six times the side AB be equal to twenty.

∞∞∞∞∞∞∞

Thus he is going to illustrate the case of a cubic term plus a linear term
equal to a constant by solving this specific equation, which is also the case
we saw 'Umar solve. As in the 'Umar selection, all terms must match up
dimensionally and are volumes in this equation; so the number 6 is really an
area and 20 a volume. We succinctly summarize his sentences by geometrical
equations, as we did with 'Umar. In this ad hoc language, where now AB^3
means $AB \bullet AB \bullet AB$, Cardano announces that he is going to solve:

$$(AB)^3 + 6 \bullet AB = 20$$

for the length AB.

There are three parts to this exposition. The first part is a demonstration,
and it starts out by assuming that there are line segments AC and BC whose
cubes differ by 20 and whose product is a third of 6. This demonstration proves
that, given AC and BC, their difference AB is the solution. The second part
is the rule, which summarizes the algorithm. The third part, which should

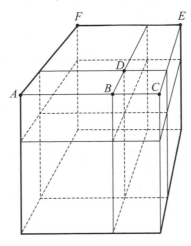

Fig. 2.6. Geometrical binomial.

connect the algorithm to the demonstration, is missing. That is, we need to know how to find AC and BC that satisfy these two conditions.

Cardano first assumes, in the demonstration, that there are two cubes, one smaller than the other, with sides of length AC and BC such that

$$AC^3 - BC^3 = 20,$$
$$AC \bullet BC = 2,$$

and then proves that their difference, $AB = AC - BC$, is the solution of

$$AB^3 + 6 \bullet AB = 20.$$

His proof is geometrical, relying on moving around the eight fundamental solids that together form the cube in Figure 2.6, which extends Cardano's original two-dimensional picture into three dimensions, as he seems to imply it should be. These solids are of four kinds: a small cube in the forefront, with a side BC; a larger cube, somewhat hidden down in the back, with sides of length AB; a flat slab with a square base of size AB^2 and a side of width BC—there are three of these; and a beam with a square cross section, like a stick of butter, with a cross section of size BC^2 and a side of length AB— also three of these. This decomposition of the cube into eight parts expresses geometrically in three dimensions the binomial theorem, which the Indians and the Chinese knew earlier [125]:

$$AC^3 = AB^3 + 3AB^2 \bullet BC + 3AB \bullet BC^2 + BC^3. \tag{2.5}$$

But the cube can be viewed another way. There is a larger slab with sides whose lengths are AB, AC and BC, combining one of the previous slabs with one of the beams. Looking closely within the figure we see that there are three of these larger slabs that can be arranged to exhaust the whole volume, except for the two cubes. Thus, the binomial theorem becomes

$$AC^3 = AB^3 + 3AB \bullet AC \bullet BC + BC^3. \tag{2.6}$$

Alternatively, the middle two terms on the right of (2.5) have merged into the middle term of (2.6); in modern notation,

$$3x^2y + 3xy^2 = 3xy\,(x+y)\,,$$

where $x + y = AC$. By assumption $AC \bullet BC = 2$, and hence (2.6) becomes

$$AC^3 - BC^3 = AB^3 + 3AB \bullet 2.$$

Therefore,

$$AB^3 + 6 \bullet AB = AC^3 - BC^3 = 20,$$

and this allows us to conclude that AB is the root.

The language of this proof, halfway between geometry and algebra, finessed much of Cardano's original proof, which is much longer. Part of what makes his proof so tedious is the need to think positively, meaning that negative numbers are lacking, which is understandable since everything is really a geometrical entity. So it is a major undertaking to transfer a term from one side of an equation to the other by changing signs. In short, modern algebra has allowed us to sidestep these difficulties.

After this proof, Cardano states his rule for solving this class of cubic equations. We quote it to give a flavor of his style, and leave its verification to Exercises 2.12 and 2.13.

∞∞∞∞∞∞∞∞∞

Cube the third part of the number of "things," to which you add the square of half the number of the equation, and take the root of the whole, that is, the square root, which you will use, in the one case adding the half of the number which you just multiplied by itself, in the other case subtracting the same half, and you will have a "binomial" and "apotome" respectively; then subtract the cube root of the apotome from the cube root of the binomial and the remainder from this is the value of the "thing."

∞∞∞∞∞∞∞∞∞

While it is exact, Cardano's method is also numerical; these two words are not antonyms. When asked to solve a cubic today, the computer package Maple apparently uses Cardano's formula to find the exact answer in terms of radicals; numerical evaluation of these then gives decimal numbers (Exercise 2.14). Until recently the formula was thought impractical; it is now a mainstay of computer algebra.

Cardano's rule produces only one root, whereas today we always expect three roots, real or complex, providing we count multiplicities. Exercises 2.15–2.19 explore this aspect of root finding. Exercise 2.20 relates Cardano's method to Euclidean constructions and Exercise 2.21 to three-dimensional perspective.

Ludovico Ferrari (1522–1565), a student of Cardano, discovered how to find the roots of fourth-degree polynomials. He related this to Cardano's work by cleverly deriving from any fourth-degree polynomial a third-degree polynomial whose roots would lead to the roots of the original equation. The result involves square roots of expressions involving cube roots of expressions with even more square roots. Cardano explains this method in Chapter 39 of [31]. See [150, p. 211] for a fuller account. After Cardano wrote his book, Viète [241] introduced letters for unknowns, extending the use already of letters for coefficients; this makes life much easier.

After these achievements, the race was on to solve fifth-degree polynomials exactly by formulas, perhaps invoking fifth roots. But there were no winners. So the problem was rephrased: can one write the roots of any polynomial as formulas in the coefficients using radicals? The celebrated answer is no. In 1826, Niels Henrik Abel (1802–1829) proved the impossibility of solving fifth-degree polynomials in this way. Evariste Galois (1811–1832) [79] developed a general theory that permutes the roots of a polynomial to explain why certain polynomials are solvable in this fashion and others are not. See [77] for a leisurely, do-it-yourself introduction to Galois theory.

From exact methods we turn now to approximate methods. The terms "exact" and "approximate" are relative to what operations are allowable; even the so-called exact methods need square and cube roots, which in general can only be approximated. The subsequent sections discuss a variety of iterative numerical methods applicable to finding nth roots, roots of polynomials, and eventually roots of rather arbitrary functions. *Iterative* means that, starting with an initial guess or two, a sequence of numbers is generated, each number calculated from previous values by some recipe. This sequence of approximations should converge to the solution. We divide the approximate methods into two camps: those where one digit at a time is determined at each stage of an algorithm, and those where two or more digits are determined. Often in the latter case, the number of digits determined in later stages increases, sometimes dramatically. Qin's method is in the former camp. Newton's and Simpson's are in the latter and came later.

The method of Qin Jiu-Shao (1202?–1261?), which we present from original source material in the next section, may have been discovered by accident. For example, to find the square root of 2 from observing that the root is between 1 and 2, compare $(1.5)^2$ with 2, and then find $(1.4)^2$, and so on by trial and error to $1.41\ldots$. Doing these calculations by hand reveals many repeated operations, at least for most of the interior multiplications. This leads to completing squares by performing only those new products needed at each step. In this way one would be led to the method of calculating square roots as a variant of long division. It is a short step to calculating cube roots similarly, and so on to higher roots. Putting all powers together would lead to Qin's algorithm, which he illustrates with a fourth-degree polynomial.

Qin's method [157] can approximate a real root of a real polynomial of any degree as accurately as needed. Its strength is its applicability to all polynomials, and its weakness is its uncertainty in readily determining the next digit. It is based on laying out the coefficients in a tableau in which the coefficients are manipulated to produce successively the digits of the answer, one by one. The discovery of numerical methods depends on having a good number system, which means having place values, a zero, and a marker to separate the integer part of a number from its fractional part. All this the Chinese had for the base 10.

A much more compact form of Qin's tableaus is the so-called Ruffini–Horner method. It was discovered, probably independently, by an Italian physician, Paolo Ruffini (1765–1822), and an Englishman, Theophilus Holdred, in 1820. William G. Horner (1786?–1837) was anticipated by Holdred [76]. This method is the same as Qin's, though with differences in layout that make it easier to explain. Several others in Europe discovered closely related algorithms earlier: Leonardo of Pisa, a.k.a. Fibonacci (c. 1180–1250) and François Viète (1540–1603) [18].

Isaac Newton (1642–1727) described a variety of iterative techniques for finding roots, mostly of polynomials, no method of which uses a derivative. In the third section, we will read his proportional method from his "Waste Book," which is what Newton called a book in which he entered some of his early mathematical ideas. These were published only long after his death and mostly since 1965. That his throwaways are thought worthy of preservation and study attests to Newton's greatness. Our selection is not "Newton's method" as we know it today, with a derivative giving the tangent, but rather a secant method, invoking only proportionality to get closer to the root. Thus we call it *Newton's proportional method*. It extrapolates new digits from estimates of the root already made. In modern language, it is based on linearizing the function locally. Newton explains his method with proportional parts in the context of a specific polynomial, with few formulas given and neither geometry nor calculus. It can nevertheless easily be rewritten strictly algebraically. If the secant is replaced by the tangent and its slope is found from the derivative, then this evolution is the formula found today in most textbooks. This latter we will call *Simpson's fluxional method,* for what is usually and erroneously called "Newton's method." In a sense, Newton's proportional method anticipates Simpson's, but it has a slower rate of convergence, as we shall see in the last section.

Without any references to prior work of others, Thomas Simpson (1710–1761) succinctly describes in words his algorithm for one function of one variable, and then goes on to systems of two functions in two unknowns [214]. He illustrates his method with five examples, but other than these, he gives neither justification for his method nor further explanation of how he arrived at it. This algorithm opened our introduction. It is the culmination of many algorithms in this chapter, and will be studied in detail in the fourth section.

In the last section we will justify and compare these various algorithms. Simpson's fluxional method belongs to a large class of numerical root finders called *fixed-point methods*. These will give deeper insight into why his method works.

Late in the eighteenth century, Joseph-Louis Lagrange (1736–1813) recognized two main strands in the studies of those who wished to find roots of equations. One strand is the theoretical, the other is the practical. On the former, he published his findings about the theoretical existence of roots and their nature as numbers in *Réflexions sur la Résolution Algébrique des Équations* (Reflections on the Algebraic Solution of Equations) [142] [150, pp. 233–247], which later led Evariste Galois (1811–1832) to explain why some polynomials have roots expressible in radicals and most others do not. On the latter, in his book *Traité de la Résolution des Équations Numériques de Tous les Degrés* (Treatise on the Solution of Numerical Equations of All Degrees) [143], published in 1798 and based on two long articles appearing in 1769 and 1770, he explores the many ways to actually find the roots numerically. In particular, Lagrange developed a general systematic algorithm for detecting, isolating and approximating, with arbitrary precision, all real and complex roots of a polynomial with real coefficients [152]. His algorithm always converges. Florian Cajori [29, p. 215] has this to say about it.

> With the publication of Lagrange's great book of 1798, containing his own rich researches and a critical summary of the work of other investigators, a brilliant period in the history of the theory of equations is drawing to its close. Seventeenth and eighteenth century mathematicians have grappled with the problem of the solution of numerical equations, wrestled with it, overcome and exhausted it for the time being. The great problems have found a solution.

Well, that is not quite true. Unlike Lagrange's algorithm, Simpson's fluxional method applies to not only polynomials but transcendental functions as well. Although highly successful in practice, Simpson's lacked a theoretical basis: did it converge, and if so, how fast? There are two kinds of convergence: local and global. Local convergence says that there is a neighborhood of a root, starting in which the algorithm always converges to that root. Global convergence means that from any starting point the algorithm converges to some root.

The proof of local convergence of Simpson's method is fairly easy, and from it readily comes a proof of its quadratic rate of convergence. Roughly, quadratic convergence doubles the number of significant digits with each iteration. These notions will be defined more precisely in the last section of this chapter.

But global convergence is not guaranteed. One likes to assume that the sequence coming out of Simpson's fluxional method converges to a root. Despite some easily constructed counterexamples exhibiting bizarre behavior, this method does seem to converge in practice. This anomaly was not explained until Stephen Smale (1930–) proved that for polynomials it can be made to converge with as large a probability as desired [216]. Although Smale's theorem guarantees convergence almost everywhere, i.e., for almost all starting values, it does not give information as to which starting points converge to which roots. In general this is rather complicated, and leads to beautiful fractal patterns such as those on the color insert.

We close this chapter by pursuing such questions as, why does a particular algorithm work, how well does it work, and what may go wrong? Does the algorithm always find a root; does it converge to the desired root and not some other? How fast does it do so, and how does its rate of convergence compare with that of other root-finders? And globally, how do the answers to these questions depend on the initial guess?

Today we can look back at these methods and see that many can be explained by expanding the function in a Taylor series, by next neglecting all terms after the linear, and then by approximating the derivative in the first term by a simple rule in order to compute the correction. For example, in Newton's proportional method, what amounts to a secant is used to approximate the derivative. In Simpson's fluxional rule the derivative is used exactly, which amounts to the tangent. Qin, al-Ṭūsī and Horner find the nearest "digit" that would give the smallest positive remainder. In a related method of Viète the remainder could be negative. For Newton and others, the correction may have more than one digit. Historically these possibilities were discovered one by one over a long period of time.

Some algorithms, such as Newton's proportional method, were first stated for polynomials, since this is how people first thought of functions, but subsequently they were applied more generally. On the other hand, some, such as Qin's, al-Ṭūsī's, and some other algorithms of Newton, are inherently confined to polynomials. For these methods we can further distinguish those that create a new polynomial by translating the argument, $P_1(x) = P_0(r + x)$, such as Qin and Horner–Ruffini, and those that only retain the remainder by completing all partial powers as we proceed, such as taking square roots and al-Ṭūsī's method.

These features just outlined may be combined in many different ways to create all sorts of new algorithms. And historically this is what happened. In fact, Newton himself sketched at least a half dozen different algorithms. For this reason, one could justify calling all these "Newton's method," but then he would certainly not be the first to have discovered it.

Whereas most chapters of this book seek out great theorems in their original settings, in this chapter we seek out great algorithms—methods of solving equations—in their original settings. Perhaps great nontheorems would be a better way to describe them. While reading them, the reader should keep in mind the historical and conceptual progression of new ideas and inventions. For example, without an electronic computer, a pocket calculator, a desk calculator, a slide rule, or even a table of logarithms, finding a numerical value for $\sqrt[5]{17}$ is nontrivial.

We must make a caveat about terms we use, such as equation, positive number, base, etc. These are relatively modern terms that did not exist in many of the periods we are writing about. Note the misnomer "equation" for early problems like duplicating the cube. Historically, terminology comes into existence only if there is a need for contrasting terms, as for example positive and negative. Although anachronistic, we must use them to be able to describe succinctly what was done.

To keep this chapter to a reasonable length, we restrict our attention largely to single equations in one unknown; for the most part we do not tackle either systems of equations or differential equations. Even so, the explanations are diverse. The reader should observe the many different ways in which mathematical ideas are conveyed. Some are verbal, others geometrical, many numerical, and eventually algebraic with the calculus. In reading the selections, be sure to contrast them with each other, in both logic and style. Enjoy their diversity.

Exercise 2.1. The point of this problem is to find bad starts deliberately. For parts (a) and (b) use the polynomial, $p(x) = x^3 - 3x$. You may need a computer programmed with an algebra package.
(a) Find a starting guess x_0 such that Simpson's fluxional method cycles with a period of 2, that is, $x_2 = x_0$ but $x_1 \neq x_0$.
(b) Find a starting guess x_0 such that Simpson's fluxional method cycles with a period of 3, that is, $x_3 = x_0$ but $x_1 \neq x_0$. Is this possible when x_0 is real? When x_0 is complex?
(c) Give an example of a specific function that has some zeros such that Simpson's fluxional method yields an increasing sequence of iterates, $x_0 < x_1 < x_2 < \cdots$, such that $\lim_{n \to \infty} x_n = \infty$. (Hint: Try $\cos x$.)

Exercise 2.2. What is a number to you, and what should it be to be useful? A positive integer, a fraction, a finite decimal, ..., a complex number, ...? (Cf. [45].)

Exercise 2.3. Can any positive real number be expressed as the sum (possibly infinite) of an integer and decreasing unit fractions? Do this: given its fractional part r between 0 and 1 subtract from it the largest unit fraction not exceeding it. With the remainder continue to subtract unit fractions as large as possible. Try representing π and e this way. This is called the *greedy* algorithm. Does it always converge to r? Why? (Hint: For this, one needs to know the δ–ε theory of convergence.)

Exercise 2.4. Prove that for any positive real number r the greedy algorithm of the preceding exercise terminates in a finite number of iterations if and only if r is rational. (Hint: Show that the numerators in the algorithm decrease if r is rational.)

Exercise 2.5. Most years consist of 365 days; every fourth year is a leap year, containing an extra day, except not every century, etc. Look up the complete list of these exceptions. What does this have to do with unit fractions? Find out the exact number of days in a year (this is not a whole number), and then use a sum of positive and unit fractions to approximate this. Write up your answers as though you were an Egyptian scribe proposing a new calendar with rules for leap years, perhaps with better subdivisions into months of maybe new lengths.

Exercise 2.6. Solve for one variable, l or w', in one equation of (2.3) and substitute it into the other equation. Solve this quadratic equation to verify (2.4) in another way. How close is this method to that of the Babylonians?

Exercise 2.7. Only the positive square root of δ^2 appears in (2.4). What happens when the negative root is used, and why wasn't it?

Exercise 2.8. Classical Greeks did not deal with equations and fractions, but rather with proportions, $a : b :: c : d$. The proportion $a : b :: c : d$ consists of two *ratios* and is to be read "a is to b as c is to d," sounding like something on an intelligence test. It is not clear from reading Euclid what exactly is meant by this [75]. Geometrically, a and b would have the same dimension, and similarly c and d. Today, for numbers, we would write this proportion as $\frac{a}{b} = \frac{c}{d}$, but appreciate that fractions are a relatively new invention. To work with proportions, formulate various properties such as the laws of *interchange*,

$$\text{if } a : b :: c : d, \text{ then } a : c :: b : d,$$

and *cancellation*,

$$\text{if } a : b :: c : d \text{ and } a = b, \text{ then } c = d.$$

Prove these two laws. Be a detective and formulate more propositions of a Euclidean nature that would back up 'Umar's argument in his own terms. For example, prove that proportions are *transitive*, meaning that the middle ratios of catenated proportions may be omitted; and prove that proportions are *multiplicative*, meaning that two proportions may be multiplied termwise.

Exercise 2.9. Solving *mean proportions* posed a challenge to the classical Greeks. *Mean* means that some of the interior quantities are the same.
(a) Realize that solving the mean proportion $a : x :: x : b$ for x is equivalent to solving $x^2 = ab$; that is, geometrically it amounts to finding a square equal in area to a given rectangle. In other words, x is the geometrical mean of a and b. Review and briefly describe how Euclid found square roots.
(b) Next consider two mean proportions that are catenated: $a : x :: x : y :: y : b$. Rediscover by algebra that solving this for x and y is equivalent to finding the intersection of any two of the conic sections, $x^2 = ay$, $xy = ab$, and $y^2 = bx$. This result goes back to Menaechmus [118, p. 100] as related by Eutocius. Further show that this is equivalent to solving the equations $x^3 = a^2b$ and $y^3 = ab^2$. Thus, being able to solve two mean proportions is equivalent to the ability to find cube roots. In particular, one can duplicate the cube, that is, solve $x^3 = 2$.

Exercise 2.10. Solve the equation $x^3 + 6x = 20$ by 'Umar's method. Just what does it mean to solve this equation? (Hint: Actually draw the circle and parabola for this equation, perhaps on graph paper, or with a computer. One way or another, find the coordinates of their point of intersection, perhaps by direct measurement.) Would 'Umar have done this, or much less the Greeks of antiquity? Compare this with Cardano's solution, given after 'Umar's method.

Exercise 2.11. Show that the roots of any cubic polynomial $x^3 + ax^2 + bx + c$ may be found at the intersections of a parabola and a hyperbola. (Hint: Introduce the variable y, related to x by the equation $y = x^2$, to reduce the original polynomial to second degree in two variables. Write this as a hyperbola $\overline{x}\,\overline{y} = C$, where $\overline{x} = x + a$ and $\overline{y} = y + b$.) By graphing these curves, find the positive root of Cardano's cubic $x^3 + 6x = 20$. Cf. Exercise 2.10.

Exercise 2.12. Today people talk about "Cardano's formula" for solving cubic equations, although del Ferro discovered it and Cardano only gave an algorithm, not a formula; it was too early in the history of algebra for that. Nevertheless, from Cardano's rule one can readily derive it. Rewrite his equation in modern notation as $x^3 + px = q$. Introduce variables u and v as his lengths AC and BC, which should satisfy $u^3 - v^3 = q$ and $3uv = p$. Introduce as well

$$\Delta \equiv \left(\frac{q}{2}\right)^2 + \left(\frac{p}{3}\right)^3.$$

Show that $u^3 = (q/2) + \sqrt{\Delta}$ and $v^3 = -(q/2) + \sqrt{\Delta}$. Prove that $u - v$ is a solution x. Relate to the geometry by showing that px is the volume of three rectangular slabs, $3AB \bullet BC \bullet AC$. Also geometrically, what are x^3 and q, and how do these pieces fit together in Figure 2.6?

Exercise 2.13. (a) Fill out the third part of the exposition that Cardano leaves out. This should connect his demonstration to his rule. From the defining equations,

$$AC^3 - BC^3 = 20 \text{ and } AC \bullet BC = 2,$$

cube each term in the second. Solve as the Babylonians might have done if they had ever considered cubic equations and negative numbers; that is, introduce a new variable δ that is the average of AC^3 and BC^3; thus $AC^3 = \delta + 10$ and $BC^3 = \delta - 10$. Substitute and solve for δ. End with a radical expression that should be a literal interpretation of Cardano's rule.
(b) Why did Cardano leave (a) out? Could he have done in his language what we just did? After all, (a) amounts to taking cube roots of square roots, a six-dimensional entity. Compare with Exercise 2.12, where the discriminant Δ would be six-dimensional by Cardano's thinking.

Exercise 2.14. Obviously 2 is a solution of Cardano's equation. How did we miss this, or did we? Use a pocket calculator to find a decimal value for this root from his expressions. Surprise! To verify this answer exactly and clarify your observation, rewrite $\sqrt{108} + 10$ as $6\sqrt{3} + 10$ and find an exact cube root of it in the form $m\sqrt{3}+n$, where m and n are integers. That is, cube $m\sqrt{3}+n$ and compare it with $6\sqrt{3}+10$. (Cardano knew that his complicated expression reduced to 2; see [31, p. 100].)

Exercise 2.15. Aren't there supposed to be three roots to any cubic, counting multiple roots? Find the other two complex roots of Cardano's equation. (Hint: Factor out a linear term using the root you found in Exercise 2.14. This will yield a quadratic polynomial as the other factor, which you can easily solve.) Check your answers directly in Cardano's equation.

Exercise 2.16. (a) Use the formula of Exercise 2.12 to find a root of the equation $x^3 = 15x + 4$, studied by both Cardano and Bombelli. (Hint: try simplifying the cube roots as you did in Exercise 2.14; now u^3 will be complex, so find its cube root as a complex number $m + in$.) Then factor out a linear term, get a quadratic equation, and solve it to find the other two roots. Graph the intersection of the curves, $y = x^3$ and $y = 15x + 4$, to see how these roots come about.
(b) In general, the other two roots, besides $u - v$, are often given as

$$\frac{v - u}{2} + \frac{u + v}{2}\sqrt{-3}, \qquad \frac{v - u}{2} - \frac{u + v}{2}\sqrt{-3}.$$

Verify that these are solutions. (Hint: to simplify, rewrite these roots as linear combinations of complex cube roots of unity: $\omega \equiv -\frac{1}{2} + i\frac{\sqrt{3}}{2}$, $\omega^2 \equiv -\frac{1}{2} - i\frac{\sqrt{3}}{2}$.)
(c) Use (b) to find all the roots of (a).
(d) Solve the equation $x^3 = x$ in two ways: first by factoring, and then by the formulas of (b). Do you see how to obtain all three roots by both methods?

As you might expect from Exercise 2.16, the formulas of Exercise 2.12 work for all cubic polynomials of the form $x^3 + px = q$, regardless of whether p and q are positive or negative, rational or irrational, or even whether they are real or complex. This shows the power of algebra over geometry. Nevertheless, geometry gives us insights that strictly algebraic methods may conceal.

Exercise 2.17. The shapes of real cubic polynomials are rather limited.
(a) Prove that any real cubic, $y = x^3 + bx^2 + cx + d$, has at least one real root and exactly one point of inflection.
(b) Move this point of inflection to the origin by translating both variables. Thus solving the cubic is reduced to solving the system $y = x^3 + px$ and $y = q$, which is essentially the form of Cardano, except that now p and q may be any real numbers.

Exercise 2.18. With Exercise 2.17, we are poised to prove the criterion for the nature of the real roots of a real cubic. Here Δ comes from Exercise 2.12; it is called the *discriminant* for obvious reasons.
If $\Delta > 0$, then the cubic has one real root and two conjugate imaginary roots.
If $\Delta = 0$, then the cubic has three real roots of which at least two are equal.
If $\Delta < 0$, then the cubic has three distinct real roots.

To prove this, find the relative maximum and minimum of the cubic curve of Exercise 2.17b, if they exist. Show that we get the three cases above according to whether q is between the maximum and minimum, at one of these, or beyond them. Which case encompasses the situation in which there are no maximum and minimum? Draw graphs to illustrate the various possibilities.

Exercise 2.19. (a) In the third case of Exercise 2.18, when there are three distinct real roots, one can work around complex numbers by the following device. Since $\Delta < 0$, show that $p < 0$ and $0 \le -\frac{27q^2}{4p^3} < 1$, and hence there must be an angle θ such that $\cos 3\theta = -\frac{3q}{2p}\sqrt{-\frac{3}{p}}$. One may assume that $0 < \theta < \frac{\pi}{3}$; verify the reasonableness of this assumption. Using the triple-angle formula for cosine, prove that the roots are

$$2\sqrt{-\frac{p}{3}}\cos\theta, \qquad 2\sqrt{-\frac{p}{3}}\cos\left(\theta + \frac{2\pi}{3}\right), \qquad 2\sqrt{-\frac{p}{3}}\cos\left(\theta + \frac{4\pi}{3}\right).$$

(In these formulas always take positive square roots.) Complete this program by trigonometry or through the complex plane.
Method 1. The equation $x^3 + px = q$ may be transformed into the equation $4y^3 - 3y = r$ by setting $x = ky$ and choosing k appropriately. Find k. The point of the expression $4y^3 - 3y$ is that $4\cos^3\theta - 3\cos\theta = \cos 3\theta$; hence, by setting $y = \cos\theta$, we obtain $\cos 3\theta = r$. If we are given r, then we may find an angle θ such that $\arccos r = 3\theta$. Trisection of this angle gives us the three solutions above.
Method 2. Plot u^3, v^3 and $-v^3$ in the complex plane, and then u, v and $u - v$. To take the cube root of a complex number, one must find the real cube root of its modulus and divide its argument by 3. Show that the cosine of the argument of u^3 is just the $\cos 3\theta$ of Method 1. Also show that there is really no cube root to compute, at least numerically, although there is still a square root left.
(b) In this way solve again the equations of Exercise 2.16.

Exercise 2.20. Gathering together the results of previous exercises, prove, within the context of Euclidean geometry, that every real cubic equation is solvable for its real roots if, and only if, one has a way of finding cube roots of positive numbers and a way of trisecting angles. By trisecting angles we do not mean dividing θ by 3, which is trivial, by rather we mean passing from an angle given geometrically to an angle a third its size. More precisely, if $\Delta \ge 0$, then only a cube rooter need be added to the classical tools of straightedge and compass; and if $\Delta < 0$, then only an angle trisector need be added.

Exercise 2.21. The annotations suppose that Cardano had in mind a three-dimensional picture (Figure 2). If so, why did he not draw it? After all, artists were now using perspective in their paintings. In fact, when was drawing in perspective discovered, and by whom? (Hint: See [133, p. 389].)

2.2 Qin Solves a Fourth-Degree Equation by Completing Powers

Chinese civilization, with a language and writing system less changed than any other, traces its culture back continuously at least three and a half thousand years. Joseph Needham, in his multivolume work [174, p. xli], states:

> The facts which have been here assembled may at first sight seem a little bewildering, but one must remember that they concern the culture of more than one-fifth of the human race, a people inhabiting for three millennia a land at least as large as Europe, and certainly no less gifted than others.

In many respects, at most times, this culture was the equal or in advance of Western civilization. On the other hand, Chinese culture has been isolated from the rest of the world for most of its history. How much, if any, in the past it has been influenced by other cultures is open to debate, and likewise how much it has influenced others is also largely unknown (Exercise 2.22).

To put some quotations into historical perspective, the Song dynasty (960–1279), in which our selection was written by Qin[4] Jiu-Shao (1202–1261), is famous for its delicate landscape paintings; the following Yuan dynasty (1297–1368), a division of the empire created by Kublai Khan, had Marco Polo as a visitor; and the next dynasty, the Ming (1368–1644), is noted for its fine monochromatic porcelain. Ulrich Libbrecht [157, p. 2] describes the period in which our selection is written:

> This later phase of the Song marks both the apogee of the development of mathematics in China and its terminal point. ... During Qin's life, China had fallen into decay; the northern part was in the hands of the Tartar Chin dynasty (1115–1234), and the western part was occupied by the Tangut dynasty of the Hsi Hsia (990–1227). Around 1230 both parts were conquered by the Mongols, who were from that time on a constant menace to the Southern Sung (1127–1279), who had their capital at Hang-chou. The empire was in a state of great unrest; nevertheless, on both sides of the demarcation line mathematics flourished.

Chinese mathematics was algebraic. Wang Ping comments [157, p. 2]:

> The achievement of Chinese mathematics up to the late Ming dynasty was certainly not inferior to that of any other contemporary civilized country. In fields such as algebra, China was even more advanced than some other countries.

---

[4] Older Roman spellings of Qin's name are Ch'in and Chhin, which give a better idea of its pronunciation.

An important early mathematical work in Chinese civilization is the *Jiu Zhang Suan Shu* (Nine Chapters on Mathematical Procedure), probably first compiled by Zhang Cang (active 165–152 B.C.E.) during the Han dynasty; it exerted enormous influence on the direction of mathematics in China, even into Qin's lifetime. During his time in the 1200s three other notable mathematicians in China wrote many treatises: Li Ye (1192–1279) wrote the *Sea Mirror of Circle Measurements* in 1248; Yang Hui wrote *A Detailed Analysis of the Arithmetical Rules in the Nine Sections* in 1261; and Zhu Shi-Jie the *Precious Mirror of the Four Elements* in 1303. Li Ye represents the North Chinese tradition (algebraic methods) and Yang Hui the South Chinese tradition (geometric methods). Zhu Shi-jie's works are synthetic: they bring together these two traditions [20].

At this time, according to [157, p. 4]:

> Mathematics of a very simple kind was one of the essential accomplishments of the post-Confucian gentleman, on the same level as propriety, music, archery, charioteering, and calligraphy. ... In other words, although mathematics was not considered a suitable livelihood for a gentleman, it was foremost of the arts of which he was encouraged to become an amateur.

For the nongentlemanly technician, practical applications mattered. Mathematics was to serve the sciences. But mathematics was separated from engineering. And writers such as Qin sneaked in problems far more involved than would occur in practice, just for the challenge.

When Genghis Khan was conquering northern China, Qin Jiu-Shao was born. In his youth Qin was an army officer, and was famous for both athletic and literary achievements. In his own words, according to [157, pp. 26–27], "In my youth I was living in the capital [Hang-chou], so that I was able to study in the Board of Astronomy; subsequently I was instructed in mathematics by a recluse scholar. ... At the time of troubles with the barbarians [the mid-1230s], I spent some years at the distant frontier; without care for my safety among the arrows and stone missiles. I endured danger and unhappiness for ten years."

Our source, *Shu Shu Jiu Zhang* (Mathematical Treatise in Nine Chapters), was written by Qin in 1247. It goes beyond what is necessary to solve the applications of the time. Although not the first to discuss polynomials, it is the first detailed account still in existence of how to find roots of polynomials of arbitrary degree by translating the unknown. For example, Yang Hui reviewed the work of Jia Xian (mid-eleventh century) on finding roots, which unfortunately is now lost.

The *Shu Shu Jiu Zhang* also contains an exposition of what today is called the Chinese remainder theorem, arising from Qin's study at the Board of Astronomy. This most likely had its origin in calendar making, to deal with the fact that different planetary bodies have different periods. However, his algorithm is tedious to follow since it is broken down into numerous cases,

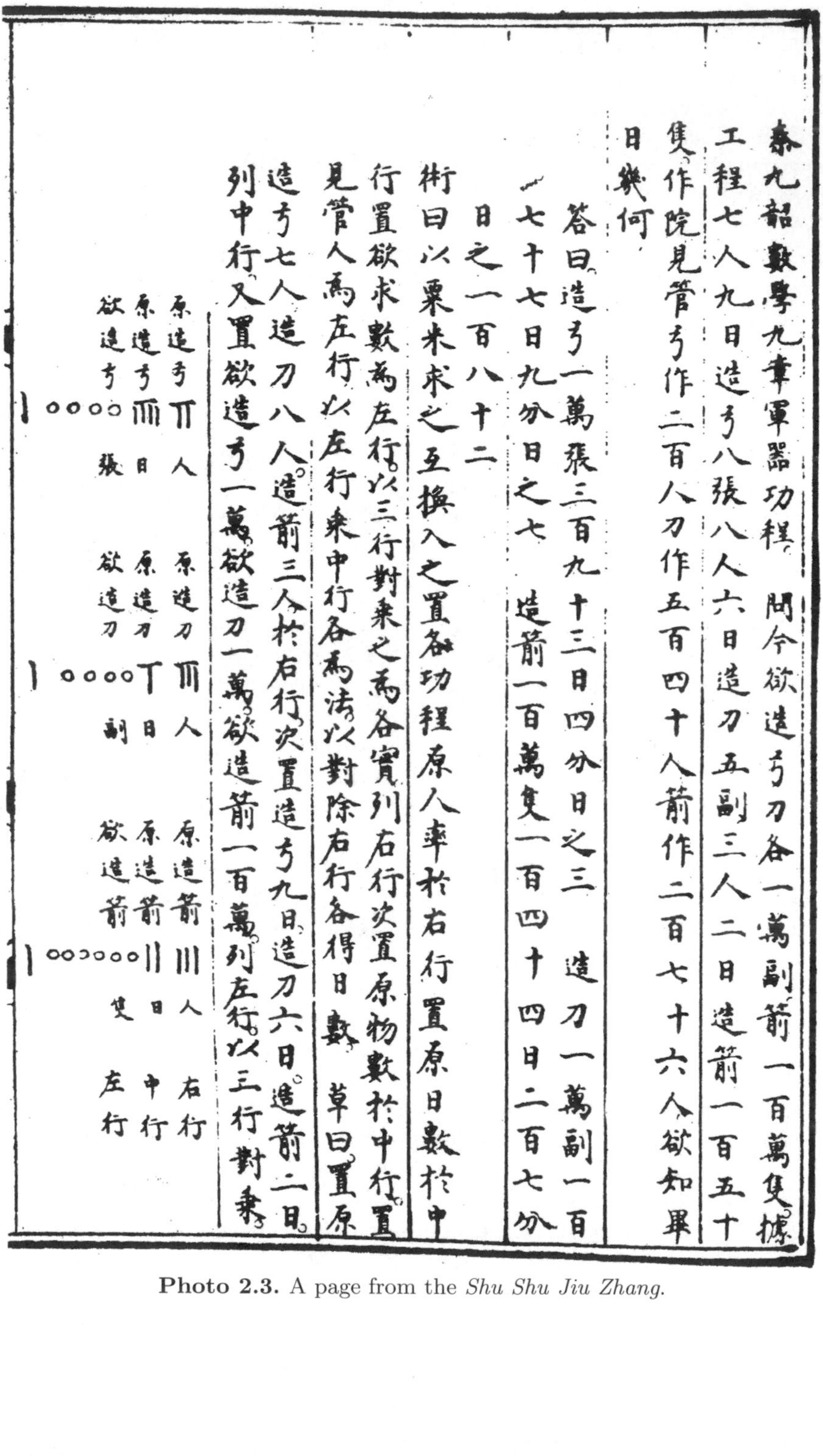

秦九韶數學九章軍器功程　問今欲造弓刀各一萬副箭一百萬隻據

工程七人九日造弓八張八人六日造刀五副三人二日造箭一百五十

隻作院見管弓作二百八刀作五百四十八箭作二百七十六人欲知畢

日幾何

答曰造弓一萬張三百九十三日四分日之三　造刀一萬副一百

一十七十七日九分日之七　造箭一百萬復一百四十四日二百七十分

日之一百八十二

術曰以粟米求之互換入之置各功程原八率於右行置原日數於中

行置欲求數為左行以三行對乘之為各實列右行次置原物數於中行置

見管人為左行乘中行各為法以對除右行各得日數草曰置原

造方七人造刀八人造箭三人於右行次置造方九日造刀六日造箭二日

列中行入置欲造方一萬欲造刀一萬欲造箭一百萬列左行以三行對乘

原造弓　　　人　　　　右行
原造弓　　　日　　　　中行
欲造弓　一〇〇〇〇　張　　左行

原造刀　　　人
原造刀　丁　日
欲造刀　一〇〇〇〇　副

原造箭　　　人
原造箭　　　日
欲造箭　一〇〇〇〇〇　隻

Photo 2.3. A page from the *Shu Shu Jiu Zhang.*

forced on the author by his inability to handle negative numbers in modular arithmetic. In addition, there are many other noteworthy topics such as the arithmetical triangle (also known as Pascal's triangle), a cubic interpolation formula, and the development of series. For more information and detail on this period of Chinese mathematics, especially Qin's work, see [125, Chapter 7], [133, pp. 199–210], [157], [174, pp. 40–48], and [235], which we have consulted and paraphrased.

Qin's method, taken from [157, pp. 181–184], bristles with numbers, but with only a few cryptic notes to explain what is going on. To explain how it works, we first look at later methods. To that end, we fall back on the Ruffini–Horner method of the nineteenth century. Both methods manipulate a polynomial in the same way and suppose a decimal representation of numbers. The basic idea is simple, but the devil is in the details.

After the first digit of the root of a polynomial P_0 is estimated, P_0 is translated into a new polynomial P_1 so that the root of P_1 is the root of P_0 with its first digit removed. The process is repeated on P_1 so that the next digit is removed from the desired root, and so on. By iterating these shifts, P_0, P_1, P_2, ... , a root of P_0 is computed to any number of decimal places, i.e., to any desired accuracy.

There are several aspects to this algorithm. Translating by linear substitutions and scaling make clear what the successive digits should be. Division of polynomials reduces the computations needed for translations, and synthetic division in the refinement of Horner–Ruffini eliminates needless copying of numbers. Qin also used synthetic division but in a different format. We tell in detail and in modern terms what is going on before presenting the original Qin in translation.

A present-day method of finding square roots is akin to long division. In long division we successively multiply new digits by the divisor and subtract the result from the current remainder, creating a new remainder. In square rooting we do something similar; we successively use a new digit to complete a square and then subtract. Since it is rarely taught these days, we illustrate it by finding $\sqrt{2}$ in Figure 2.7; the letters on the right are usually absent in practice. We pair the trailing zeros after the decimal point since the exponent is 2. The decimal point is subsequently neglected. At the end of each step a completed square has been subtracted from 2: first 1^2, then $(1.4)^2$, $(1.41)^2$, etc. We could have found this sequence by trial and error, squaring numbers as needed to see how close they are to 2; but as necessity is the mother of invention, the algorithm avoids continually redoing interior multiplications, and so is more efficient. The binomial theorem makes clear what is happening: $(a + d)^2 = a^2 + (2a + d)\, d$. The accumulator a takes on closer and closer approximations to the root: 1, 1.4, 1.41, etc.; and d is successively the new digits 4, 1, 4, etc. (multiplied by appropriate negative powers of 10). We estimate each new digit d by doubling the accumulator and dividing it into the current remainder. We complete the previous square to a new accuracy

$$
\begin{array}{r}
1.4\ 1\ \dots \\
\sqrt{2.00\ 00\ 00} \\
\end{array}
$$

$$
\begin{array}{rl}
& -1 \qquad\qquad a^2 \\
& \overline{1\,00} \\
24 \quad & -96 \qquad (2a+d)\,d \\
& \overline{4\,00} \\
281 \quad & -2\,81 \qquad (2a+d)\,d \\
& \overline{1\,19\,00} \\
& \qquad\vdots
\end{array}
$$

Fig. 2.7. Square rooting.

by multiplying $2a + d$ by d (Exercise 2.23). Now compute two more decimal places (Exercise 2.24(a)).

Although this brief explanation suffices for explaining the square root algorithm, the method for finding roots of higher-degree polynomials requires more. Especially, not only must squares be completed, but all powers appearing in the polynomial must be completed (Exercise 2.25). In preparation for that, we analyze algebraically in detail what we have done, which helps greatly in approaching Qin's method.

The polynomial whose root $\sqrt{2}$ we wish to find will be $2 - x_0^2$, which will be called $P_0(x_0)$ since it will be continually modified, creating a new sequence of translated polynomials: $P_1(x_1)$, $P_2(x_2)$, $P_3(x_3)$, ..., whose formation in Figure 2.8 will be explained in a moment. Along with these will also be created a sequence of approximations to the root: $r_0 = 1$, $r_1 = 1.4$, $r_2 = 1.41$, ..., as well as the increments: $d_0 = 1$, $d_1 = .4$, $d_2 = .01$, ... that refine the r_n. That is, $r_{n+1} = r_n + d_{n+1}$, or equivalently, $r_n = d_0 + d_1 + d_2 + \cdots + d_n$. The notation d_n is chosen to suggest "digit," but we incorporate into it its corresponding power of 10, in order to make clear what is taking place as we move from one step to the next. After approximating the root with first "digit" d_0, we replace x_0 in $P_0(x_0)$ by $d_0 + x_1$. This has the effect of cutting out the most significant part d_0 of what the root should be and creating a new variable x_1 and a new polynomial $P_1(x_1)$ whose root will be the remainder of the root of P_0 after d_0 is subtracted. In other words, we see a "root" of P_0 but only approximate it by its correct first digit d_0. By translating out this digit, we can then see the second digit in P_1. The root of P_1 will be of significantly smaller magnitude than that of P_0. And so on . . . . In general, to obtain $P_{n+1}(x_{n+1})$ from $P_n(x_n)$, we replace x_n in $P_n(x_n)$ by $d_n + x_{n+1}$, that is, $P_{n+1}(x_{n+1}) = P_n(d_n + x_{n+1})$. For example, $P_1(x_1) = P_0(1 + x_1) = 2 - (1 + x_1)^2 = 1 - 2x_1 - x_1^2$ and $P_2(x_2) = P_1(.4 + x_2) = .04 - 2.8x_2 - x_2^2$. Observe that the numbers appearing in the table of Figure 2.8 also appeared in the earlier method for finding square roots.

The crucial part of each iteration is determining the new "digit" d_{n+1}. We illustrate this by finding d_2, which should be close to a root of $P_2(x_2)$. If it were

n	$P_n(x_n)$	d_n	r_n
0	$2 - x_0^2$	1	1
1	$1 - 2x_1 - x_1^2$	.4	1.4
2	$.04 - 2.8x_2 - x_2^2$	.01	1.41
3	$.0119 - 2.82x_3 - x_3^2$	.004	1.414
4	$.000604 - 2.828x_4 - x_4^2$	.0002	1.4142
$\vdots$	$\vdots$	$\vdots$	$\vdots$

Fig. 2.8. Iterating $\sqrt{2}$.

to be exactly a root, then we would be finished. But we must settle for d_2 to be a "digit," that is, at this iteration it should be one of $.00, .01, .02, \ldots, .09$. To determine it, we realize that d_n^2 will be small compared to the other terms of P_2; therefore, we truncate P_2 to its linear term and constant, $-2.8x_2 + .04$, and set to 0 to find that $x_2 = .0142\ldots$, and round down to .01. Why round *down*? That's so that the following "digit" will be nonnegative; this is equivalent in this example to ensuring that the constant term of the next polynomial P_3 is positive. Of course, the quadratic term $-(.01)^2$ in P_2 might have overpowered the constant $P_2(0)$, pushing P_2 negative. In such a case we would have had to try a new estimate for d_2, something less than the previous.

This is a real problem with polynomials of high degree: a large number of nonlinear terms with varied signs and coefficients can make it difficult to figure the new "digit." Typically, however, after one perseveres for several iterations, the later polynomials become more and more linear in the sense that the quadratic terms and higher eventually contribute very little to determining the d_n. Intuitively, think of what happens with a graphing calculator. Calculating a new P_n moves the Y-axis. Searching for new and smaller d_n is like blowing up the graph of the polynomial and zooming in, making the graph almost linear, i.e., a straight line.

The astute reader may have noticed that the first step in computing $\sqrt{2}$ is different from the remaining steps. This is not accidental. For polynomials in general, one has to have a good initial estimate. Without this, any converging may veer off to a wrong root. After all, most polynomials have many roots, but often there is only one particular root that one is interested in. One must first isolate the roots with a small and well-defined interval about each. This is an important and difficult area, isolating the possible solutions of an equation, with an extensive literature; but we will not pursue it [29, 143, 152, 238].

We now turn to Qin's polynomial and its origin in a geometrical problem of determining an area given the lengths of the sides of the symmetrical quadrilateral in Figure 2.9 ([133, p. 233]). The known sides are $a = 39$, $b = 25$, and $c = 30$; and the unknown is the total area x. With some work (Exercise 2.26), one finds that the unknown must satisfy the equation $-x^4 + 763200x^2 - 4064256\,0000 = 0$, the left side of which we call "Qin's polynomial."

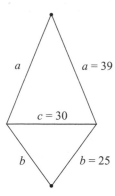

Fig. 2.9. Qin's field.

Mental arithmetic tells us that 8 or something nearby will work for the first digit. The reason is that in this case, there is a simple way to estimate the solution from the picture by "guesstimating" the area. If this were a rectangle with the same sides, the area would be $ab = 39 \cdot 25 = 975$, the real area being somewhat less since the sides are skew. If we should try 9 as the first digit, we will realize that it is too large; so we decrease it to 8, where a sign change occurs. Some trial and error is typical of finding the first digit.

When the polynomial does not come from an application, there is still a way to estimate the solution. Using only the first two terms set to zero and solving for x, we find that $x^4 = 76\,3200x^2$ and hence $x \approx 800$. The last term of Qin's polynomial is overwhelmed, i.e., it is of insignificant magnitude compared to the first two terms evaluated at $x = 800$; if it could not be ignored, then one would have to balance all three terms, approximately. But only when the first transformation is completed can we be sure the first digit is correct. At each transformation, guessing the next digit is the tricky part, but it becomes less murky as the algorithm proceeds, with linearization emerging and attention eventually shifting to the low-order terms (Exercise 2.27).

As an aside, if we should use the lower-order terms of Qin's polynomial to compute a first digit, then, upon completing his algorithm, we will have computed another root. Of course, in a fourth-degree polynomial there are also two more (Exercise 2.28).

We will ease into Qin's solution since it appears rather opaque; but after exploring several related ways to solve his equation, his method should become transparent. We will adapt the method just exhibited for solving quadratics to this higher-degree polynomial. We successively transform it by (1) ridding ourselves of excessive multiplications, when translating it, by dividing instead by simple linear divisors; (2) simplifying long division with synthetic division, arriving at the Horner–Ruffini method; and (3) refining the synthetic divisions into Qin's tableaus. As we build up to Qin's method, the reader should peek ahead and start identifying numbers there with numbers we are about to compute.

n	$P_n(x_n)$	d_n	r_n
0	$-x_0^4 + 76\,3200x_0^2 - 406\,4256\,0000$	800	800
1	$-x_1^4 - 3200x_1^3 - 307\,6800x_1^2 - 8\,2688\,0000x_1 + 382\,0544\,0000$	40	840
2	$-x_2^4 - 3360x_2^3 - 347\,0400x_2^2 - 10\,8864\,0000x_2$	0	Finished

Fig. 2.10. Solving Qin's equation.

Having explained the square-rooting algorithm in algebraic terms earlier, we have a scheme to find the roots of any polynomial, in particular Qin's polynomial P_0 in Figure 2.10. Recall that $P_{n+1}(x_{n+1}) = P_n(x_{n+1} + d_n)$ and $r_n = d_0 + d_1 + \cdots + d_n$. The working assumption is that each "digit" d_n is significantly smaller than the previous one, being attached to a lesser power of 10. We piece the root together with these successive corrections found from transformed polynomials (Exercise 2.29). The crucial observation is that the constant term is the value of the original polynomial in the latest approximation. Thus, if at any stage this constant term $P_n(0)$ is 0, then we are finished and we have an exact solution. Why? Otherwise, these constant terms, which are measuring an error, should decrease in value and approach 0.

Since the constant term of P_2 has vanished, clearly 0 is a root of it, and there is no more to do. How neat! Except that computing the coefficients of these successive polynomials is horrendous. Try it!

So the order of the day is to try to reduce these computations. Perhaps a historical note is in order. Before electronic computers and calculators, all computations had to be done by hand, perhaps with the help of tables or simple physical devices (Exercise 2.30). These were nowhere near as powerful as today's computers, nor as automated. So great store was set in finding the most efficient modes of computation. The Qin and Horner–Ruffini methods are complementary in this respect and illustrate how the most efficient mode depends on whether one uses counting rods or pen and paper. In hand computations, one avoids multiplying two multidigit numbers at any cost. In these methods one of the multipliers is always a single digit. In the larger picture, a computer takes the same time whether it multiplies single-digit numbers or multidigit ones; so the advantage of Qin is lost. It is an example of a fossil algorithm.

Recall that the method starts off by in effect substituting $x_1 + d_0$ for x_0 in $P_0(x_0)$, yielding a translated polynomial $P_1(x_1)$. We could just plough through the algebra (Exercise 2.31), but it is more efficient to proceed differently. Dividing $P_0(x_0)$ by $x_0 - d_0$ gives

$$P_0(x_0) = (x_0 - d_0)Q(x_0) + P_0(d_0),$$

and the remainder $P_0(d_0)$ becomes the constant term for the new P_1, as we will discuss later. Usually, dividing polynomials is performed by something akin to long division; with our polynomial, we calculate as in Figure 2.11, where the top line is $Q(x_0)$ and the lower right-hand number is $P_0(d_0)$.

$$
\begin{array}{r}
-x_0^3 \ -800x_0^2 \ +12\,3200x_0 \ +9856\,0000 \\
\hline
x_0-800)\,\overline{-x_0^4 \quad +0x_0^3 \ +76\,3200x_0^2 \qquad\qquad\quad +0x_0 \ -406\,4256\,0000} \\
-x_0^4 \ +800x_0^3 \\
\hline
-800x_0^3 \ +76\,3200x_0^2 \\
-800x_0^3 \quad 64\,0000x_0^2 \\
\hline
12\,3200x_0^2 \qquad\qquad +0x_0 \\
12\,3200x_0^2 \ -9856\,0000x_0 \\
\hline
9856\,0000x_0 \ -406\,4256\,0000 \\
9856\,0000x_0 \ -788\,4800\,0000 \\
\hline
382\,0544\,0000
\end{array}
$$

Fig. 2.11. Dividing polynomials.

When the divisor is of the particular form $x_0 - d_0$, we can condense this long division with its many redundancies by eliminating all mention of x_0, by being careful to fill in missing powers with zeros, and by multiplying by d_0 rather than $-d_0$, which allows one to add rather than subtract. Knowing full well that the first difference in each line will be 0, we need not take up space with these. With these abridgements long division becomes *synthetic division* [238] in Figure 2.12. Here 800 $(= d_0)$ multiplies each entry in the bottom line and the result is added to the top entry of the next column, giving the next entry in the bottom line, e.g., $800 \times (-800) + 76\,3200 = 12\,3200$, which is seen to be the coefficient of x_0 in $Q(x_0)$. By comparing the computations in the long division with those in the synthetic division, we see that the last line gives all the coefficients of $Q(x_0)$; its coefficient -1 of x_0^3 is of course the first entry on the bottom line, that is, $-x_0^4/x_0 = -x_0^3$.

To obtain the other coefficients of the translated polynomial P_1, perform a similar division on $Q(x_0)$, and so on through each of what Qin calls "transformations":

$$
\begin{aligned}
P_0(x_0) &= (x_0 - d_0)\,Q(x_0) + P_0(d_0) \\
&= (x_0 - d_0)\,\{(x_0 - d_0)\,R(x_0) + Q(d_0)\} + P_0(d_0) \\
&= (x_0 - d_0)\,\{(x_0 - d_0)\,[(x_0 - d_0)\,S(x_0) + R(d_0)] + Q(d_0)\} + P_0(d_0) \\
&= (x_0 - d_0)\,\{(x_0 - d_0)\,[(x_0 - d_0)\,((x_0 - d_0)\,T(x_0) + S(d_0)) + R(d_0)] \\
&\quad + Q(d_0)\} + P_0(d_0) \\
&= P_1(x_0 - d_0).
\end{aligned}
$$

Notice that the degrees of the polynomials P, Q, R, S, T are decreasing one by one, so T is actually a constant polynomial. Upon multiplying this out

$$
\begin{array}{r}
800\,)\ -1 \quad\ 0 \quad\ 76\,3200 \qquad\qquad 0 \ -406\,4256\,0000 \\
-800 \ -64\,0000 \ 9856\,0000 \quad 788\,4800\,0000 \\
\hline
-1 \ -800 \quad 12\,3200 \ 9856\,0000 \quad 382\,0544\,0000
\end{array}
$$

Fig. 2.12. Synthetic division.

Line	c_4	c_3	c_2	c_1	c_0	Qin
1	−1	0	76 3200	0	−406 4256 0000	Tableau 1
2		−800	−64 0000	9856 0000	788 4800 0000	
3	−1	−800	12 3200	9856 0000	382 0544 0000	Tableau 8
4		−800	−128 0000	−92544 0000		
5	−1	−1600	−115 6800	−82688 0000		Tableau 13
6		−800	−192 0000			
7	−1	−2400	−307 6800			Tableau 15
8		−800				
9	−1	−3200				Tableau 16
10						
11	−1					

Fig. 2.13. Sequence of synthetic divisions.

and setting $x_1 = x_0 - d_0$, we see that $P_0(d_0)$, $Q(d_0)$, $R(d_0)$, ... become the coefficients of $P_1(x_1)$. In other words, as before, $P_0(x_1 + d_0) = P_1(x_1)$. With synthetic division we perform these calculations, which are presented in Figure 2.13; the last column is added in anticipation of explaining Qin's method.

The unlined entries in the odd-numbered lines of Figure 2.13 give the coefficients of P_0, Q, R, S and T; and the even-numbered are intermediate calculations. For example, from line 5, $R(x_0) = -x_0^2 - 1600x_0 - 115\,6800$; and from line 9, $T(x_0) = -1$. The doubly underlined entries are the remainders, which give the coefficients of the translated polynomial P_1, that is, $P_1(x_1) = -x_1^4 - 3200x_1^3 - 307\,6800x_1^2 - 8\,2688\,0000x_1 + 382\,0544\,0000$, which is also the second line in Figure 2.10. The translated polynomial P_1 could also be found by direct substitution or by a Taylor expansion (Exercise 2.32).

We are now ready to find the second digit of the root. Proceeding as in Figure 2.13 above, we create a new Figure 2.14. We estimate the next digit by dividing c_0 by c_1 (coefficients in P_1) and taking the negative, i.e., we solve the equation $c_1 x + c_0 = 0$. In other words, we linearize, which, as observed before, we can usually do after finding the first few digits (Exercise 2.33). This transformation need go no further with more synthetic divisions since $c_0 = 0$ in Line 3 of Figure 2.14. This means that the algorithm has terminated; so the answer is $800 + 40$.

There is one aspect to the Horner–Ruffini method, as it was used in practice, that we omitted to avoid muddying the waters even further. At the

	c_4	c_3	c_2	c_1	c_0	Qin
Line 1	−1	−3200	−307 6800	−8 2688 0000	382 0544 0000	Tableau 17
2		−40	−12 9600	−1 2825 6000	−382 0544 0000	
3	−1	−3240	−320 6400	−9 5513 6000	0	Tableau 21

Fig. 2.14. The second digit.

$$
\begin{array}{lllll}
-1\,oooo\,oooo & 0oo\,oooo & 76\,3200\,oooo & 0oo & -406\,4256\,0000 \\
& -8\,00oo\,oooo & -64\,0000\,oooo & 98\,5600\,00oo & 788\,4800\,0000 \\
\hline
-1\,oooo\,oooo & -8\,00oo\,oooo & 12\,3200\,oooo & 98\,5600\,00oo & \underline{382\,0544\,0000}
\end{array}
$$

Fig. 2.15. Scaling by powers of ten.

beginning and between each synthetic division, the variable x is scaled by a power of 10 in order to better estimate the next digit. For that reason we replace x by $100x$ in Qin's polynomial; this has the effect of multiplying the coefficients by powers of 100 and helps us to compare the sizes of terms. Carrying this through the first synthetic division would create Figure 2.15 out of the earlier Figure 2.12.

Clearly, the last four zeros of each number might be omitted everywhere.

Qin's method for finding roots appears to be a massive array of confusing computations, without rhyme or reason (see the tableaus below); actually the motivation is rather simple. He describes it by working through his fourth-degree polynomial on a counting board. In the columns on the counting board one places configurations of short sticks that represent numerals arranged as the coefficients of the polynomial and its subsequent translates. We use Hindu-Arabic equivalents:

$$
\begin{array}{cccccccccc}
| & || & ||| & |||| & \underline{\quad} & | & \underline{\quad} & || & ||| & |||| & \bigcirc \\
1 & 2 & 3 & 4 & 5 & 6 & & 7 & 8 & 9 & 0
\end{array}
$$

The symbol $\bigcirc$ would not actually appear on the counting board, but only in the illustration of the method; it simply means no sticks in that position. An experienced user of the counting board could move these sticks about very quickly, changing from one tableau to the next. Thus the calculations in the sequence of tableaus presented below actually took place on just one counting board, and only when replicated on the printed page does it seem so complicated. The Chinese, at this time, did not have a decimal point, but something quite close to it. Since the problems were typically applied, under the digit in the units position, the unit of measurement was placed, e.g., day, inch, etc. (The use of measures before most numbers, where they would not be considered necessary in English, is a characteristic feature of the Chinese language.)

Now the method, with the details and explanations interpolated between the tableaus. In the first tableau, the first column contains the Arabic numerals corresponding to the numerals represented by the Chinese counting sticks, the second column has a free translation of the Chinese terms describing these numbers, and the third does not appear in the Chinese scheme, but has our modern equivalent for these terms. The directions to the right of each tableau are a very free translation of the Chinese.

Qin's method is essentially the same as the much later Horner–Ruffini method with some differences in the layout that made it easier to explain the latter first. The coefficients run horizontally in Horner–Ruffini and vertically in Qin. Both algorithms proceed to find the root one digit at a time, the polynomial being continually transformed by scaling so that each new root of a transformation can be approximated by a whole number from 0 to 9. Scaling is accomplished differently in the two methods. The extra zeros added to numbers in the Horner–Ruffini method are effected in Qin by shifting the numbers to the left in Tableaus 1–3. These right indentations line up the coefficients in Qin so that one easily sees how the various powers of the first digit will cancel out to make the polynomial close to 0; similarly in Horner the same scaling makes the coefficients about the same magnitude.

The Horner–Ruffini method appears more compact in that one line of it captures several tableaus of Qin, but the same arithmetical operations on the same numbers are performed in both cases. For example, the synthetic division of Figure 2.12 corresponds to Tableaus 3–8, called the first transformation by Qin. Tableaus 4 and 7 carry intermediate calculations: each of these tableaus corresponds in Figure 2.12 to multiplying an entry in Line 3 by 800 and adding it the entry in Lines 1 in the next column. Tableau 8 gives the coefficients of $Q(x)$.

The second transformation of Qin (not shown) corresponds to the second synthetic division, lines 3–5 in Figure 2.13. And so on. The rest is mechanical.

Here is the detailed explanation of the first tableau of Qin. The first column sets up the polynomial with its coefficients originally designated by rods as explained earlier. The 0 in its first line means that root finding has yet to start. In the second column are free translations of the Chinese terms where, of course, variables and their powers x, x^2, etc. were not available at the time and had to be expressed by Chinese characters. The third column contains our modern symbolic equivalents, with which we carry on. At the bottom of the tableau are the directions for operations to be performed that will carry us onto the next tableau. In this instance, these directions scale the polynomial, that is, substitute $10x$ for x, so that the coefficients are beginning to be comparable in value.

0	quotient Q		00 Q	
$-406\,4256\,0000$	constant term c_0		$-406\,4256\,0000$	c_0
0	coefficient of x c_1		0	c_1
$+76\,3200$	coefficient of x^2 c_2		$7632\,00$	c_2
0	coefficient of x^3 c_3		0	c_3
-1	coefficient of x^4 c_4		-1	c_4
Move Q 1 column.			Move Q left 1 column.	
Move c_2 2 columns.			Move c_2 2 columns.	
Move c_4 4 columns.			Move c_4 4 columns.	
			Find first digit of Q	

Tableau 1   Tableau 2

To go from Tableau 2 to Tableau 3, we scale again by powers of 10. Notice that now the coefficients are roughly aligned to the left. It looks as though another scaling might be in order, but this would push the leading coefficient too far to the left. With the coefficients roughly aligned, one may now find the first digit by estimating powers of it multiplied by the coefficients and added up algebraically. In this instance it suffices to look only at the leading digits of c_2 and c_4, as explained earlier; that is, $\sqrt{76/1} \approx 8$. The constant c_0 may be ignored since it is overwhelmed by the leading terms: roughly, and looking at only leading digits, $c_4 8^4 + c_2 8^2 \approx |-1 \cdot 4096 + 76 \cdot 64| = 778$, which is greater than $406 \approx |c_0|$.

800 Q	800 Q	800 Q
$-406\,4256\,0000$ c_0	$-406\,4256\,0000$ c_0	$-406\,4256\,0000$ c_0
0 c_1	0 c_1	0 c_1
76 3200 c_2	76 3200 c_2	12 3200 c_2
	$-64\,0000$ c_3'	
0 c_3	$-8\,00$ c_3	$-8\,00$ c_3
-1 c_4	-1 c_4	-1 c_4
Multiply c_4 by Q; add to c_3.	Multiply c_3 by Q. Subtract from c_2.	Multiply c_2 by Q. Add to c_1.

Tableau 3   Tableau 4   Tableau 5

In Tableau 3 we are directed to create a new c_3 by multiplying Q by c_4 and adding the result to c_3 to create Tableau 4. This corresponds precisely to the first step of synthetic division in Figure 2.13. Notice that a new intermediary line, $c_3' = c_3 \times Q$, is created in Tableau 4 when it is too complicated to do the calculation in one step. Although Qin could easily handle negative quantities, he directs us to "subtract" where we would say "add". This is the second step of synthetic division, and it results in c_2 becoming 12 3200 both in Tableau 5 and in Figure 2.13. We move right along through the first synthetic division, the arithmetic steps being exactly the same in Qin and Horner–Ruffini.

Tableau 8 finishes the "first transformation," in other words, the synthetic division of Figure 2.12. Qin now proceeds to the second transformation, which corresponds to the second synthetic division in Figure 2.13, whose beginning instructions are at the bottom of Tableau 8. The reader should now be able to recreate the remaining tableaus of Qin, which are to be found in [157].

We pause at the end of the fourth transformation at his Tableau 16. This corresponds to finishing Figure 2.13. Are you ready for the next digit? Since it will have a decimal weight of 10, whereas the first digit 8 had a weight of 100, shift back by a factor of 10 to arrive at Tableau 17. As in the Horner–Ruffini method, the leading digits of the linear and constant terms, $382/82 \approx 4$, estimate the next digit. With a few more tableaus, corresponding to

800 Q	
$-406\,4256\,0000$	c_0
$98\,5600\,00$	c_1
$12\,3200$	c_2
$-8\,00$	c_3
-1	c_4

Multiply c_1 by Q.
The result is a
positive number c_1'.
Add to c_0.

800 Q	
$-406\,4256\,0000$	c_0
$788\,4800\,0000$	c_1'
$98\,5600\,00$	c_1
$12\,3200$	c_2
$-8\,00$	c_3
-1	c_4

To the negative c_0
add the positive c_1'.
The remainder is a
positive constant
term.

800 Q	
$+382\,0544\,0000$	c_0
$98\,5600\,00$	c_1
$12\,3200$	c_2
$-8\,00$	c_3
-1	c_4

First transforma-
tion is completed.
Multiply c_4 by Q.
Add to c_3.

Tableau 6   Tableau 7   Tableau 8

Figure 2.14, Qin finds that he has an exact solution. For practice, do Exercises 2.24(b) and 2.34.

800 Q	
$382\,0544\,0000$	c_0
$-826\,8800\,00$	c_1
$-307\,6800$	c_2
$-32\,00$	c_3
-1	c_4

End of fourth transformation
Move c_1 back 1 column.
c_2, 2 columns.
c_3, 3 columns.
c_4, 4 columns.
In Q, set up the next figure.

800 Q	
$382\,0544\,0000$	c_0
$-82\,6880\,000$	c_1
$-3\,0768\,00$	c_2
$-320\,0$	c_3
-1	c_4

Divide c_0 by c_1;
as the next figure in Q take 40.
Multiply 40 by c_4 $(40 = Q')$.
Add to c_3.

Tableau 16   Tableau 17

In its day, the method of Qin and Horner–Ruffini was important and widely used. Many improvements were advanced (Exercise 2.24(c)). However, today it has been superseded by Simpson's fluxional method, which is more direct to program on a computer and more efficient to execute. Most importantly, Simpson's works not only with polynomials but also with the much wider class of transcendental functions (Exercise 2.35).

Exercise 2.22. Several centuries after Qin's book appeared there was a decline in science and mathematics in China. In contrast there was the Renaissance in Europe. Using Needham's short book [175], explain why China, which earlier was more advanced, did not go on to develop its counterpart of European science.

Exercise 2.23. Imagine you are a logic designer assigned the task of creating a square rooter for a modern computer. Today computers work in binary. Write out an algorithm, taking full advantage of the binary base. It will allow you to avoid entirely any guesswork about the next bit. Also no multiplication or division is needed. Try it out by computing $\sqrt{2}$ in binary, i.e., $\sqrt{10_2}$, to six bit places. Convert your answer to decimal and compare with $\sqrt{2_{10}}$.

Exercise 2.24. (a) Compute $\sqrt{2}$ to four decimal places by the square root algorithm.
(b) Recompute $\sqrt{2}$ by Qin's method to two decimal places, solving the equation, $x^2 - 2 = 0$, with a first digit $Q = 1$. Compare the intermediate numbers with those found by the square root algorithm.
(c) In all the methods developed in this section, there is a way to substantially increase the accuracy with only a little more work. One is estimating the next digit by linearization using division. This suggests that, rather than settling on just one new digit, one should compute several digits of the division. Of course, this works only at the last step. Why? Try this out on part (a) or (b). Observe that the number of significant digits roughly doubles with this modification. This phenomenon of the accuracy doubling is rather universal, and will be explored more thoroughly in the sequel.

Exercise 2.25. Compute $\sqrt[3]{2}$ analogously to the way $\sqrt{2}$ was computed. (Hint: At each step complete the cube by $\left(3a^2 + 3ad + d^2\right)d$.)

Exercise 2.26. Derive Qin's polynomial from Figure 2.9. (See [133, p. 233].)

Exercise 2.27. Try to find all four roots of $x^4 - 139\,4500x^2 + 4860\,8784\,0000$ using the Horner–Ruffini or Qin method. (Hints: The first digit cannot be estimated from just two terms. Two roots are very close together.)

Exercise 2.28. Find another positive solution to Qin's equation by estimating the leading digit from the two lower-order terms. Use either the Horner–Ruffini or Qin method. Notice that this comes from a variant of the field in Figure 2.9 by flipping the smaller, lower triangle over on top of the larger triangle. Finally, what are the other two solutions of this fourth-degree polynomial? (Hint: It's an even function.)

Exercise 2.29. Graph Qin's polynomial P_0. Locate r_0 and r_1 on the X-axis. Also find where $x_0 = 0$, $x_1 = 0$, and $x_2 = 0$. This is like shifting the Y-axis. With respect to these new Y-axes, where are P_1 and P_2? Discuss the other places where P_0 crosses the X-axis, i.e., how would you find its other roots? Why does the constant term of these polynomials change sign in going from P_0 to P_1 and then become 0 in P_2?

Exercise 2.30. If you had to find a root of a polynomial by hand, would you prefer Horner's method or Simpson's method, which you learned in calculus as Newton's method? Be careful how you answer: each multiplication or division of multidigit numbers takes time.

Exercise 2.31. (a) By expanding with the binomial theorem and gathering together common powers of y, show that with the substitution $x = d + y$, the polynomial, $p(x) = c_4 x^4 + c_3 x^3 + c_2 x^2 + c_1 x + c_0$, becomes

$$c_4 y^4 + (4c_4 d + c_3) y^3 + (6c_4 d^2 + 3c_3 d + c_2) y^2$$
$$+ (4c_4 d^3 + 3c_3 d^2 + 2c_2 d + c_1) y + (c_4 d^4 + c_3 d^3 + c_2 d^2 + c_1 d + c_0).$$

(b) Recast algebraically Qin's method up to the first transformation so that we achieve (a). To accomplish this, rewrite each step in the form $c_4 Q + c_3 \to c_3'$ (Tableau 3 to Tableau 4, using primes differently from Qin). Be wary of the word "subtract" after Tableau 4, and be sure to accumulate primes. Compound these substitutions. The final c_4'''', c_3'''', etc., at Tableau 16 should be the coefficients of the powers of y found in (a). But notice how Qin's method, with its clever sequencing of multiplications and additions, implicitly mimics binomial coefficients while explicitly avoiding them.

Exercise 2.32. (a) Substitute $x_1 + 800$ into Qin's polynomial P_0 and thereby obtain P_1.
(b) Expand Qin's polynomial P_0 in a Taylor series about $d = 800$ and obtain P_1.

Exercise 2.33. Estimating the initial or even subsequent digits of a root of a general polynomial is tricky. Suppose, after having computed a new translation, you suspect your choice of a new digit was wrong. How do you know whether to add or subtract 1 from your suspect digit to create a new trial digit? Formulate a rule to determine this by using the signs of the coefficients c_0 and c_1 of the new polynomial. Recall that c_0 is its value and c_1 its derivative at the new estimate of the root.

Exercise 2.34. Find a root of $x^3 - 130x^2 - 250x - 1848$ by Qin's method.

Exercise 2.35. Reflect on the differences between the method of Qin for finding roots of polynomials and modern methods, such as pocket calculators and computer packages, e.g., Maple and Mathematica. In particular, find out in detail how a particular modern method actually does it.

2.3 Newton's Proportional Method

Here is how John Fauvel opens the book *Let Newton Be!* [72].

> In April 1727, the French writer Voltaire viewed with astonishment the preparations for the funeral of Sir Isaac Newton. The late President of the Royal Society lay in state in Westminster Abbey for the week preceding the funeral on 4 April. At the ceremony, his pall was borne in a ceremonious pageant by two dukes, three earls, and the Lord Chancellor. "He was buried," Voltaire observed, "like a king who had done well by his subjects." No scientist before had been so revered. Few since have been interred with such dignity and high honour.

Since there are many fine books about Newton (1642–1727) and his extraordinary contributions to science, let us just note that he lived in tumultuous times. He was born at Woolsthorpe in Lincolnshire, England, when the British Civil War was beginning. He came of age at the restoration of Charles II, who, although his rule was relaxed in many ways, received secret financial help from Louis XIV so that England might become again an absolute monarchy like France. Newton studied at Cambridge University for three years, which then closed because of the bubonic plague then raging. He returned to his family manor from 1664 to 1666, where the magic apple presumably fell and he discovered the universal law of gravitation. At that time he also demonstrated that all colors of the rainbow compose white light, and he began to develop the calculus. His lifetime was the time in England of the writers John Milton, John Dryden and John Bunyan, the composers Henry and Daniel Purcell, and fellow scientists Edmund Halley, Robert Hooke, Robert Boyle and William Harvey.

Photo 2.4. Newton.

"Newton's proportional method" [178, pp. 489–491] was probably written in 1665. Not surprisingly, as just a sketch in his "Waste Book," already described in the introduction of this chapter, his informal exposition has several annoying but minor errors, as well as a few inconsistencies, most of which we correct. Newton had read Wallis's (1616–1703) *Arithmetica Infinitorum*, by which time, through the efforts of Simon Steven (1548–1620), Rafael Bombelli (1526–1572) and François Viète (1540–1603), algebraic notation had evolved close to its present form. Consult [93] for details about the history of the Waste Book.

In preparation for reading the selection of Newton in its original language, we make several preliminary remarks about his first paragraph.[5] The comma serves as a decimal point in Europe. Parenthetically, Newton lists various ways of obtaining a first estimate, which is needed to start iterating. "Geometrically by description of lines" would mean something like the method 'Umar used in the introduction; and "an instrument ... of numbers made to slide by one another" would be a slide rule [120], invented by William Oughtred and Edmund Wingate about 1630, and superseded by the pocket calculator in the 1970s. Then Newton splits his first estimate 2,2 into $g = 2$ and $y = 0,2$. At each step in the algorithm these are updated, although g and y are never mentioned again. Maybe this split is indicated by Newton initially writing 2,2 as 2|2 in the original manuscript. In other words, 2 and 2,2 are the first two estimates for the root. From these he will compute a better third estimate, and a final and significantly better fourth. Beware that his letters are slippery: they are recycled with new values, over and over again.

◯◯◯◯◯◯◯◯◯◯◯

Newton, from

The resolusion of y^e affected Equation $x^3 + pxx + qx = r$.
Or $x^3 + 10x^2 - 7x = 44$.

First having found two or 3 of y^e first figures of y^e desired roote viz 2,2 (wch may bee done either by rationall or Logarithmical tryalls as Mr Oughtred hath tought, or Geometrically by descriptions of lines, or by an instrument consisting of 4 or 5 or more lines of numbers made to slide by one another wch may be oblong but better circular) this knowne pte of y^e root I call g, y^e other unknowne pte I call y then is $g + y = x$. Then I prosecute y^e Resolution after this manner (making $x + p$ in $x = a$. $a + q$ in $x = b$. $b - r$ in $x = c$. &c.)

◯◯◯◯◯◯◯◯◯◯◯

---

[5] As an aside, we explain the origin of the curious word "y^e" occurring here, and meaning simply "the." The "y" was sometimes used by printers to mimic an old runic character called "thorn," which looked somewhat like a "y." But "y^e" should not be confused with the archaic English pronoun "ye," as used in the Christmas carol *God rest ye merry Gentlemen*.

We interrupt the flow to make several points. To follow Newton's sketch, one could rewrite his equation as a polynomial set to zero: $x^3 + pxx + qx - r = x^3 + 10x^2 - 7x - 44 = 0$.[6] This is helpful in comparing his algorithm to the other algorithms of this chapter. But it seems best for now to follow Newton as he wrote it; then we will be finding out how much the left side deviates from 44.

There are two parts to his algorithm: computing the values that his polynomial $x^3 + 10x^2 - 7x$ achieves at the estimates, which he does in the arrays below; and calculating a new estimate from a natural proportion created from four numbers: the previous two estimates and their polynomial values, i.e., by linear extrapolation a new estimate is found.[7]

Values of the polynomial are going to be calculated at 2 and 2,2; the results are both designated by b and these subtracted from r are designated by h and k, respectively, which are the deviations from 44, and which should be approaching 0. Newton evaluates a polynomial, $P(x) = x^3 + px^2 + qx$, as $((x + p)x + q)x$, a form computer programmers rediscovered as minimizing the number of multiplications (this is what his parenthetical phrase above explains). In the arrays below, $a = (x + p)x$ and $b = (a + q)x$; in short, multiply 12 by 2 to get 24, etc.

$\infty\infty\infty\infty\infty\infty$

$$\begin{array}{|ll|}\hline 12 = x + p & \times \\ \hline 24 = a & x = 2 \\ \hline\end{array} \cdot \begin{array}{|ll|}\hline a + q = 17 & \times \\ \hline b = 34 & 2 = x \\ \hline\end{array} \cdot$$

$r - b = 10 = h$. by supposing $x = 2$.

$\infty\infty\infty\infty\infty\infty$

This is essentially synthetic division, as described in the last section, although here the g at which the polynomial is evaluated may be longer than one digit, for example, 2,2. When $g = 2$, we have, in the language of that section, that $P(x) = (x - 2)Q(x) + P(2) = (x - 2)(x^2 + 12x + 17) + 34$, calculated in the synthetic division

$$\begin{array}{r} 2\,)\,1\ 10\ -7\ \ 0 \\ \underline{2\ \ 24\ 34} \\ 1\ 12\ \ 17\ 34 \end{array}$$

These coefficients are readily read off from Newton's calculations. But Newton does not complete the sequence of synthetic divisions to translate the polynomial. At the next iteration he uses the original polynomial, increasing his

---

[6] Actually Newton starts out his sketch with the equation $x^3 + pxx + qx + r = 0$ and then shortly shifts his r to the right side without changing its sign! But we have altered his original in order to treat r consistently throughout.

[7] This method of finding a real root by approaching it from both sides is called *Regula falsi* or the *method of false position* [123, article 301c]. It is natural, old, and has appeared in many different cultures [133]. Newton's contribution was to find an efficient way to evaluate polynomials.

2.3 Newton's Proportional Method

labors unnecessarily (but Joseph Raphson (1648–1715) did complete the substitution to obtain a new polynomial, although not in the efficient manner of Qin, Horner, et al. [29, pp. 193–194]). Here are more of Newton's calculations, where he distributes the multiplication by 2,2 over 2 and ,2, and adds up the partial products.

∞∞∞∞∞∞∞∞∞∞

Againe supposing $x = 2,2$.

$x + p = 12,2$	$\times$		$a + q = 19,84$	$\times$
2,44	,2		3,968	,2
24,4	2,		39,68	2,
26,84 $= a$			43,648 $= b$	

$r - b = 0,352 = k.\ h - k = 9,648.$

∞∞∞∞∞∞∞∞∞∞

In preparation for creating a proportion, differences in the value of the polynomial are being calculated, corresponding to differences in the argument. In geometrical terms, the heights of similar right triangles are being figured, 9,648 and 0,352; the bases will be 0,2 and y, if the graph of the polynomial is visualized (Exercise 2.36). Corresponding sides of the similar triangles lead to the proportion, as Newton explains next and which is at the heart of the method.

∞∞∞∞∞∞∞∞∞∞

That is y^e

$\left\{ \begin{array}{l} \text{latter } r - b \text{ substracted from the former } r - h \text{ there remaines} \\ \text{difference twixt this \& } y^e \text{ formor valor of } r - b \text{ is} \end{array} \right\}$ 9,648.

& y^e difference twixt this & y^e former valor of x is 0,2. Therefore make

$$9,648 : 0,2 :: 0,352 : y.$$

Then is $y = \frac{0,0704}{9,648} = 0,00728$ &c. the first figure of wch being added to y^e last valor of x makes $2,207 = x$.

∞∞∞∞∞∞∞∞∞∞

Now iterate the process (Exercise 2.37).

∞∞∞∞∞∞∞∞∞∞

Then wth this valor of x prosecuting y^e over$\overline{acon}$ as before tis

$x + p = 12,207$	$\times$		$a + q = 19,94084$	$\times$
0,085449	,007		0,13958588	,007
2,44140	,20		3,9881680	,20
24,414	2,		39,88168	2,
26,9408 49	$= a$		44,00943388	$= b$

$r - b = -0{,}00943\,388$. wch valor of $r - b$ substracted from y^{e} precedent valor of $r - b$ y^{e} diff: is $+0{,}36143\,388$. Also y^{e} diff twixt this & y^{e} precedent valor of x is $0{,}007$. Therefore I make

$$0{,}36143\,388 : 0{,}007 :: -0{,}00943\,388 : y.$$

That is

$$y = \frac{-0{,}00005\,903716}{0{,}36143\,388} = -0{,}0001633\&\text{c}.$$

2 figures of w^{ch} (because negative) I substract from y^{e} former value of x & there rests $x = 2{,}20684$. And so might y^{e} Resolution be prosecuted.

∞∞∞∞∞∞

The final answer is slightly wrong. The leading 59 in the numerator of y should really be 66; thus $y = \frac{-0{,}00006\,603716}{0{,}36143\,388} = -0{,}0001827$, and thus $x = 2{,}20682$.

Stimulated by the need to solve equations arising from the birth of modern science, Newton and others created diverse algorithms (Exercises 2.38, 2.39 and 2.40).

Exercise 2.36. Graph the polynomial $x^3 + 10x^2 - 7x$ carefully, perhaps with a pocket calculator, together with the horizontal, $y = 44$. Plot all the points, quantities, and similar triangles that enter into the first iteration of Newton's example. Emphasize the secant. Notice that the first correction is positive, but the next is negative. Why is this? What sign do you expect the next corrections to have?

Exercise 2.37. (a) For Newton's proportional method derive the iterative formula,

$$x_{n+1} = x_n - \frac{y_n}{\frac{y_n - y_{n-1}}{x_n - x_{n-1}}} \qquad (n = 1, 2, 3, \ldots),$$

where x_0 and x_1 are initial estimates, $y_n = f(x_n)$, and f is any continuous function whose root we desire. Why is the fraction in the denominator close to the derivative, if it exists? How close are we to the classical "Newton's" method.

(b) Simplify the formula of (a) to

$$x_{n+1} = \frac{x_{n-1}y_n - x_n y_{n-1}}{y_n - y_{n-1}}.$$

Exercise 2.38. Rather than guessing the new root proportionally, that is, by linear extrapolation, try to speed up convergence by using quadratic extrapolation, i.e., by fitting a quadratic curve to the old estimates. Work out the details of this by resolving the equation, $x^3 + 10x^2 - 7x = 44$, of the selection in this new way. You will need three original guesses x, so add 2.1 to 2.0 and

2.2. Use these to plot three points on the cubic curve, and then solve three linear equations to find the a, b, c that fit $ax^2 + bx + c$ to the cubic polynomial at these three values of x; that is, make the quadratic pass through the three points. Then find where the quadratic crosses the line $y = 44$, and use this for a new estimate of x. How many iterations do you need to match Newton's final accuracy? Edmund Halley (1656?–1743) proposed quadratic extrapolation in 1694. (See [29, pp. 191–195] for the history of this method and many other proposed variants of Newton's and Simpson's methods.)

Exercise 2.39. Newton had another way of finding a root of a polynomial, which should first be read in [179, pp. 328–340], and which is sometimes cited as evidence for Newton discovering "Newton's method." But to do this, one must conjecture what Newton was thinking. He applied this method only to specific polynomials where the coefficients were numbers, with the arithmetical operations leaving no trace of any algebra, and even obscuring the presence of a derivative. So the general form is not obvious. The reader is asked to create the theory from his specific example.
(a) Explain how close it is to Qin's and the Horner–Ruffini methods, excepting that Newton's layout is radically different and he allows the answer to increase more than one decimal place at a time.
(b) When Newton estimates the next few digits, he is in effect computing a derivative. Show numerically that in his polynomial, $f(y) = y^3 - 2y - 5$, when he substitutes $y = 2 + p$ and uses the linear and constant terms to estimate the next few digits, he has computed $-\frac{f(2)}{f'(2)}$.
(c) Algebraically, retrace (b) for an arbitrary polynomial with literals as coefficients: make a linear substitution, use the binomial theorem to expand, throw out all terms higher than the first power, and show that indeed the correction term has the renowned form of Simpson's fluxional method [29, pp. 191–194].

Exercise 2.40. [179, p. 326] Although Newton did not discover the method named after him, he did invent many remarkable techniques. In particular, he was fond of taking strictly numerical methods and applying them algebraically. As an example of this technique, you are to expand $\sqrt{a^2 + x}$ in a power series in x.
(a) By completing powers in complete analogy with the numerical method of calculating square roots at the beginning of Section 2, find $\sqrt{a^2 + x}$ by starting with a trial divisor of a into $a^2 + x$. Continue through the fourth power of x.
(b) Find $(a^2 + x)^{1/2}$ by the binomial theorem, and compare with (a). They should be the same.
(c) Find $\sqrt{5}$ by taking $a = 2$ and $x = 1$, using the series of (b) through the fourth power. How accurate is the answer? (Hint: Partial sums of an alternating series with the terms decreasing in size bracket the answer.)

2.4 Simpson's Fluxional Method

Thomas J. Simpson (1710–1761) gave the first account describing how to find roots of functions that explicitly uses the calculus. Hence, it is also the first general method to work for transcendental functions.

In reading about this new method, bear in mind the times. Culturally we have moved from the Restoration in England to the Age of Reason. The beginning of Simpson's life overlaps the end of Newton's; both were country boys. England was predominantly rural, with industry scattered throughout small villages.

> The curfew tolls the knell of parting day,
> > The lowing herd wind slowly o'er the lea.
> The plowman homeward plods his weary way,
> > And leaves the world to darkness and to me.
>
> Thomas Gray (1716–1771),
> > *Elegy Written in a Country Churchyard*

T. J. Simpson was a writer of mathematical textbooks in eighteenth-century England. He was colorful, self-taught, pugnacious, and trained as a weaver at his father's insistence. Francis Clarke [39, Chapter II] sets the tone.

> The children of the working class were expected to gain the training that was needed for their social and economic role under the system of apprenticeship. The boy who was to become a weaver entered the home and employ of a master weaver and learned from him his art. ... The education offered to Thomas Simpson was no exception to the general rule. Thomas was only taught to read English, but possessing a rare intellect he showed an interest in study in general. He educated himself through reading the literature with which he came in contact. He taught himself to write, and made opportunities to associate with people from whom he might learn.
>
> During this period he continued his work of self-education, and acquired such a knowledge of arithmetic, algebra, and geometry as to permit him to understand certain questions proposed in the *Ladies' Diary*. Through this magazine he learned of a still higher branch of mathematics which was much talked of at the time. This was the method of fluxions.

Clarke continues in his Chapter III.

> Upon his arrival in London Simpson settled in Spitalfields, a Huguenot weaving community, and while establishing himself as a weaver, taught mathematics during his spare time. In this settlement he not only found opportunity for self-advancement but also a great demand for his teaching. Although engaged in trades and ordinary occupations, the Huguenots had carried with them into Great Britain

their love of intellectual activities .... This is seen in their musical societies, their debating clubs, and, what is still more remarkable, in their scientific associations.

Simpson mastered the mathematics of the day and wrote a number of widely read textbooks. Besides his writing and his method, he also is known today for several rules for integration and some ways of solving differential equations.

Written in 1740, the selection below has no references, which is typical of textbook writers. Still one would like to know whether Simpson picked this up somewhere or whether he invented it himself.

With effort we recognize his Case I as the algorithm appearing in calculus textbooks. Remarkably he extends this from one function in one variable in Case I to two functions in two variables in Case II; this is much less intuitive. It would be of some interest for a historian to explore how Simpson found this. He introduces his method in the preface to his collection [214] of thirteen unrelated papers.

ᘛᐧᐧᐧᐧᐧᐧᘚ

Simpson, from
Essays on Several Curious and Useful Subjects, ...

The Sixth [paper], contains a new Method for the Solution of all Kinds of Algebraical Equations in Numbers; which, as it is more general than any hitherto given, cannot but be of considerable Use, though it perhaps may be objected, that the Method of Fluxions, whereon it is founded, being a more exalted Branch of the Mathematicks, cannot be so properly applied to what belongs to common Algebra.

ᘛᐧᐧᐧᐧᐧᐧᘚ

Before beginning to read the selection, we should say something about the controversial notion of fluxions. Simpson is following in the tradition of Newtonian calculus rather than that of Leibniz, which had a different notation. In Newton's calculus, the fluxion of any variable existed, often with respect to time, otherwise with respect to another variable or parameter, but more often than not with respect to something not specified, that is, somehow a fluxion measured how much a variable is changing (how much a fluent is flowing). As such one should not equate fluxions with derivatives as we know them. Simpson, in another work [213], defines a fluxion:

ᘛᐧᐧᐧᐧᐧᐧᘚ

The magnitude by which any flowing quantity would be uniformly increased in a given portion of time with the generating celerity at any proposed position or instant (was it from thence to continue invariable) is the fluxion of the said quantity at that position or instant.

ᘛᐧᐧᐧᐧᐧᐧᘚ

E S S A Y S

ON SEVERAL

Curious and Useful SUBJECTS,

In SPECULATIVE and MIX'D

MATHEMATICKS.

Illuftrated by a Variety of EXAMPLES.

By *THOMAS SIMPSON,*

L O N D O N:

Printed by H. WOODFALL, *jun.* for J. NOURSE, at the *Lamb* without *Temple-Bar.*

M.DCC.XL.

Photo 2.5. Simpson's *Essays.*

Such haziness, typical of contemporary attempts to define fluxions and derivatives, was ably attacked earlier by Bishop George Berkeley (1685–1753) [14] (Exercise 2.41).

Here are several comments about the first paragraph of Simpson's Case I. Today, instead of "equation" we would more carefully say "function." The fluxion of x is $\dot{x}$, and the fluxion of a function f would be $\dot{f}$. The quotient A is thus $\dot{f}/\dot{x}$, which today we would simplify in Leibniz's notation to df/dx. By "Error" is meant the error in the functional values, just $f(x)$ since $f(r) = 0$ when r is the root sought. If x_0 is the first estimate, then the last sentence tells how to compute a new value x_1,

$$x_1 = x_0 - \frac{f(x_0)}{\frac{\dot{f}(x_0)}{\dot{x}(x_0)}},$$

and we have essentially the first formula that opened this chapter (Exercise 2.42).

∞∞∞∞∞∞∞∞∞∞

A new Method for the Solution of Equations in Numbers.
CASE I
When only one Equation is given, and one Quantity (x) to be determined.

Take the Fluxion of the given Equation (be it what it will) supposing, x, the unknown, to be the variable Quantity; and having divided the whole by $\dot{x}$, let the Quotient be represented by A. Estimate the Value of x pretty near the Truth, substituting the same in the Equation, as also in the Value of A, and let the Error, or resulting Number in the former, be divided by this numerical Value of A, and the Quotient be subtracted from the said former Value of x; and from thence will arise a new Value of that Quantity much nearer to the Truth than the former, wherewith proceeding as before, another new Value may be had, and so another, etc. 'till we arrive to any Degree of Accuracy desired.

∞∞∞∞∞∞∞∞∞∞

Before proceeding to Case II, to which Simpson immediately jumps, the reader may first want to work through Examples I and II, which follow Case II.

∞∞∞∞∞∞∞∞∞∞

CASE II.
When there are two Equations given, and as many Quantities $(x$ and $y)$ to be determined.

Take the Fluxions of both the Equations, considering x and y as variable, and in the former collect all the Terms, affected with $\dot{x}$, under their proper Signs, and having divided by $\dot{x}$, put the Quotient $= A$; and let the remaining Terms, divided by $\dot{y}$, be represented by B: In like manner, having divided the Terms in the latter, affected with $\dot{x}$, by $\dot{x}$, let the Quotient be put $= a$, and the rest, divided by $\dot{y}$, $= b$. Assume the Values of x and y pretty near the Truth, and substitute in both the Equations, marking the Error in each, and let these Errors, whether positive or negative, be signified by R and r respectively: Substitute likewise in the Values of A, B, a, b, and let $\frac{Br-bR}{Ab-aB}$ and $\frac{aR-Ar}{Ab-aB}$ be converted into Numbers, and respectively added to the former Values of x and y; and thereby new Values of those Quantities will be obtained; from whence, by repeating the Operations, the true Values may be approximated *ad libitum*.

◯◯◯◯◯◯◯◯◯

To understand what Simpson studies in his Case II, suppose that F and f are the two functions in the equations that are to be simultaneously solved:

$$\begin{cases} F(x,y) = 0, \\ f(x,y) = 0. \end{cases}$$

Using a modern notation for partial derivatives, we would calculate the fluxions:

$$\begin{cases} \dot{F} = F_x \dot{x} + F_y \dot{y}, \\ \dot{f} = f_x \dot{x} + f_y \dot{y}. \end{cases}$$

Thus $A = F_x$, $B = F_y$, $a = f_x$, and $b = f_y$. If x_0 and y_0 are initial estimates with $R = F(x_0, y_0)$ and $r = f(x_0, y_0)$, then $x_1 = x_0 + \frac{Br-bR}{Ab-aB}$, and so forth (Exercise 2.43).

More revealing is a geometrical interpretation. The function F describes a surface, $z = F(x, y)$, in three-dimensional space. On this there is a plane Π tangent to the point (x_0, y_0, Z_0), where $Z_0 = F(x_0, y_0)$. Likewise, the other function f describes a surface with its tangent plane π at the point (x_0, y_0, z_0), where $z_0 = f(x_0, y_0)$. There is also the XY-plane given by $z = 0$. These three planes, Π, π, and the XY-plane, generally intersect in one point $(x_1, y_1, 0)$, yielding the first iterate (x_1, y_1) (Exercise 2.44).

Simpson concludes the theory with three observations. In the last one he observes what today we would call quadratic convergence. But he also notes that in some cases the rate of convergence may be much slower. How is his condition couched in fluxions related to our derivatives today (Exercise 2.45)?

◯◯◯◯◯◯◯◯◯

Note, 1. That every Equation is first to be so reduced by Transposition, that the Whole may be equal to Nothing.

2. That, if after the first Operation, the value of x or y be not found to come out pretty nearly as assumed, such Value is not to be depended on, but a new Estimation made, and the Operation begun again.

3. That, the above Method, for the general part, when x and y are near the Truth, doubles the Number of Places at each Operation, and only converges slowly, when the Divisor A, $Ab - aB$, at the same time converges to nothing.

EXAMPLE I.

Let $300x - x^3 - 1000$ be given $= 0$; to find a Value of x. From $300\dot{x} - 3x^2\dot{x}$, the Fluxion of the given Equation, having expunged $\dot{x}$, (Case I.) there will be $300 - 3xx = A$: And, because it appears by Inspection, that the Quantity

$300x - x^3$, when x is $= 3$, will be less, and when $x = 4$, greater than 1000, I estimate x at 3.5, and substitute instead thereof, both in the Equation and in the Value of A, finding the Error in the former $= 7.125$, and the Value of the latter $= 263.25$: Wherefore, by taking $\frac{7.125}{263.25} = .027$ from 3.5 there will remain 3.473 for a new Value of x; with which proceeding as before, the next Error, and the next Value of A, will come out $.00962518$, and 263.815 respectively; and from thence the third Value of $x = 3.47296351$; which is true, at least, to 7 or 8 Places.

EXAMPLE II.

Let $\sqrt{1-x} + \sqrt{1-2xx} + \sqrt{1-3x^3} - 2 = 0$. This in Fluxions will be $\frac{-\dot{x}}{2\sqrt{1-x}} - \frac{2x\dot{x}}{\sqrt{1-2xx}} - \frac{9x^2\dot{x}}{2\sqrt{1-3x^3}}$, and therefore A, here, $= -\frac{1}{2\sqrt{1-x}} - \frac{2x}{\sqrt{1-2xx}} - \frac{9x^2}{2\sqrt{1-3x^3}}$: wherefore if x be supposed $= .5$, it will become -3.545: And, by substituting 0.5 instead of x in the given Equation, the Error will be found $.204$; therefore $\frac{.204}{-3.545}$ (equal $-.057$) subtracted from $.5$ gives $.557$ for the next Value of x; from which, by proceeding as before, the next following will be found $.5516$, &c.

◯◯◯◯◯◯◯◯◯◯◯

Exercises 2.46 and 2.47 give additional practice in Case I. Omitted are two examples of Case II, which the reader may find in [214].

◯◯◯◯◯◯◯◯◯◯◯

EXAMPLE V

Let $x^x + y^y - 1000 - 0$, and $x^y + y^x - 100 = 0$. Here we shall have $A = \overline{1 + L : x} \times x^x$, B equal $\overline{1 + L : y} \times y^y$, $a = \frac{y}{x} \times x^y + y^x L : y$, and b equal $\frac{x}{y} \times y^x + x^y L : x$. Now, it appearing from the first Equation, that the greatest of the two required Quantities cannot be lesser than 4, nor greater than 5; and from the first and second together, that the Difference of x and y must be pretty large; otherwise $x^x + y^y$ could not be 10 times as great as $x^y + y^x$: I therefore take x (which I suppose the greater Number) equal 4.5, and y equal 2.5; and then by a Table of Logarithms, or otherwise, find the next Values of these Quantities to be 4.55 and 2.45; and the next following 4.5519, &c. and 2.4405, &c. respectively.

◯◯◯◯◯◯◯◯◯◯◯

In Example V the vinculum, for example in $\overline{x+y}$, is used for grouping where today we would use parentheses, $(x + y)$. The notation $L : x$ means $\ln x$. Exercise 2.48 gives additional practice with Case II.

Simpson's fluxional method is often popularly called "Newton's method." But this is wrong. Simpson's has an appreciably faster rate of convergence than Newton's proportional method. So we distinguish them. Geometrically, we can see the secant in Newton's and the tangent in Simpson's. It is true that Newton gave a variety of methods for finding roots of polynomials, some of which would grow into the method of this section if algebra and calculus were used. But Newton, one of the inventors of the calculus, never employed his fluxions in any of his known examples of solving equations. And in the one example of finding a zero of a transcendental function, where the method usually named after him would have quickly given the answer, Newton goes through a convoluted argument based on the origin of the problem in planetary orbits to eventually find a value. So, barring further evidence to the contrary, to Simpson should go the credit for the fluxional method. "The immense popularity of Fourier's writings led to the universal adoption of the name 'Newton's method'" [29, pp. 194–195]. To explore this viewpoint further, consult also [37, 139, 259] and do Exercise 2.49.

But what else has been done? The literature is quite large. We have neglected the problem of choosing a starting value, of making an initial guess. A significant difficulty with these root-finding methods is how to pick an initial value that will lead to the right root. If the problem is physical, as many early problems were, then it should be obvious what a ballpark figure would be. But if not, it is even more pressing to know the distribution of the roots and their multiplicities before one begins iterating. An extensive literature developed in the nineteenth century to answer these questions. To read further, see one of the old texts on the theory of equations, such as [28, 47, 238]. Also the method of this section has been extended to functions of any number of variables and to spaces more general than Euclidean [8] (Exercises 2.50 and 2.51). In the next section we advance to contemporary developments of the twentieth century such as where algorithms converge and how fast. To what functions could they apply (polynomial, differentiable, continuous)? How robust might they be: how do small errors along the way affect them? And what could go wrong to prevent convergence?

Exercise 2.41. Define a fluxion. Offer more detail than Simpson provides.

Exercise 2.42. Look up an exposition of the so-called one-variable Newton's method that is written today, say from your calculus textbook. Compare with Simpson's exposition. Illustrate how much they are the same and how they differ, in general tenor, style and detail.

Exercise 2.43. Formulate Simpson's fluxional method in modern mathematical language using derivatives rather than fluxions. Use matrices and determinants in his Case II, $F(x, y) = 0$ and $f(x, y) = 0$, and rewrite his method as

$$\begin{bmatrix} F_x & F_y \\ f_x & f_y \end{bmatrix} \left(\begin{bmatrix} x_1 \\ y_1 \end{bmatrix} - \begin{bmatrix} x_0 \\ y_0 \end{bmatrix} \right) = - \begin{bmatrix} F \\ f \end{bmatrix},$$

where the functions and their derivatives are all evaluated at (x_0, y_0). Here Cramer's rule is useful, but not necessary. Rewrite Case I similarly.

Exercise 2.44. Show that Simpson's Case II is really supported by the geometrical argument given in the text. That is, derive from equations for the tangent planes his formulas with fluxions, or the modern formulas with derivatives.

Exercise 2.45. Simpson's Note 3 (just before Example I) is couched in the language of fluxions. What does it amount to in the language of derivatives?

Exercise 2.46. (a) A very old, popular and natural method for computing square roots is the following. For example, to find the square root of 2, take an initial estimate of $x_0 = 1$, and divide 2 by it. Average this quotient with x_0 to get the next iterate, $x_1 = 1.5$, and so on; work through x_4. Notice the quadratic convergence: the number of correct digits seems to double at each step.
(b) Try to find a corresponding algorithm to find the cube root of 2. Is it converging to the answer? Quadratically? If not, weight the average.

Exercise 2.47. (a) For computing $\sqrt{A}$, show that Simpson's fluxional method reduces to

$$x_{n+1} = \frac{1}{2}\frac{A}{x_n} + \frac{1}{2}x_n.$$

(Hint: Let $f(x) = A - x^2$.) Compare with Exercise 2.46a.
(b) For computing $\sqrt[3]{A}$, show that the method reduces to

$$x_{n+1} = \frac{1}{3}\frac{A}{x_n^2} + \frac{2}{3}x_n.$$

Compare with Exercise 2.46b.
(c) For computing A^s ($s \neq 0$), discover and verify an averaging formula anal-ogous to parts (a) and (b) to which Simpson's fluxional method reduces.
(d) Use (c) to evaluate inverses ($s = -1$) while avoiding divisions! Calculate 3^{-1} starting with $x_0 = 0.1$.

Exercise 2.48. Return to the Babylonian problem of Section 1 after it is put into standard form:

$$xy = 210,$$
$$x + y = 29.$$

Solve this anew by Case II of this section starting with guesses, $x_0 = 10$ and $y_0 = 20$.

Exercise 2.49. Why did Newton not discover his method?

Exercise 2.50. For differentiable functions from the complex numbers to complex numbers, there are at least two possible algorithms, patterned after Simpson's. The first would use Case I with complex differentiation. The second would use Case II by changing the complex function into two real functions, each of two arguments. Work out both algorithms. Do you end up with the same algorithm? (Hint: first learn about the Cauchy–Riemann equations [38].)

Exercise 2.51. Formulate Simpson's fluxional method geometrically for three functions each in three variables. Tangent planes now become tangent "hy-perplanes." Derive his formulas or their modern counterparts.

2.5 Smale Solves Simpson

One might think that a field almost four thousand years old would now be exhausted. On the contrary, the study of numerical methods is alive and well today. Finding roots of functions is one small corner of a discipline called numerical analysis, which can trace its ancestry back to the ancient Egyptians, but which got its start as a distinct field of study only with the invention of electronic computers; previous contributions tended to be subordinated to other scientific concerns. However, it is an important corner, for careful programming of computers requires a thorough examination of the underlying algorithms. Computation done by hand allows a perceptive mathematician to sense how well the computation is going; it naturally uncovers bizarre behavior that a computer might mask. Accordingly, theoretical questions of convergence tend to be neglected. However, a generic computer program running on an anonymous computer has to have all the necessary safeguards built into it to sense when sufficient accuracy has been achieved, or otherwise, to sense when convergence is not possible, or at least not guaranteed. A program that crashes may crash an airplane.

It might be good to pause to realize how much the methods of the past have permeated much of what we do today, and yet also to realize how much electronic computers have changed how we do it. Only sixty years ago computation by hand dominated, for there was no other way. Of course, there were tables of logarithms, slide rules and desk calculators. These were invented in the 1600s; and over time they evolved with increasing accuracy and speed. Tables of logarithmic, exponential and trigonometric functions grew bigger and easier to use. Slide rules added more and more special scales. But such improvements were limited by the capabilities of paper, wood and metal, and came slowly.

In what was an unexpected breakthrough, by the standards of the time, the Friden company in 1952 introduced an electromechanical desk calculator that could extract a square root at the touch of a single key and within a few seconds! Unfortunately, it was undercut by the quickly increasing sales of the first commercially available, but expensive, electronic computer, the Univac, already introduced in 1947. These two events dramatize the shift from electromechanical machines to electronic computers, and underscore the beginning of the dramatic increase in computer power that is still continuing today.

However, all the devices before the advent of electronic computers only provided the basic arithmetical operations and a few transcendental ones. There was no way to automate iteration. Each step had to be carried through on paper with the help of these auxiliary devices. Up until fifty years ago, science and engineering required only limited accuracy, rarely over five significant digits. Greater accuracy taxed one's ingenuity. One milked each digit for all it was worth. Also one had a limited vision of the number line; numbers were neither exceptionally large nor exceptionally small. To give a flavor of such computation as late as the 1950s, here is the homely advice of Cyrus C. MacDuffee [161, pp. 64–67].

> The roots of an equation are said to be *isolated* when intervals have been determined each of which contains just one root. To be useful these intervals should be small, usually between consecutive integers, or better, between consecutive tenths. A refinement is impossible until we have decided which root it is that we wish to approximate. ... It is impossible to state in general how many decimal places to keep, for this differs from equation to equation. Usually each repetition of the Newton process just about doubles the number of significant figures in the approximation. The last digit may not be correct, but it may be better to keep it than to replace it with 0.

Let us take stock of the many algorithms and their numerous variants for finding roots, and put them into a broader framework. They fall into two broad classes: exact methods in the first section of this chapter and inexact methods in the middle three sections. This classification is somewhat arbitrary. The algebraic method presented by Cardano to find the roots of a third-degree polynomial requires the extraction of square and cube roots, which at that time was nontrivial, and in any case, they can only be approximated in general.

In discussing the algorithms presented in the last three sections, both in the expositions and the exercises, we have encountered a plethora of names, both proper and descriptive, that are sometimes inaccurate and often confused with one another. Add to this some apparently self-perpetuating historical myths. Writers use such phrases as "essentially the same," "equivalent," and "a special case of" to lump together algorithms that we have distinguished. Of course, this depends on what is meant by "equivalent," and how coarse or fine this equivalence is. Some would view the methods of Qin, Horner, Newton and Simpson as all cut from the same cloth. As all are iterative processes using linear interpolation or extrapolation to compute a correction term, in that sense they are all close in spirit. For they all depend, once they are in the neighborhood of a root, on linearizing or extrapolating in some way what still needs to be computed: by omitting terms higher than the first, by proportional juggling, or by using the derivative to find a line tangent to the graph of the function. But with more discrimination, we shall see that we should consider them as not equivalent since their speeds of convergence are different: linear for Qin, quadratic for Simpson, and in between for Newton's proportional method. Whether an algorithm always converges to a root, how fast it does so, and whether it corrects small errors, these are also important characteristics that should distinguish different algorithms. This section discusses in turn each of these important aspects of a computation, makes them precise, and uses them to discriminate among the algorithms we have studied. Many of the techniques and results we use and engage in are part of the modern theory called "real analysis."

We start off with the general theory of fixed points, which will help us to answer some of these specific questions; and, in particular, make it trivial to derive the local convergence of Simpson's fluxional method. We will find that there is a trade-off between speed of convergence and sensitivity to initial conditions. The most reliable routines are the slowest. Some algorithms, such as the bisection method (Exercise 2.55), always converge to a root, but slowly, whereas Simpson's fluxional method is quite fast—when near a root it doubles the number of significant digits at each step—but it may fail to converge, or worse yet, converge to a root different from that expected.

Although Simpson's fluxional method has proven quite reliable in practice, there was no rigorous justification for its success. In fact, one can easily construct counterexamples showing how it can fail in many different ways. Toward the end of this section we will address this; it leads to the celebrated theorem of Stephen Smale, which explains what is going on, at least for polynomials. Colorful fractal patterns illustrate this root-finding by tracing its dependence on initial guesses.

What happens when a pen makes a slip or a computer loses an electron? Is all lost? Not necessarily. A *robust* algorithm is one that corrects errors as long as they are small and there are not too many of them. Of course, convergence may be delayed. The methods of Newton and Simpson are robust, whereas that of Qin is not. Any mistake along the way may doom the algorithm to a wrong answer. There are also subtle problems with computer arithmetic that complicate life and which we will ignore [36, Chapter 2].

In practice, and beyond the scope of this chapter, one often uses up to three different routines in a successful root-finder: first one finds a ballpark estimate of where the root lies; then a reliable, general routine will narrow the root to within maybe 10% of its true value; and finally a much faster, more narrowly focused routine will complete the calculation. So, how do we begin? While it is often difficult to guess the first digit and since many algorithms cough and sputter at the beginning, let us assume at first that somehow a good initial estimate of the root has been made.

LOCAL BEHAVIOR: FIXED POINTS, CONVERGENCE, AND SPEED. As opposed to his proportional method, Newton already noticed that in his polynomial method (Exercise 2.39) the number of significant digits doubles at each step. This is characteristic also of Simpson's fluxional method, whose precise rate of convergence will be worked out shortly. It is typical of the numerical algorithms of this chapter that the number of significant digits eventually increases by a certain amount at each iteration, with the amount depending on the particular algorithm. Algorithms that find one digit at a time, such as described by Qin, al-Ṭūsī, Viète, Ruffini and Horner, are called *linear*. For a particular method, these digits are calculated in a particular base b (where $b > 1$). For the algorithms just mentioned the base is 10, but it may be otherwise, such as 2 in the bisection method (Exercise 2.55). More than one digit in the base b could be calculated at a time, say n of them, but then this would be the same as calculating one digit at a time in the base b^n.

We shall show that Simpson's fluxional method is generally *quadratic*. By this is meant that, if r is the root of a function and $x_0, x_1, x_2, \ldots$ are successive approximations to r, then there is a positive constant C such that for all n,

$$|x_{n+1} - r| \leq C |x_n - r|^2. \qquad (2.7)$$

Alternatively, if we designate the error as $\varepsilon_n = |x_n - r|$, then $\varepsilon_{n+1} \leq C\varepsilon_n^2$. These ε_ns are called *absolute* errors.

To see the connection of quadratic convergence to the doubling of significant digits, it is necessary to introduce the *relative* error: $\rho_n = \varepsilon_n / |r|$. This measures as a fraction of r how far the current approximation is from the true value of the root. The smaller ρ_n is, the larger the number of significant digits. Quantitatively, the number σ_n of significant digits of x_n in its approximation to r is measured roughly by the negative of the logarithm (to the base 10) of ρ_n, i.e., $\sigma_n = -\log_{10} \rho_n$ (look at the number of leading zeros in ρ_n after the decimal point and ignore the fractional part of the log). If there is quadratic convergence, then (2.7) becomes

$$\rho_{n+1} = \frac{|x_{n+1} - r|}{|r|} \leq C |r| \frac{|x_n - r|^2}{|r|^2} = C |r| \rho_n^2.$$

Taking logarithms and negating, we obtain

$$-\log_{10} \rho_{n+1} \geq -\log_{10} \left(C |r| \rho_n^2 \right) = 2 \left(-\log_{10} \rho_n \right) - \log_{10} \left(C |r| \right).$$

Equivalently,

$$\sigma_{n+1} \geq 2\sigma_n \quad C',$$

where $C' = \log_{10}(C|r|)$. Thus the number σ_n of significant digits of the iterates x_n tends to double as n grows large, although, because of the presence of the constant C', they may not do so early on (Exercise 2.52).

For other methods of finding roots, the exponent in (2.7) may be different. Even better than Simpson's method, using quadratic extrapolation at each step gives cubic convergence, with the exponent of (2.7) changed from 2 to 3, but the extra computations at each iteration nullify its advantages (Exercise 2.38). Newton's proportional method has an exponent less than 2; curiously, this exponent is the golden section: $\phi = \left(\sqrt{5} + 1 \right) / 2 = 1.618\ldots$; a proof of this is outlined in Exercise 2.53. Since the analysis is easier for Simpson's fluxional method, we do it thoroughly in the text.

The exponent in (2.7) might also be 1. But care is needed here if the desire is to capture linearity as defined earlier when one is adding one digit at a time. The problem is that there may be in the root a string of successive digits that are all 0, in which case the successive errors are not decreasing for a while. So linearity may not be defined locally by comparing successive approximates, but must be defined globally. Using the notation now developed, we may define it more generally by saying that an algorithm is *linear* if there is a constant C between 0 and 1 and another positive constant D such that for all n,

$$\varepsilon_n < C^n D. \tag{2.8}$$

Some of the algorithms of this chapter belong to a larger class of numerical root finders called fixed-point methods. We study these first to gain a deeper insight into how these algorithms work, e.g., Simpson's method and Smale's modification of it. For simplicity, let us assume that the function f is analytic in some interval about the root r, that is, developable in a power series (Taylor series) about r such that the series converges to f, although often only a continuous second derivative is needed. When we talk about an interval $[a, b]$ being about a root r, we mean that $a < r < b$.

Definition 2.1. By a *fixed point* of the function φ is meant a number x such that

$$\varphi(x) = x.$$

The *fixed-point algorithm* for solving such an equation is the iterative scheme

$$\begin{cases} x_0 = \text{initial guess}, \\ x_{n+1} = \varphi(x_n). \end{cases}$$

Our exposition of the elementary theory of fixed points follows roughly that of [242, Chapter 17]. Fixed points and roots of functions are intimately related.

Proposition 2.2. *If two functions φ and ψ are so related that $\varphi(x) = \psi(x) + x$, then φ has a fixed point at r if, and only if, ψ has a root at r.*

So, searching for roots of ψ is equivalent to searching for fixed points of φ. To view Simpson's fluxional method in terms of the fixed-point algorithm, let $\varphi(x) = x - f(x)/f'(x)$ and $\psi(x) = -f(x)/f'(x)$. Assuming $f'(x) \neq 0$, clearly $\psi(r) = 0$ iff $f(r) = 0$; so f has the root r iff φ has the fixed point r. Thus, the iterates in Simpson's method are the same as those in the corresponding fixed-point algorithm for φ.

Figure 2.16 illustrates the fixed-point algorithm of the next theorem. Sought is a fixed point of the function, $\varphi(x) = \text{arccot}\, x$, i.e., an angle x (in radians) equal to its cotangent. From the preceding, this corresponds to a root of $\text{arccot}\, x - x$. The dashed diagonal is the graph of $y = x$, so where it intersects the curve $y = \text{arccot}\, x$ is the fixed point: $\varphi(r) = r$. In the figure, by sending the functional value $\text{arccot}\, x_0$ over to the diagonal and up or down to the curve, we obtain the next functional value $\text{arccot}\, x_1$, and so on. Since $-1 < \varphi'(x) < 0$ when $x > 0$, the hypothesis of the next theorem is satisfied and therefore the fixed-point algorithm converges to $r = 0.86033\ldots$. In the proofs to come, some standard theorems of calculus will be used; see [227] for example.

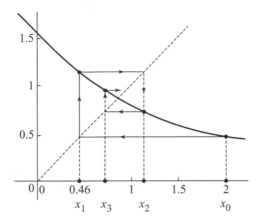

Fig. 2.16. Convergence to a fixed point of arccot.

Theorem 2.3. *If $|\varphi'(x)| < 1$ throughout an interval $[a, b]$ containing a fixed point r of φ, and the initial guess x_0 is in this interval, then the fixed point algorithm converges to r.*

Proof. Let x_0, x_1, x_2, ... be successive iterates, and let ε_0, ε_1, ε_2, ... be the absolute errors in these approximations: $\varepsilon_n \equiv |x_n - r|$. Recall that $x_{n+1} = \varphi(x_n)$ and $r = \varphi(r)$. By the mean value theorem, for each n there is an intermediate value ξ_n between x_n and r such that

$$(x_n - r)\,\varphi'(\xi_n) = \varphi(x_n) - \varphi(r).$$

But $\varphi(x_n) - \varphi(r) = x_{n+1} - r = \pm\varepsilon_{n+1}$. Thus $\varepsilon_n\,|\varphi'(\xi_n)| = \varepsilon_{n+1}$. Since φ' is continuous throughout $[a, b]$, by the extreme value theorem the function $|\varphi'(x)|$ must attain its maximum value M on $[a, b]$. So $|\varphi'(x)| \leq M$ throughout $[a, b]$, with $M < 1$ since M is a value of $|\varphi'(x)|$ in $[a, b]$ and $|\varphi'(x)| < 1$ for all x in $[a, b]$. Consequently, from the preceding, $\varepsilon_{n+1} = \varepsilon_n\,|\varphi'(\xi_n)| \leq M\varepsilon_n$; hence $\varepsilon_n \leq M^n\varepsilon_0$ for all n. Now $\lim_{n \to \infty} M^n = 0$ since $|M| < 1$. Therefore, as $n \to \infty$, so $\varepsilon_n \to 0$. We conclude that $x_n \to r$.

This proof shows that the fixed-point algorithm converges linearly (where $M = C$ in (2.8)). The statement of the theorem also implies that the fixed point is unique within the specified interval. Some functions, such as $y = x^2$, may have more than one fixed point overall. The reader should investigate its fixed points and how the algorithm works or doesn't work in the vicinity of each fixed point (Exercise 2.54). For more about the theory of fixed points, see [208].

Let us apply this theorem to Simpson's fluxional method.

Corollary 2.4. *If $|f(x)\,f''(x)| < |f'(x)|^2$ throughout a closed interval that contains a root of f, then Simpson's fluxional method converges in that interval to the root.*

Proof. Let $\varphi(x) \equiv x - \frac{f(x)}{f'(x)}$. Observe by the hypothesis that $f'(x) \neq 0$ on the interval. Then $\varphi'(x) = \frac{f(x)f''(x)}{f'(x)^2} < 1$.

Augustin-Louis Cauchy (1789–1857) first gave the criterion in this corollary for convergence, but he did not use fixed points to prove it. (See [34, pp. 210–216].)

Next we examine the rate of convergence of the fixed-point method in a special case, and then apply this to showing that Simpson's fluxional method converges quadratically in general.

Theorem 2.5. *If r is a fixed point of φ, and if $\varphi'(r) = 0$, then the fixed-point algorithm converges quadratically in an interval about the fixed point r.*

Proof. Since $\varphi'(r) = 0$, there is an interval around r such that $|\varphi'(x)| < 1$ in this interval. So now we can use the notation from Theorem 2.3 and its proof. By the mean value theorem, for each x in the interval there is a ξ between x and r such that

$$\varphi''(\xi)(x - r) = \varphi'(x) - \varphi'(r) = \varphi'(x).$$

In particular, for each ξ_n as in the previous proof, there is another $\widehat{\xi}_n$ between ξ_n and r such that

$$\varphi''(\widehat{\xi}_n)(\xi_n - r) = \varphi'(\xi_n).$$

By the extreme value theorem, $|\varphi''(x)|$ attains its maximum C in the interval. Hence

$$|\varphi'(\xi_n)| = |\varphi''(\widehat{\xi}_n)| \, |\xi_n - r| \leq C \, |x_n - r| = C\varepsilon_n,$$

since ξ_n is between x_n and r. Also from the proof of Theorem 2.3, we know that $\varepsilon_{n+1} = \varepsilon_n |\varphi'(\xi_n)|$. We conclude that $\varepsilon_{n+1} \leq C\varepsilon_n^2$, which is quadratic convergence.

Corollary 2.6. *If $f(r) = 0$ and $f'(r) \neq 0$, then there is an interval about r such that Simpson's fluxional method converges quadratically in it to the root r.*

Proof. As usual, set $\varphi(x) = x - \frac{f(x)}{f'(x)}$. Since $f(r) = 0$, it follows that $\varphi'(r) = \frac{f(r)f''(r)}{f'(r)^2} = 0$.

Even though we expect Simpson's fluxional method to double the number of digits at each step, realize that this doubling will be delayed if the constant C in the proof of Theorem 2.5 is large. Also multiple roots impede the convergence, that is, if $f'(r) = 0$ at the root r, then the convergence is only linear.[8] However, there are ways to overcome this, providing one knows ahead of time that the root is multiple (Exercise 2.58) [36, p. 89].

---

[8] If the multiple zero has order m, then the constant C of linearity in (2.8) is $(m - 1)/m$. So the algorithm converges very slowly when m is large.

GLOBAL BEHAVIOR: SMALE'S THEOREM. The final two topics are contemporary and reflect in different ways the impact of high-speed electronic computers on mathematics. The first is Smale's justification of Simpson's method; the second is about fractals that illustrate the dependence of the solution on the initial guess. First a few words about Smale.

Stephen Smale was born in Flint, Michigan, July 15, 1930, and educated at the University of Michigan, receiving the Ph.D. in 1957. He was a professor at the University of California at Berkeley until he retired in 1994. In 1966 he received the Fields medal, the highest research award in mathematics, sometimes compared to the Nobel prize, which has no award for mathematics. René Thom gave the address honoring Smale for his original research in differential geometry: "... if the work of Smale does not possess perhaps the formal perfection of definitive work, it is because Smale is a pioneer who takes risks with tranquil courage" [237]. The field of differential geometry traces its origins to the notion of curvature, whose history we relive in another of our chapters. In 1988 Smale received the Chauvenet prize for expository writing from the Mathematical Association of America for his paper *On the Efficiency of Algorithms of Analysis* [209], a sequel to the paper that we are shortly going to read a part of. And there are many more prizes he has received. For more detail, see [13] and Smale's web site.

Smale protested the Vietnam war vigorously, denounced both the United States and the Soviet Union on the steps of Moscow State University a day after he received the Fields medal, and was subsequently taken for a fast ride through the streets of Moscow by the Soviet authorities before being released [217].

Photo 2.6. Smale.

Our excerpt of the first two pages of [216] summarizes succinctly the essence of Smale's theorem justifying the use of Simpson's fluxional method in practice, which Smale calls Newton's method, and which by this time had become widespread and accepted as fast and reliable, although there was no proof guaranteeing global convergence. We encourage the reader to delve further into the original article. As you read it, you should become aware of three complementary undercurrents bearing ideas and techniques into it: the extraordinary rise of mathematical, and more generally scientific, activity in the United States; the internationalization of mathematics, evidenced by the variety of countries in the references; and finally the close collaboration of Smale with three of his colleagues at Berkeley: Gerard Debreau, Morris W. Hirsch, and Richard Karp. See [218] for further development of some of these ideas.

To help in reading the excerpt, we make several preliminary annotations.

- The coefficients may now be complex numbers, and the iterates may be complex, including the initial value z_0.
- In the third paragraph, the Bieberbach conjecture is this: if a complex-valued function is representable by a power series, $f(z) = z + \sum_{n=2}^{\infty} a_n z^n$, for all complex numbers z such that $|z| < 1$, and if it is one-to-one there, then $|a_n| \leq n$ for all n, with equality holding only for Koebe's functions, $f(z) = z/(1 - cz)^2$ ($|c| = 1$). Louis DeBranges settled this conjecture in 1985, building on the results of many mathematicians [123, article 438 C].
- In the fifth paragraph, an example of such a search method is the bisection method of Exercise 2.55.

∞∞∞∞∞∞∞∞∞∞

Stephen Smale, from

The Fundamental Theorem of Algebra and Complexity Theory

Part I

1. The main goal of this account is to show that a classical algorithm, Newton's method, with a standard modification, is a tractable method for finding a zero of a complex polynomial. Here, by "tractable" I mean that the cost of finding a zero doesn't grow exponentially with the degree, in a certain statistical sense. This result, our main theorem, gives some theoretical explanation of why certain "fast" methods of equation solving are indeed fast. Also this work has the effect of helping bring the discrete mathematics of complexity theory of computer science closer to classical calculus and geometry.

A second goal is to give the background of the various areas of mathematics, pure and applied, which motivate and give the environment for our problem. These areas are parts of (a) Algebra, the "Fundamental theorem of algebra," (b) Numerical analysis, (c) Economic equilibrium theory and (d) Complexity theory of computer science.

An interesting feature of this tractability theorem is the apparent need for use of the mathematics connected to the Bieberbach conjecture, elimination theory of algebraic geometry, and the use of integral geometry.

Before stating the main result, we note that the practice of numerical analysis for solving nonlinear equations, or systems of such, is intimately connected to variants of Newton's method; these are iterative methods and are called fast methods and generally speaking, they are fast in practice. The theory of these methods has a couple of components; one, proof of convergence and two, asymptotically, the speed of convergence. But, not usually included is the total cost of convergence.

On the other hand, there is an extensive theory of search methods of solution finding. This means that a region where a solution is known to exist is broken up into small subsets and these are tested in turn by evaluation; the process is repeated. Here it is simpler to count the number of required steps and one has a good knowledge of the global speed of convergence. But, generally speaking, these are slower methods which are not used by the practicing numerical analyst.

The contrast between the theory and practice of these methods, in my mind, has to do with the fact that search methods work inexorably and the analysis of cost goes by studying the worst case; but in contrast the Newton type methods fail in principle for certain degenerate cases. And near the degenerate cases, these methods are very slow. This motivates a statistical theory of cost, i.e., one which applies to most problems in the sense of a probabilistic measure on the set of problems (or data). There seems to be a trade off between speed and certainty, and a question is how to make that precise.

One clue can be taken from the problem of complexity in the discrete mathematics of theoretical computer science. The complexity of an algorithm is a measure of its cost of implementation. In these terms, problems (or algorithms) which depend on a "size" are said to be tractable provided the cost of solution does not increase exponentially as their size increases. The famous $P = NP$ problem of Cook and Karp lies in this framework.

∞∞∞∞∞∞∞∞∞

Here P means that a class of problems is solvable on a Turing machine in polynomial time, meaning that the number of machine steps required is bounded by a polynomial whose argument is the size of a particular problem in the class. A *Turing machine* is an idealized model of the modern digital computer. On the other hand, NP has a similar meaning, but the computation is to be done on a nondeterministic Turing machine. By *nondeterministic* is meant that at each step there are several alternatives to choose from, and from each alternative even more choices, and so on. If some sequence of choices for all the steps leads to the answer, then the computation is considered successful, but a priori one would not know which path to choose through all the alternatives. Needless to say, because the number of possibilities generally increases exponentially, nondeterministic Turing machines are not practical [123, article 71E]. In the next paragraph, notice Smale restricting himself to polynomials, while ignoring transcendental functions. And in the paragraph after that, z^* is the desired root.

∞∞∞∞∞∞∞∞∞∞

In the case of a single polynomial the obvious "size" is the degree d. So these considerations pose the problem. Given $\mu > 0$, an allowable probability of failure, does the cost of computation via the modified Newton's method for polynomials in some set of probability measure $1 - \mu$, grow at most as a polynomial in d? Moreover, one can ask that as μ varies, this cost be bounded by a polynomial in $1/\mu$. I was able to provide an affirmative answer to these questions.

Let me be more precise. The problem is to solve $f(z^*) = 0$ where $f(z) = \sum_{i=0}^{d} a_i z^i$, $a_i \in \mathbf{C}$ and $a_d = 1$. The algorithm is the modified Newton's method given by: let $z_0 \in \mathbf{C}$ and define inductively $z_n = T_h(z_{n-1})$ where $T_h(z) = z - hf(z)/f'(z)$ for some h, $0 < h < 1$. If $h = 1$, this is exactly Newton's method.

∞∞∞∞∞∞∞∞∞∞

The importance of h is that by taking it smaller the algorithm has less chance to go wild, and either not converge or converge to an unexpected root. But this is at the cost of slowing it down (Exercises 2.56, 2.57, 2.58).

Here are a few preparatory remarks for the next paragraph. The sequence z_n will fail to be defined completely if $f'(z_n) = 0$ for some n. Even if $f'(z_n) \neq 0$, worse things can happen (Exercise 2.1).

The inequality at the end of the same paragraph means that convergence is linear, adding at least one bit at a step. We have already seen how the algorithm may fool around, even when it is going to converge, before it finally takes off. An "approximate zero" z_0 means that we are in the home stretch, at least linearly to the base 2.

This requires some explanation. It is important to notice that the error is now measured by functional values: $|f(z_n)| = |f(z^*) - f(z_n)|$, rather than arguments: $\varepsilon_n = |z^* - z_n|$. What is their relationship? If f is expressed as a Taylor series expanded about z^*, then $f(z_n) \approx f'(z^*)(z_n - z^*)$, where higher powers of $z_n - z^*$ are usually negligible in a neighborhood of z^*. Thus $\varepsilon_n/\varepsilon_{n-1} \approx |f(z_n)/f(z_{n-1})| < \frac{1}{2}$. So the error in the argument also shrinks.

Even though Smale expects only linear convergence, usually the speed of convergence eventually becomes quadratic. But observe what happens with $f(z) = z^2$, where these approximations converge exactly linearly. To see this, calculate that $z_{n+1} = \frac{1}{2}z_n$ (cf. Corollary 2.6 and Exercise 2.58).

∞∞∞∞∞∞∞∞∞∞

We will say that z_0 is an *approximate zero* provided if taking $h = 1$, the sequence z_n is well defined for all n, z_n converges to z^* as $n \to \infty$, with $f(z^*) = 0$ and $|f(z_n)/f(z_{n-1})| < \frac{1}{2}$ for all $n = 1, 2, \ldots$.

Practically and theoretically this is a reasonable definition. One could say that in this case, z_0 is in a strong Newton Sink.

Let $\mathcal{P}_d$ be the space of polynomials f, $f(z) = \sum_{i=0}^{d} a_i z^i$, $a_d = 1$. Thus $\mathcal{P}_d$ can be identified with $\mathbf{C}^d$, with coordinates $(a_0, \ldots, a_{d-1}) = a \in \mathbf{C}^d$.

Define

$$P_1 = \{f \in \mathcal{P}_d \mid |a_i| < 1, i = 0, \ldots, d - 1\}$$

and use normalized Lebesgue measure on P_1, for a probability measure.

∞∞∞∞∞∞∞∞

Henri Lebesgue (1875–1941) generalized Riemann's theory of integration to cover exotic functions that were not integrable in Riemann's original sense; Lebesgue's notion of measure also satisfied some properties not shared by Riemann's. Roughly, to explain the phrase "normalized Lebesgue measure," one should think of $\mathcal{P}_d$ as a Euclidean space of d dimensions, but with complex coordinates. P_1 is a bounded subset of $\mathcal{P}_d$ with a finite volume. Each coordinate is confined to a disk of radius 1 in the complex plane, and therefore has area, $\pi r^2 = \pi$. We multiply areas from different coordinates to get the total volume π^d of P_1. Divide by this to normalize the volume of P_1 to 1. Thus, when $d = 3$ the volume of the set of cubic polynomials in P_1 with coefficients whose real parts are positive is $\frac{1}{8}\pi^3$, and its measure is $\frac{1}{8}$.

Subtly the viewpoint changes in the Main Theorem. Instead of fixing a polynomial and varying the starting point to observe whether and how the method converges, Smale turns the tables on us, and fixes the starting point while varying the polynomial.

∞∞∞∞∞∞∞∞

MAIN THEOREM. *There is a universal polynomial $S(d, 1/\mu)$, and a function $h = h(d, \mu)$ such that for degree d and μ, $0 < \mu < 1$, the following is true with probability $1 - \mu$. Let $x_0 = 0$. Then $x_n = T_h(x_{n-1})$ is well defined for all $n > 0$ and x_s is an approximate zero for f where $s = S(d, 1/\mu)$.*

∞∞∞∞∞∞∞∞

The last equality should really be $\geq$ since the universal polynomial S may give nonintegral values. In the next paragraph, one possibility for S is given; later in his article, in section 3, beyond this extract, Smale defines h.

∞∞∞∞∞∞∞∞

More specifically we can say, if $s \geq [100(d+2)]^9/\mu^7$, then with probability $1 - \mu$, x_s is well defined by the algorithm for suitable h and x_s is an approximate zero of f.

Note especially that h and s do not depend on the coefficients.

The use of probability is made more precise in the following very brief idea of the proof. There is a certain $W_* \subset \mathcal{P}_d$ such that $f \in W_*$, $z_0 = 0$ is a "worst case" for the algorithm "in the limit" $h \to 0$. We don't expect the algorithm to work in this case, no matter how small h is taken. But if $f \notin W_*$ the algorithm will converge for sufficiently small h.

∞∞∞∞∞∞∞∞

For example, the polynomial $z^2 + 1$ will belong to this exceptional set W_* since a real polynomial starting out at a real z_0 can never lead to an imaginary root.

FRACTALS. Although Smale's theorem guarantees convergence almost everywhere, it does not give information as to which starting points converge to which roots. Such investigations go back to at least Pierre Fatou [70] and Gaston Julia [126] in the early twentieth century. In general this is rather complicated, and leads to the beautiful fractal patterns of the color insert at the beginning of this chapter.

What we see in the first image "Fractal with Basins" is the coding of the three attractor basins for $f(z) = z^3 - 1$ by different colors, to show which points in the complex plane iterate to each of the three roots of unity under Simpson's method. Rather surprisingly, where any two regions meet, so does the third! This picture also illustrates how some points arbitrarily far away from the roots but arbitrarily close together may end up at different roots, and this can happen in a rather complicated way.

Nevertheless, there are simple observations to be made, but not proven here. Each basin of attraction is open, in the sense that about each point P of the basin a small disk may be drawn totally within the basin. Also, the three basins together are dense, i.e., every point of the complex plane has a basin point arbitrarily near to it. In other words, the points that fail to converge (which are invisible in the picture) are numerous but scattered. And yet in another sense they are very sparse, occupying no measurable space (this is called having measure zero) [21, pp. 263–68]. Returning to Figure 2.2 of the first section, we observe that the points failing to converge to 1, advancing to the left along the X-axis, are just the places where all three basins meet on the real axis in "Fractal with Basins". See [46, 95, 164, 184, 185, 255] for many more such pictures and do Exercises 2.59 and 2.60 for further activity.

The second color image "Fractal with Iterations" illustrates how the number of iterations needed in order to approach the root 1 increases as one moves away from the root within its attractor basin. The other two basins of attraction are now colored a background yellow; the origin is in the center of the picture. One must first choose a radius of tolerance ε circumscribing the root 1; in this particular picture it is taken to be 0.05. This means that when a sequence of iterations falls within ε of 1, it is considered to have arrived at the root, and is terminated. The large blue blob on the right contains those points that are already within ε of 1 and also those further points that need one iteration to get within this tolerance of 1. Those in the green region surrounding the blue need two, those next in the aqua three, etc. But notice that this sequence of eight colors starts repeating to the left of their first appearance, with the new blue containing points requiring nine iterations, and so on. Of special interest is the much smaller blue blob on the negative real axis; points in it are iterated just once for them to jump over to within ε of 1. (Why aren't there points in it needing none?)

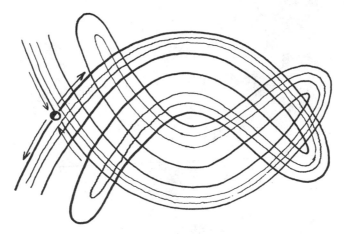

Fig. 2.17. Smale horseshoe.

Fractals relate to the branch of chaos theory called *strange attractors*. Earlier in his career, before settling the practicality of Simpson's method, Smale investigated the global behavior of nonlinear differential equations. His equations for what is now called the Smale horseshoe [95, p. 51] are in three variables x, y and t, and their derivatives $\dot{x}$, $\dot{y}$, $\ddot{x}$, $\ddot{y}$, etc. He represents solutions of his horseshoe by curves in the xy-plane (Figure 2.17). The characteristic feature of this whirl of curves is that trajectories that start close together may diverge and end up far apart, the so-called butterfly effect [1, p. 95], [2, p. 125], [95, p. 9ff]. Another such dynamical system, discovered ten years later by Edward Lorenz from biological considerations, is the Lorenz attractor of Figure 2.18; see [1, p. 61], [2, p. 128], [95, facing p. 114] and [184, p. 2].

The similarity of nonlinear dynamical systems and fractal patterns is this: solutions with initial conditions that are near each other may separate and go their own way. The difference is that t is a continuous parameter in a dynamical system coming from differential equations, whereas in Simpson's fluxional method it is discrete: $n = 0, 1, 2, \ldots$. Note that his formula (2.1) is also nonlinear: the derivative occurs in the denominator.

For those who would like to delve further into the past, we recommend the compendiums of original sources of algorithms by Goldstine [97] and Chabert et al. [34]. Do also Exercises 2.61 and 2.62.

Exercise 2.52. We want to solve numerically the equation $100x^2 + x = 0$.
(a) With 1 as the initial value, use Simpson's fluxional method to compute the first four iterations.
(b) Why is the convergence apparently so slow and not quadratic? Hint: look at the first derivative.
(c) When might the iterates start converging quadratically?

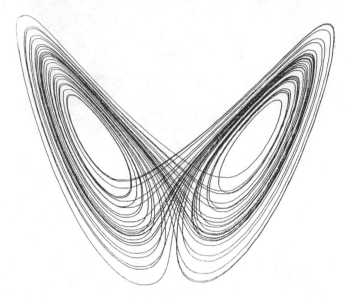

Fig. 2.18. Lorenz attractor.

Exercise 2.53. Newton wished to find a root of

$$f(x) = x^3 + 10x^2 - 7x - 44.$$

The initial data are $x_0 = 2$ and $x_1 = 2.2$. The iteration step (Exercise (2.37) is

$$x_{n+1} = \frac{x_{n-1} f(x_n) - x_n f(x_{n-1})}{f(x_n) - f(x_{n-1})}.$$

(a) Pursue this for several steps, using Maple, Mathematica, or some other computer package to make a table of x_n for n from 0 to 6, going farther and more accurately than Newton did. Newton appears to be adding two significant figures at each step. This would appear to be linear convergence, adding significant digits twice as fast as Qin's algorithm. Is this justified, and why? This exercise answers these questions. As you can see in your computed values of x_n, the significant digits appear to be increasing faster than linear but less than quadratic. This is a general phenomenon with Newton's proportional method, and the rate of increase of significant digits approaches a limit that can be discovered analytically; this limit is independent of the function, subject to certain differentiability conditions.

We will do a rough analysis, without invoking δ's and ε's to justify our assertions. We assume the iterates converge to a limit, $\lim_{n\to\infty} x_n = r$. It is convenient to assume that $r = 0$. We have shifted the x-axis and the error at any step is now just x_n. Without loss of generality we may believe that $|x_n| < 1$. The function f is assumed to be twice differentiable, and also its first and second derivatives are not zero at r. Thus, near the root r, by Taylor's theorem,

$$f(x) = c_1 x + c_2 x^2, \text{ approximately, with } c_1 \neq 0 \neq c_2.$$

(b) Establish that $x_{n+1} \approx \frac{c_2 x_n x_{n-1}}{c_1}$, as x_n becomes small. Now take the negative of logarithms, setting $u_n = -\log|x_n|$ and $d = -\log\left|\frac{c_2}{c_1}\right|$. The last line should become

$$u_{n+1} \approx d + u_n + u_{n-1}. \tag{2.9}$$

Since $|x_n| < 1$, the u_n are positive and increasing in magnitude. Thus d may be eventually ignored.

(c) Verify that we are left with an iterative step that is the same as for the Fibonacci numbers: $F_{n+1} = F_n + F_{n-1}$. To get started finding a solution, assume that it has the form $u_n = Kb^n$. This is the standard approach for solving linear difference equations [23], which is quite similar to that for solving linear differential equations. Into (2.9), with d removed, substitute $u_n = Kb^n$. Simplify to a quadratic equation in b, and find its roots. Note that any two solutions of the difference equation may be added to obtain a new solution. Show that as n grows large, the solution tends to

$$u_n \approx K\phi^n,$$

where ϕ is the golden section.

(d) Verify that K is $\frac{1}{\sqrt{5}}$ for the Fibonacci numbers F_n when $F_0 = 0$ and $F_1 = 1$; but in general it depends on the initial conditions. Make a table of $\frac{\phi^n}{\sqrt{5}}$ and u_n for n from 0 to 6, and notice that the nearest integers to the $\frac{\phi^n}{\sqrt{5}}$ are just the F_n.

(e) Now let's return to the original problem. The logarithmic error u_n measures the number of significant digits; thus it grows as the Fibonacci numbers grow. Verify this from your tables. You may first wish to subtract the best estimate of r from the iterates. By your first table, you can see that Newton was right in his choice of significant digits, but the stages after his allow for more. Did Newton know this? See [36, pp. 97–101] for a related derivation.

Exercise 2.54. Explore the use of the fixed-point algorithm to find the fixed points of the function $\varphi(x) = x^2$. Try several initial points, lesser and greater than each fixed point, and iterate from each to see what happens. Such fixed points are called *attractors* and *repellers*.

Exercise 2.55. Perhaps the simplest and most general algorithm for finding roots is the bisection method, which requires only one evaluation of a continuous function at each step, and makes minimal demands on memory. It is also the slowest. Let f be a function continuous on the interval $[x_0, x_1]$; we will find a sequence of arguments $x_0, x_1, x_2, \ldots$, and corresponding values of the function $y_0 = f(x_0), y_1 = f(x_1), y_2 = f(x_2), \ldots$. To calculate a root of f one starts with two values of the argument, x_0 and x_1, such that y_0 and y_1

have opposite signs, say $y_0 < 0$ and $0 < y_1$. Take the midpoint $\frac{x_0 + x_1}{2}$ as a new argument x_2 and evaluate f at it. Choose y_0 or y_1 to have opposite sign from y_2. Keep repeating the process to get a sequence of iterates: $x_0, x_1, x_2, \ldots$.
(a) Prove that this sequence always converges to a root r of f in $[x_0, x_1]$.
(b) Prove that the rate of convergence is bound by

$$|x_n - r| \le |x_1 - x_0| \, 2^{-(n-1)};$$

that is, it is linear in the sense that at least one binary bit of significance is added to the answer at each step.
(c) There is a curious feature of the bisection method: the errors, $\varepsilon_n = |x_n - r|$, do not necessarily decrease at each step. To observe this phenomenon, calculate the first nine approximates x_n of the root r of the function $f(x) = x - \pi$, starting with initial values of 3 and 4.

Exercise 2.56. This exercise examines precisely how varying Smale's h, coming from the "modified Newton's method" in his text, changes the rate of convergence for the simplest nontrivial polynomial: $f(x) = a_1 z + a_0$.
(a) Prove by induction that $z_n = (1 - h)^n \left(z_0 + \frac{a_0}{a_1} \right) - \frac{a_0}{a_1}$.
(b) How many iterations are needed to obtain the root when $h = 1$?
(c) Show that $f(z_n) = (1 - h)^n (a_1 z_0 + a_0)$.
(d) For which h will z_0 be an approximate zero? (Allow h to be different from 1 in Smale's definition of an approximate zero.)
(e) Does the algorithm ever fail to converge to a root of f?

Exercise 2.57. This exercise explores what Smale calls the modified Newton's method.
(a) Show graphically why this method (when $h < 1$) slows down the rate of convergence.
(b) Calculate $\varphi'(x)$ for the fixed-point function φ associated with this method. Write it as

$$\varphi'(x) = 1 - h \left(1 - \frac{f(x) f''(x)}{f'(x)} \right).$$

Using Theorem 2.3, illustrate why decreasing h generally increases the region of convergence of this method.

Exercise 2.58. The modified Newton's method is sometimes useful even when $h > 1$. For example, when the root is not simple, i.e., $f'(r) = 0$, then convergence (when $h = 1$) is no longer quadratic, but only linear. The remedy is to set $h = 2$.
(a) Show graphically why this generally increases the rate of convergence.
(b) Using Theorem 2.5, demonstrate that the rate of convergence is quadratic, assuming $f''(r) \ne 0$. (Hint: Let f have a Taylor series expansion about r, and use L'Hospital's rule to evaluate $\varphi'(r)$.)

Exercise 2.59. To learn more, search the World Wide Web under the term *fractal*, for example [33].

Exercise 2.60. Find an application of fractals and report on it to the class. Is this application a bona fide use of mathematics? In other words, do you believe this application will actually work in practice?

Exercise 2.61. At the time of Stephen Smale's research into the cost of computing zeros, all of the mathematical background that he used was fairly well known. Why was it Smale, and not somebody else, who discovered this theorem?

Exercise 2.62. In this chapter, many episodes were about earlier mathematics in a particular part of the world, apparently isolated from other cultures. By way of contrast, find evidence of the international nature of mathematics today in the makeup of a contemporary conference, the reports in *Mathematical Reviews* on the Internet (www.ams.org/mathscinet), or elsewhere.

3

Curvature and the Notion of Space

3.1 Introduction

On June 10, 1854, at the University of Göttingen, in a lecture that nearly did not occur, Georg Friedrich Bernhard Riemann (1826–1866) proposed a visionary concept for the study of space [223, pp. 132–133]. To obtain the position of an unsalaried lecturer (Privatdozent) in the German university system, Riemann was required to submit an inaugural paper (Habilitationsschrift) as well as to present an inaugural lecture (Habilitationsvortrag). The topic of the lecture was selected from a list of three provided by the candidate, with tradition suggesting that the first would be chosen. The most prominent member of the faculty at Göttingen and arguably the preeminent mathematician of his time, Carl Friedrich Gauss (1777–1855) passed over Riemann's first two topics (concerning his recent investigations into complex functions and trigonometric series), and chose the third as the subject of the lecture: *Über die Hypothesen, welche der Geometrie zu Grunde liegen* (On the Hypotheses That Lie at the Foundations of Geometry) [173, p. 22]. Gauss's decision, undoubtedly motivated by his own unpublished work on non-Euclidean geometry, elicited a lecture that changed the course of differential geometry. Some years prior to Riemann's lecture, Gauss had developed a consistent system of geometry in which the Euclidean parallel postulate (see below) does not hold, but wishing to avoid controversy, he did not publish these results [101]. Riemann, however, did not present a lecture tied to the tenets of a particular geometry (Euclidean, hyperbolic, or otherwise), but offered a new paradigm for the study of mathematical space with his notion of an n-dimensional manifold. His ideas remain the standard for the classification of space today. Although many modern textbooks on geometry and topology offer a rather technical definition of a manifold, the ultimate goal of this chapter is to present Riemann's own lucid description of what space ought to be. What developments in mathematics helped to precipitate Riemann's lecture? What mathematical concepts are needed for an appreciation of the ideas therein? Why have his thoughts endured the test of time?

The title of Riemann's lecture, *On the Hypotheses That Lie at the Foundations of Geometry* [193], suggests immediately that the essay concerns fundamental principles (axioms) of geometry, which the author regards not as given, as in most treatises on the subject, but as the hypotheses of an empirical science. This is indeed the case, with the key axiom in question being Euclid's fifth postulate, the parallel postulate, which in modern parlance states, Given a line L and a point P not on L, then there is a unique line M through P parallel to L. For nearly two millennia mathematicians and philosophers had tried to prove the parallel postulate from Euclid's first four axioms, an activity that reflected the fundamental belief that space is Euclidean, and that this fact must follow logically from more basic ideas of geometry.

In his *Critique of Pure Reason* (1781), Immanuel Kant (1724–1804) espouses the idea of Euclidean geometry as a philosophical necessity. Compare the following two viewpoints, the first by Kant, the second by Riemann:

> Space is not a conception which has been derived from outward experiences. ... the representation of space must already exist as a foundation. Consequently, the representation of space cannot be borrowed from the relations of external phenomena through experience [129, p. 23].

> Thus arises the problem of seeking out the simplest data from which the metric relations of Space can be determined ... the most important system is that laid down as a foundation of geometry by Euclid. These data are—like all data—not necessary, but only of empirical certainty, they are hypotheses ... [166, p. 269].

Riemann's use of "metric relations" above refers to the determination of the length of a segment or an arc, which is then used to determine the nature of space. This stands in direct opposition to Kant's "the representation of space cannot be borrowed from the relations of external phenomena through experience."

Riemann's treatment of space does not involve a study of the axioms of geometry, but instead the inauguration of a new concept for thinking about space. A detailed study of the axiomatic geometry that results from replacing the parallel postulate by a particular case of its negation was undertaken by János Bolyai (1802–1860), Nikolai Lobachevsky (1792–1856), Gauss, and others, for which the reader is referred to [17, 150, 198]. Although pioneering in its spirit, the bold new geometry of Bolyai and Lobachevsky, today called hyperbolic geometry, suffered from a key drawback: neither author provided an example of hyperbolic geometry. Riemann's notion of a manifold offers a setting that encompasses not only Euclidean and hyperbolic geometry, but also many other new geometries, and proved essential for the study of relativity and space-time in the work of Albert Einstein (1879–1955) [58] and Hermann Minkowski (1864–1909) [172].

As the reader will discover, a manifold, in Riemann's words, is a continuous transition of an instance, and need not be contained in two- or even three-dimensional Euclidean space. Moreover, manifolds may have any dimension, finite or infinite. The crucial feature of manifold theory that allows non-Euclidean geometries is curvature. Whereas the Euclidean plane is flat, the surface of a sphere or the saddle $z = x^2 - y^2$ are both curved and provide two examples of manifolds. To determine what alternative to the parallel postulate holds on a curved space, the idea of line must be generalized to an arc of shortest distance between two points (a geodesic). On a sphere, such arcs form great circles, and given a great circle L and a point P not on L, there are no great circles through P parallel to L. In short, there are no parallel "lines" on a sphere, where "line" must be interpreted as a great circle.

Spheres have constant positive curvature and provide a setting for what is known today as elliptic geometry. A surface of constant negative curvature (Exercise 3.26) is a model for hyperbolic geometry, where given a "line" L and a point P not on L, there are many "lines" through P parallel to L. Notice how the determination of "lines" on a surface (two-dimensional manifold) provides the proper version of the parallel postulate for that surface. Of course, there must be a method to determine arc length on a surface (or within a manifold) in order to identify the geodesics. "Thus," as Riemann states "arises the problem of seeking out the simplest data from which the metric relations of Space can be determined ... ." These metric relations, as the visionary genius claims, are determined by the curvature of the manifold.

What is curvature and how is it computed? Through a sequence of selected original sources, answers to these questions will be provided. The goal of the chapter is not to conclude with what today is called the Riemann curvature tensor, an advanced topic [166, 223], but to tell the story of curvature through the work of pioneers such as Christiaan Huygens (1629–1695), Isaac Newton (1642–1727), Leonhard Euler (1707–1783), and Carl Friedrich Gauss (1777–1855). For a surface, Riemann's notion of curvature is simply Gaussian curvature, and to discuss the curvature of higher-dimensional manifolds, Riemann considers the Gaussian curvature of certain two-dimensional surfaces within the manifold.

Although the story of curvature could begin with the work of Apollonius (250–175 B.C.E.) on normals to a plane curve and the envelope such normals form when drawn to a conic section [5], the goal of Apollonius's *Conics* appears to be the study and applications of conic sections and not the description of how a plane curve is bending. To construct the envelope of a plane curve at the point P, consider another point P' on the curve very close to P and draw perpendiculars to the curve through the points P and P'. Suppose that the two perpendiculars intersect at the point Q on one side of the curve. The limiting position of Q as P' approaches P is designated as a point on the envelope, with the envelope itself being the set of all such limiting points as different locations are chosen for P to begin the process. The envelope to a general plane curve, not just a conic section, was systematically studied by Huygens in his work on pendulum clocks, and such an envelope he called an evolute [256] (see below). The curvature of a given curve at some point P

is the reciprocal of the length of the normal drawn from P to the evolute. Huygens's construction is general enough that it can be applied to any curve in the plane (with a continuous second derivative). His method, taken from the *Horologium Oscillatorium* (Pendulum Clock) (1673) [121] is presented in Section two.

The *Horologium Oscillatorium* finds its motivation in a rather applied problem: the need, during the Age of Exploration, for ships to determine longitude when navigating the round earth [221]. If a perfect timekeeper could be built, then longitude could be determined at sea by first setting the clock to read noon when the sun is at its highest point at the port of embarkment. A reading of the clock at sea when the sun is again at its highest point would yield a discrepancy from noon, depending on how many degrees of longitude the ship had progressed from embarkment, with one hour corresponding to $360°/24 = 15°$. Although in 1659 Huygens did construct a chronometer that theoretically keeps perfect time, his device did not perform reliably on the high seas [256]. The history of science credits the Dutch inventor with constructing the first working pendulum clock,[1] while physics is indebted to Huygens for the isochronous pendulum, a special type of pendulum with the mathematical property to keep perfect time. So acute was the need to determine longitude at sea that the British government issued the Longitude Act on July 8, 1714, which offered £20,000 for a method to determine longitude to an accuracy of half a degree of a great circle. The prize, after much haggling with the Board of Longitude, was awarded nearly in full to John Harrison (1693–1776) for his maritime clock known as H-4 [22, 221].

Huygens's isochronous pendulum, although it did not solve the longitude problem, employs certain techniques that soon became standard for the study of curvature of plane curves. The first of these is the osculating circle, and the second is the radius of curvature. Given a curve in the plane, to find its curvature at some point B, construct a circle that best matches the curve at B. This is the osculating circle; its radius is the radius of curvature at B, and the measure of curvature is the reciprocal of the radius. The locus of the centers of the osculating circles as the point B moves along the given curve forms what Huygens calls the evolute of the original curve. The reader is invited to witness how the ideas of the osculating circle and radius of curvature arise in the original work of Huygens presented in Section two. Huygens's scientific legacy as portrayed in modern physics texts is touched on in Exercise 3.8.

Astonishingly, Huygens arrived at his description of the radius of curvature before the development of the differential or integral calculus. His results are stated in geometric terms without the use of derivatives or even equations. Moreover, the term osculating circle was coined by Gottfried Wilhelm Leibniz (1646–1716), who spent the years 1672–1676 in Paris, where he met the renowned Huygens and received a copy of the *Horologium Oscillatorium*

---

[1] The idea for the use of a pendulum as a regulating device in a clock goes back to Galileo, but he never built such a clock [221, p. 37].

[117]. An analytic expression for the radius of curvature was found by Isaac Newton (1642–1727) and appears in his *De methodis serierum et fluxionum* (Methods of Series and Fluxions), published in 1736. The reading selection of Section three is precisely Newton's solution to what he states as "To find the curvature of any given curve at a given point" [178, p. 150]. With his newly developed calculus of fluxions (differential calculus), Newton arrived at an equation for curvature that can easily be implemented and is equivalent to expressions for curvature found in modern calculus texts. The geometry behind Newton's construction, however, is strikingly similar to that of Huygens. It is profitable to compare Huygens's construction of the evolute in Figure 3.5 with Newton's derivation of the radius of curvature in Figure 3.8. Newton has essentially assigned coordinates and their associated fluxions to the geometry behind the osculating circle.

This concludes the discussion on the curvature of plane curves, which is necessary for an understanding of the curvature of other objects. Curves in three-dimensional (Euclidean) space had been studied by Alexis Clairaut (1713–1765) [230, p. 100], and described as curves of double curvature in his 1731 text *Recherches sur les courbes à double courbure* (Researches on curves of double curvature), a topic not pursued here. See [223, p. 38] for further details. The chapter instead moves forward with the study of surfaces (two-dimensional manifolds) in Euclidean three-space with emphasis on their curvature. In this regard, we turn to the contributions of the prolific Leonhard Euler (1707–1783).

In a paper presented to the St. Petersburg Academy of Science in 1775, *De repraesentatione superficiei sphaericae super plano* (On Representations of a Spherical Surface on the Plane) [66, v. 28, pp. 248–275], Euler proved what cartographers had long suspected, namely the impossibility of constructing a flat map of the round world so that all distances on the globe are proportional (by the same constant of proportionality) to the corresponding distances on the map. In a preceding paper (1770), *De solidis quorum superficiem in planum explicare licet* (On Solids Whose Surfaces Can Be Developed in the Plane) [66, v. 28, pp. 161–186], Euler had studied the problem of describing all surfaces that can be mapped to the plane. In doing so he introduced two techniques that would become standard tools in differential geometry. The first is the use of two parameters to describe points on the surface, an idea used again by Gauss and extended by Riemann to higher-dimensional manifolds. The second is the use of a line element, i.e., the "metric data" needed to compute arc length on a surface. Gauss would later re-prove Euler's result on map projections [84] as a special case of a more general theorem in which the problem of mapping one surface onto another (not necessarily a plane) is reduced to knowing the curvature of both surfaces. If a distance-preserving map between two surfaces exists, then both surfaces must have the same value of

Gaussian curvature at corresponding points, a theorem that Gauss christens the *theorema egregium*[2] (remarkable theorem).

How exactly is the curvature of surfaces computed? The reading selection of Section four offers Euler's answer to this question from his 1760 essay *Recherches sur la courbure des surfaces* (Researches on the Curvature of Surfaces) [66, v. 28, pp. 1–22]. He begins by considering a planar cross section of the surface, and then determines the curvature of the curve formed by the intersection of the plane with the surface. At a given point P of the surface, Euler further restricts his attention to those planes that are perpendicular to the surface, and identifies two "principal" cross sections at P, one with maximum curvature and one with minimum curvature. Moreover, any other perpendicular cross section at P has a value for its curvature that can be expressed in terms of these maximum and minimum values via a simple formula. In this way Euler reduces the curvature of surfaces to that of curves. The Euler Archive [136, Eneström 333] offers an English translation of his original proof of this result.

The calculational genius begins his paper thus (translated from the original French):

> In order to know the curvature of curved lines, the determination of the radius of the osculating circle offers the proper method .... But for ... surfaces, one would not even know how to compare the curvature of the surface with that of a sphere, as one can always compare the curvature of a curved line with that of a circle [66, v. 28, p. 1].

The idea expressed here, that the curvature of a surface might be computed in terms of an osculating sphere, much as the curvature of a plane curve is expressed in terms of an osculating circle, is not realized. (See the conclusion of Exercise 3.17 for the quadratic surface that best matches a given surface at a given point.) Carl Friedrich Gauss, however, does make incisive use of an auxiliary sphere to compute the curvature of surfaces (see below), although this sphere is not, strictly speaking, the two-dimensional analogue of the osculating circle. Foreshadowing Gauss's deep results, Oline Rodrigues (1794–1851) [230, p. 116] had studied the ratio of a small area on a surface and the corresponding area on an auxiliary sphere [197], but did not develop this idea to the extent of Gauss. Furthermore, Sophie Germain (1776–1831) had introduced the notion of a referent sphere to a surface and proposed that this sphere have curvature given by the mean (average) of the maximum and minimum cross-sectional curvatures found by Euler [88, 89]. She does not, however, offer a construction that would show how the referent sphere would arise out of geometric considerations (such as the construction of an osculating circle). Nonetheless, Germain's mean curvature proved to be essential in her work on elasticity [24], and later became a key tool in the study of minimal surfaces (surfaces of least area with prescribed boundary [224]).

---

[2] The converse of the *theorema egregium*, at least for surfaces of constant curvature, was studied by Ferdinand Minding (1806–1885) [170].

In a deep and highly polished essay, *Disquisitiones generales circa super-ficies curvas* (General Investigations of Curved Surfaces) (1827) [84], Carl Friedrich Gauss (1777–1855) introduced his own concept for the measure of curvature of a surface at a point. Unlike Euler, who had considered planar cross sections to a surface, Gauss begins by considering vectors normal (perpendicular) to the surface, and then transports these vectors to an auxiliary sphere. By making an adroit comparison between the area of a small triangle on the surface around some point P and the corresponding area of the triangle on the auxiliary sphere, Gauss develops a formula for the curvature of the surface at P in terms of a single value. In an elegant and unifying theorem, the preeminent geometer proves that this value, today known as Gaussian curvature, is equal to the product of the maximum and minimum cross-sectional curvatures found by Euler. The essay continues with a proof of the *theorema egregium*, that one surface can be mapped onto another in a distance-preserving fashion only if the surfaces have the same value of Gaussian curvature at corresponding points. This is followed by a study of geodesics (arcs of shortest length) on surfaces, and a detailed analysis of angles in a geodesic triangle (a triangle, all sides of which are geodesics).

In Euclidean geometry all triangles have angle sum 180°, a result that is itself logically equivalent to the parallel postulate. Gauss proves, however, that on a surface of negative curvature, a geodesic triangle has angle sum less than 180°, and on a surface of positive curvature, a geodesic triangle has angle sum greater than 180°. In particular, "The excess of the angles of a triangle formed by shortest paths over two right angles is equal to the total curvature of the triangle" [51, p. 90]. The total curvature here refers to the integral of the Gaussian curvature over the triangle. Notice how vividly curvature enters into a result that transcends a basic tenet of Euclidean geometry. The curvature of the Euclidean plane is, of course, zero. Gauss's own unpublished work on hyperbolic geometry served in part to motivate these results [51]. Before the authorship of *Disquisitiones superficies* or even its 1825 draft, the German master had been fully aware of the logical basis for a geometry satisfying Euclid's first four axioms, but not the fifth, yet he shared with few his work on non-Euclidean geometry [17, 101]. Primarily to avoid controversy, Gauss did not publish his results on hyperbolic geometry, and in a letter concerning János Bolyai's discovery in this field, Gauss wrote to János's father on March 6, 1832: "My intention was, in regard to my own work [on non-Euclidean geometry] of which very little up to the present has been published, not to allow it to become known during my lifetime" [249, p. 52]. Nowhere in *Disquisitiones superficies* is there mention of the parallel postulate.

Further inspiration for *Disquisitiones superficies* may have been drawn from Gauss's own field work as surveyor of the Kingdom of Hanover (now a German state) during the years 1821–1825 [51, p. 129]. Following his geodetic survey during the summer of 1825, he writes to a friend Christian Schumacher on November 21 of that year:

Recently I have taken up again a part of the general investigations on curved surfaces which are to form the basis of my projected essay on advanced geodesy. It is a subject which is as rich as it is difficult, and it takes me from accomplishing anything else [85, p. 400].

The reading selection from *Disquisitiones superficies* in Section five includes a derivation for Gaussian curvature of a surface in terms of its partial derivatives, as well as the description of a surface in terms of two parameters p, q. The curious reader is encouraged to consult Gauss's original tract [87] for a proof of the *theorema egregium*, which relies on the equations for the "metric data" (identified as E, F, G in Gauss's notation) needed to compute the length of curves in terms of the two-parameter coordinate system. The proof is too lengthy to be reproduced in this chapter.

The publication of *Disquisitiones superficies* precipitated the study of surfaces with specific properties that could then be stated in terms of Gaussian curvature, such as surfaces of constant curvature [192]. In addition to this the search for minimal surfaces entered a new era of growth around 1830. Before then the only (nontrivial) examples of minimal surfaces had been discovered by Euler and Jean-Baptiste Meusnier (1754–1793), namely the catenoid and the helicoid[3] [192]. The final section of the chapter, however, moves forward with Riemann's essay *On the Hypotheses That Lie at the Foundations of Geometry*, where the notion of manifold offers a new paradigm for the study of space. Generalizing from the idea of a curved surface, Riemann's notion of extended quantity encompasses objects of any dimension, and, moreover, objects that do not necessarily exist in a three-dimensional Euclidean world. (For a two-dimensional surface that cannot be constructed in an ambient three-dimensional world, see Exercise 3.31.) Riemann clearly intended that the metric data needed to compute the length of a curve in a manifold be given by generalizing the ideas of Gauss to higher dimensions. In a striking result, the visionary Riemann claims that the curvature of the manifold determines the metric: "If the curvature is given in $\frac{1}{2} n(n-1)$ surface directions at every point, then the metric relations of the manifold may be determined" [166, p. 274]. The surfaces referred to here are certain two-dimensional surfaces within the n-dimensional manifold, and curvature refers to the Gaussian curvature of these surfaces. Thus, curvature determines the metric, which in turn determines the geodesics on the manifold, and these provide us with the version of the parallel postulate that holds in a given manifold. In this sense, curvature determines the nature of space.

Riemann's lecture was meant for a general scholarly audience, and as such contains virtually no formulas for the metric relations or curvature. In a subsequent paper (1861) [194, pp. 391–404] he does introduce specific formulas for these, although it is not the goal of the chapter to present this material. All of the claims in his 1854 lecture can be substantiated; for proofs, the reader is referred to the specialized texts [166, 223]. In the one and one-half

---

[3] The mathematical properties of these surfaces are discussed in [224].

centuries following his ground-breaking address, Riemann's ideas have proven very fertile, with an entire subject, Riemannian geometry [50, 199], having its roots in this single source. In the twentieth-century work of Einstein and Minkowski, manifold theory proved to be essential in the study of relativity and four-dimensional space-time [58, 172].

Since Riemann and Einstein, the classification of manifolds has continued in all dimensions, with recent spectacular progress in the fourth dimension [188]. In particular, by the work of Simon Donaldson (1957–), the notion of derivative has special interpretations in dimension four that do not occur in any other dimension [9, pp. 3–6]. The most elusive dimension, however, remains the third, with an outstanding problem being the classification of three-dimensional manifolds that can be continuously deformed into the three-dimensional sphere:

$$S^3 = \left\{ (x_1,\, x_2,\, x_3,\, x_4) \in \mathbf{R}^4 \mid \sum_{i=1}^{4} x_i^2 = 1 \right\}.$$

This problem, known as the three-dimensional Poincaré conjecture, has been solved for every dimension except possibly the third, with even the herculean task of dimension three apparently also solved as this text goes into print. The four-dimensional Poincaré conjecture was only recently proved (in 1982) by Michael Freedman (1951–) [168, pp. 13–15].

Our story of curvature bears witness to some very applied problems, such as constructing an accurate timekeeper, and mapping the round globe onto a flat plane, whose solutions or attempted solutions resulted in key concepts for an understanding of space. The development of relativity and space-time provided a particular impetus for the study of manifolds, with the curvature of space-time being a key feature that distinguishes it from Euclidean space. The reader is invited to see [181] for a delightful informal discussion of curvature, and [222, 223] for a more advanced description of manifold theory. This chapter closes with the open problem of the classification of space, space in the name of mathematics and philosophy, space in the name of physics and astronomy, space in the name of curiosity and the imagination.

3.2 Huygens Discovers the Isochrone

Holland during the seventeenth century was a center of culture, art, trade, and religious tolerance, nurturing the likes of Harmenszoon van Rijn Rembrandt (1606–1669), Johannes Vermeer (1632–1675), Benedict de Spinoza (1632–1677), and René Descartes (1596–1650). Moreover, the country was the premier center of book publishing in Europe during this time, with printing presses in Amsterdam, Rotterdam, Leiden, the Hague, and Utrecht, all publishing in various languages, classical and contemporary [108, p. 88]. Into this environment was born Christiaan Huygens (1629–1695), son of a prominent statesman and diplomat.

The young Huygens showed an interest in astronomy, developed improved methods of grinding and polishing lenses for telescopes, and made notable discoveries about the rings of Saturn and the length of the Martian day [212, p. 801]. During a visit to Paris in 1655, the Dutchman began to study probability, and authored the book *De Ratiociniis in Aleae Ludo* (On the Calculations in Games of Chance), published in 1657 [133, p. 456]. At the invitation of Jean-Baptiste Colbert (1619–1683), minister of King Louis XIV (1638–1715), Huygens moved to Paris in 1666 as a member of the newly established *Académie des Sciences*, where he resided for the next 15 years. Aside from his work on pendulum clocks (discussed below), he formulated a principle for the conservation of energy for an elastic collision of two bodies, and correctly identified the centripetal force of an object moving in circular motion. Newton held the work of Huygens in high regard, and used the Dutch scholar's results in some of his own investigations [212, p. 802]. the reader is encouraged to compare the work of these two intellectual giants on the derivation of the radius of curvature in this and the following section. Alas, growing religious intolerance for Prostestants in Paris prompted Huygens to return to the Hague in 1681.

Photo 3.1. Huygens.

Later in life, he launched a study of microscopy in loose connection with
Anton van Leeuwenhoek (1632–1723), and developed highly original ideas in
protozoology. In 1690 Huygens published his *Traité de la Lumière* (Treatise
on Light), in which he proposed a wave theory of light. His final publication,
Cosmotheoros, appeared posthumously, and contains a summary of what was
known about the universe at the time. We turn now to the master's work on
horology.

In a burst of inspired creativity during 1659, Christiaan Huygens developed a pendulum clock that theoretically keeps perfect time [121, 256]. In the
years prior to his landmark discovery, Huygens had studied the simple pendulum, which consisted of a bob attached by a thread to a fixed point. The bob
then oscillated in a circular arc. As a timekeeper, the simple pendulum is not
entirely accurate, since the time required to complete one oscillation depends
on the amplitude of the swing. The greater the swing, the more time is needed
for an oscillation. Huygens's genius was to discover a curve for which the time
of an oscillation is independent of the swing amplitude, an idea that at first
glance seems a virtual impossibility.

Such a curve is described either as isochronous or as tautochronous, both
terms referring to the "same-time" property at which the bob reaches its lowest point, regardless of the amplitude. Astonishingly, Huygens showed that
the shape of the tautochrone is given by a curve that had been studied intensely and independently during the seventeenth century, namely a cycloid.
Consider a point P on the circumference of a wheel and suppose that the
wheel begins to roll along a flat surface. The curve traced by the point P is
called a cycloid (Figure 3.1). For use in the pendulum, this curve could simply
be turned upside down (inverted), which would then serve as the path of the
bob. The cycloid had already occupied the minds of great mathematicians and
scientists such as Galileo, Torricelli, Mersenne, Roberval, Fermat, Descartes,
Pascal, and others [18], yet none of them discovered its isochronous property.

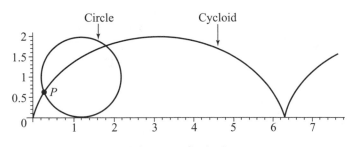

Fig. 3.1. Cycloid.

Of course, once the shape of the tautochrone had been determined, the
problem of forcing a pendulum bob to oscillate along such a curve remained.
This the Dutch scholar solved by placing two curved metal or wooden plates

at the fulcrum of the pendulum (Figure 3.2, II). As the bob swings upward, the thread winds along the plates, forcing the bob away from the path of a perfect circle, and as the bob swings downward, the thread unwinds. This leads then to another problem in what today would be called mathematical physics: what should be the shape of the metal plates? Huygens called the curve for the plates an evolute of the cycloid, or *evolutus* (unrolled) in the original Latin, and went on to discuss the mathematical theory of evolutes for general curves, not just cycloids. The key idea for the construction of the evolute is this: Suppose (Figure 3.3) that the thread leaves the plate at point A, the bob is at B, and segment AB is taut. Although B is no longer traversing a circle, the bob is instantaneously being forced around a circle whose center is A and radius is AB. To find A and AB, simply determine the circle that best matches the cycloid at point B. The length of AB became known as the radius of curvature of the cycloid at B, while A became known as the center of curvature. Finding the evolute of the cycloid is then reduced to finding the locus of centers of curvature, a locus that Huygens demonstrated to be another congruent cycloid shifted so that its cusp lies at the fulcrum of the pendulum. The construction of a perfect timekeeper (assuming no friction) is thus accomplished by attaching metal jaws in the shape of a cycloid to the top of the pendulum.

The Dutch scientist published his findings in the magnum opus *Horologium oscillatorium sive de motu pendulorum ad horologia aptato demonstrationes geometricae* (The Pendulum Clock or Geometrical Demonstrations Concerning the Motion of Pendulums as Applied to Clocks) in Paris in 1673, with license and approval from King Louis XIV. In Huygens's own words [121, p. 11]:

> For the simple pendulum does not naturally provide an accurate and equal measure of time, since its wider motions are observed to be slower than its narrower motions. But by a geometrical method we have found a different and previously unknown way to suspend the pendulum and have discovered a line[4] whose curvature is marvelously and quite rationally suited to give the required equality to the pendulum. After applying this line to clocks, we have found that their motion is so accurate on both land and sea, it is now obvious that they are very useful for investigations in astronomy and the art of navigation. ... The geometers of the present age have called this line a cycloid ... .

Huygens is a bit too optimistic, with further trials revealing that a pendulum clock behaved unreliably at sea. Nonetheless, the idea of the radius of curvature in Huygens's work would become central for the study of the

---

[4] Today, one would write "curve" instead of "line."

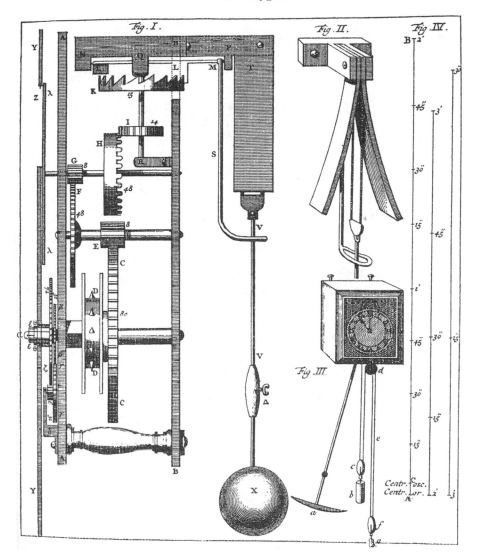

Fig. 3.2. Huygens's pendulum.

bending of curves. Let's continue with a discussion of pendular motion before reading a passage from the *Horologium oscillatorium*.

To briefly describe the physics of the simple pendulum in modern dressing, consider a bob B of mass m suspended on a thread of length L (Figure 3.4). Suppose further that the thread forms an angle θ with the vertical. The force due to gravity acting on the bob is $\mathbf{F} = m\mathbf{g}$, and is directed downward. Now, $\mathbf{F}$ can be written as the sum of two forces, $\mathbf{F}_t$ and $\mathbf{F}_p$, where $\mathbf{F}_t$ is the component of $\mathbf{F}$ tangent to the path of the bob, and $\mathbf{F}_p$ is the component of

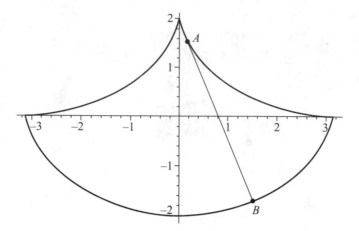

Fig. 3.3. The isochronous pendulum.

$\mathbf{F}$ perpendicular to $\mathbf{F}_t$. The magnitude of $\mathbf{F}_t$ is $mg\sin\theta$, while the magnitude of $\mathbf{F}_p$ is $mg\cos\theta$, where g is the scalar value of acceleration due to gravity. Since the thread and $\mathbf{F}_p$ lie on the same line, the tension in the thread and $\mathbf{F}_p$ sum to a vector that lies on this line. The tangential component of the *net force* acting on the bob is thus $\mathbf{F}_t$.

Letting a denote the tangential acceleration of the bob, we have

$$ma = -mg\sin\theta,$$
$$a = -g\sin\theta.$$

The negative sign is used since $\mathbf{F}_t$ points to the left for positive values of θ and to the right for negative values of θ. (Equivalently, $\mathbf{F}_t$ always points in the direction of decreasing magnitude for θ.) Tangential acceleration is the first derivative of speed, v, with respect to time,

$$a = \frac{dv}{dt},$$

while speed is the derivative of the distance traveled, i.e., the arc length of the path from O. Letting s denote this arc length, we have

$$v = \frac{ds}{dt},$$
$$\frac{d^2s}{dt^2} = -g\sin\theta.$$

Note that the arc length from O to B is $s = L\theta$, and thus

$$\frac{d^2s}{dt^2} = -g\sin(s/L)$$

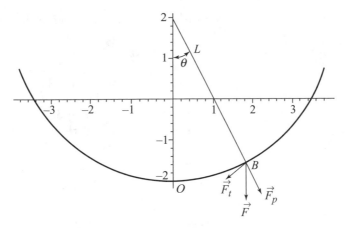

Fig. 3.4. The simple pendulum.

is the equation of motion for the simple pendulum. Huygens knew that this type of pendulum is very accurate for small values of θ, for which $\sin\theta \approx \theta$, and hence

$$\frac{d^2s}{dt^2} \approx \frac{-gs}{L}.$$

Letting $k = \frac{g}{L}$, then

$$\frac{d^2s}{dt^2} \approx -ks$$

rather accurately models small displacements of the simple pendulum. Huygens's insight was to find a curve for which $\frac{d^2s}{dt^2}$ is exactly, not just approximately, equal to $-ks$ for all values of s. A modern derivation of his results can be found in Exercises 3.2 through 3.5. An examination of the improved accuracy of the isochronous pendulum over the simple pendulum is presented in Exercise 3.6.

The Dutch clockmaker's discovery is perhaps all the more striking, since he arrived at his results before the advent of the calculus of Newton and Leibniz. Huygens did, however, make ready avail of the geometric idea of a tangent line, which was part of the mathematical culture at the time, and he exploited what today would be called the constant acceleration of a body in free fall. Since acceleration was not yet articulated as a separate concept, Huygens expressed constant acceleration as "In equal times equal amounts of velocity are added to a falling body, and in equal times the distances crossed by a body falling from rest are successively increased by an equal amount" [121, p. 43]. The latter idea goes back to Galileo's *Discorsi e dimostrazioni matematiche intorno a duo nuove scienze* (1638), which is the original Italian title for *Dialogues Concerning the Two New Sciences* [78] [256, p. 9]. Huygens's mathematical work is very geometric, with statements written verbally, and not couched in formulas. Imagine finding the equation of motion for even the simple pendulum, not to mention the isochronous pendulum, without the use of derivatives or integrals.

Let's read then a passage from the *Horologium oscillatorium* [121, pp. 94–96] in which Huygens isolates a geometric quantity that would become the radius of curvature. Proposition XI in part III of the text is concerned with the construction of the evolute of a given curve as well as finding the arc length of the evolute. Recall that a tangent to a circle at some point B is perpendicular to the radius drawn from the center of the circle to B, a fact that Huygens used liberally in his description of the circle that best matches a given curve at a given point.

∞∞∞∞∞∞∞∞

Huygens, from
The Pendulum Clock

PROPOSITION XI

Given a curved line, find another curve whose evolution describes it. Show that for any geometrical curve, there exists another geometrical curve for which an equal straight line can be given.[5]

Let ABF (Figure 3.5) be any curved line, or part thereof, which is curved in one direction. And let KL be a straight line to which all points are referred. We are required to find another curve, for example DE, whose evolution will describe ABF.

Assume that such a line has already been found. Now since all the tangents to the curve DE must meet at right angles with the line ABF, which is described by evolution, it is also clear in the reverse relation that lines which are perpendicular to ABF, for example, BD and FE, will be tangents to the evolute CDE.[6]

Next select the points B and F, which are close to each other. Now if the evolution begins from A, and if F is more distant than B from A, then the contact point E will also be more distant than D from A. And the intersection of the lines BD and FE, which is G, will fall beyond the point D on the line BD. For BD and FE must intersect since they are perpendicular to the curve BF on its concave side.

---

[5] The term "equal straight line" is used to refer to arc length. In 1658 Christopher Wren computed the arc length of the cycloid in response to a challenge posed by Pascal [18, 256] Given a curve C and a point P on C, let Q be the point on the evolute corresponding to P. Then the arc length of the evolute from a given endpoint to Q is equal the length of the radius of curvature of C at P.

[6] The point C, missing in Figure 3.5, should be in the lower right-hand side of the diagram, along the right most curve, just above the point D. The point C appears in Huygens's original *Horologium oscillatorium*.

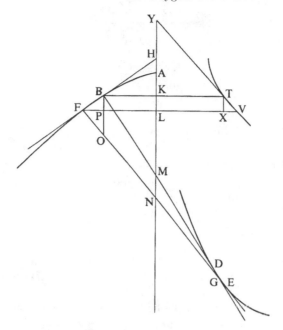

Fig. 3.5. Construction of the evolute.

Moreover, insofar as the point F is closer to B, to the same extent the points D, G, and E will also appear to come together. And if the interval BF is taken to be infinitely small, these three points can be treated as one. As a result the line BH, after having been drawn, is tangent to the curve at B and also can be thought of as tangent at F. Let BO be parallel to KL, and let BK and FL be perpendiculars to KL. FL cuts the line BO at P, and let M and N be the points where the lines BD and FE meet KL. Since the ratio of BG to GM is the same as that of BO to MN, then when the latter is given, so is the former. And when the line BM is given in length and in position, so is the point G on the extension of BM, and also D on the curve CDE, since we have taken G and D to be one. But the ratio of BO to MN is already known both in the case of the cycloid, which we investigated first and found to be 2 to 1, and in the case of the other curves which we have examined so far where we found it to be the [composition] of two given ratios. Now since the ratio of BO to MN is composed of the ratio of BO to BP or of NH to LH [and[7]] the ratio of BP or KL to MN, it is clear that if either[8] of these latter are given, then the ratio of BO to MN, which is composed of them, will also be given. It will be clear in what follows that the former are given for all geometrical curves. And as a result

---

[7] $BO/MN = (BO/BP)(BP/MN)$.
[8] Either BO/BP or NH/LH and either BP/MN or KL/MN.

by their use it is always possible to designate curves by whose evolution the given curves are described.[9]

∞∞∞∞∞∞∞∞∞

In the construction of the evolute, the quantity BG becomes the radius of curvature of the curve ABF at the point B, and Exercise 3.1 describes how Huygens's geometric methods can be employed to compute BG in a specific case, while Exercise 3.7 outlines a derivation of the modern equation for BG. The circle with center G and radius BG would be called the osculating circle (circulus osculans) by Gottfried Wilhelm Leibniz (1646–1716) in his 1686 paper *Meditatio nova de natura anguli contactus et osculi* (New Mediations on the Nature of Contact and Osculation Angles) [154]. The German-born Leibniz spent the years 1672–1676 in Paris, where he met the renowned Huygens, and became his pupil [117]:

> Huygens came to like the studious and intelligent young German more and more, gave him a copy of the *Horologium* as a present and talked to him about this latest work of his, the fruit of ten years of study, of the deep theoretical research to which he had been led in connection with the problem of pendular motion. [117, pp. 47–48].

Exercise 3.1. In this exercise the radius of curvature of $y = 4 - x^2$ is estimated at the point $B(-1, 3)$ using the geometric ideas of Huygens. In modern terminology, let the y-axis be placed along the line HL (Figure 3.5), and suppose that the x-axis is parallel to the line FL, so that B has coordinates $(-1, 3)$. What is the y-coordinate of the point K? Using a modern equation for the slope of the tangent line to $y = 4 - x^2$ at B, find the y-coordinate of H. Find the y-coordinate of M from the equation of the line perpendicular to $y = 4 - x^2$ at B. Let $F(-1.1, 2.79)$ be another point on the parabola $y = 4 - x^2$, close to B. From F, determine the y-coordinates of L and N. From the equations

$$BG = BM + MG, \qquad \frac{BG}{MG} = \frac{HN}{HL}\frac{KL}{MN}$$

estimate BG, the radius of curvature of the parabola at B. Repeat the construction using the same point B, now considering F as $(-1.01, 2.9799)$.

Exercise 3.2. The goal of this exercise is to develop the equation of the tautochrone by using a few modern techniques from physics and integral calculus. Recall that the equation of motion for the simple pendulum is approximately given by

---

[9] Once the length BG and the position of G can be determined, then the evolute of the curve ABF can be constructed by allowing the point B to vary, and repeating the construction for BG. Further details can be found in Exercise 3.7

$$\frac{d^2s}{dt^2} \approx -ks$$

for small values of the arc length s. Consider instead a particle of mass m moving along the x-axis subject to a net force $F = -kx$, where k is a positive constant. Let $x(t)$ denote the position of the mass at time t, $v(t)$ its velocity, and $a(t)$ its acceleration. Suppose that at time $t = 0$, the mass is released at the position $x(0) = A_0$ with initial velocity $v(0) = 0$. Still assuming that $F = -kx$, find an equation for $x(t)$. The reader may wish to review the ideas of simple harmonic motion in a physics or calculus text. Show that the particle oscillates about the origin with maximum displacement A_0. Show that the time required for the particle to travel from A_0 to the origin is independent of the value of A_0. What is the significance of this finding?

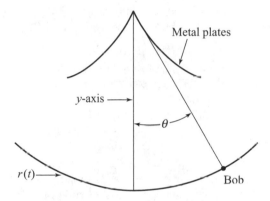

Fig. 3.6. The pendulum bob.

Suppose now that the particle is a pendulum bob swinging on a thread, and that the thread itself is constrained by metal plates (Figure 3.6). Let the y-axis be placed along the vertical position of the thread when the bob is at rest and at its lowest position. We wish to find a path $r(t)$ such that gravity and tension in the thread combine to produce a net force whose tangential component has magnitude kms. Here s denotes the arc length of $r(t)$, measured from the bob's lowest position. From the above assumption about the tangential component, show that for the curve $r(t)$, we have (which holds for the isochronous pendulum, but not the simple pendulum)

$$\frac{d^2s}{dt^2} = -ks.$$

Letting θ denote the angle formed by the taut thread and the y-axis, show that

$$\frac{d^2s}{dt^2} = -g \sin \theta$$

as well, assuming that the net force on the bob points in the direction of decreasing magnitude of θ. Equating the two expressions for $\frac{d^2s}{dt^2}$, find an equation for s, and then compute $\frac{ds}{dt}$. For

$$r(t) = \big(x(t),\, y(t)\big),$$

explain how $\frac{dx}{dt}$ and $\frac{dy}{dt}$ are related to $\frac{ds}{dt}$ and θ. Integrate the resulting expressions to find parametric equations for x and y in terms of θ. Apply the initial conditions $x = 0$ and $y = -\frac{g}{2k}$, both when $\theta = 0$, to conclude that

$$x(\theta) = \frac{g}{4k}\Big(2\theta + \sin 2\theta\Big),$$

$$y(\theta) = -\frac{g}{4k}\Big(1 + \cos 2\theta\Big).$$

Finally, graph this set of parametric equations in the xy-plane in the special case $\frac{g}{4k} = 1$. Be sure to notice the negative sign for $y(\theta)$.

Exercise 3.3. In this exercise the shape of the tautochrone studied above is identified. Consider a circle of radius R and center C rolling to the left, below and tangent to the x-axis (see Figure 3.7). Let P be a point on the circumference of the circle and let α denote the angle between CP and the lower half of the vertical diameter. We wish to find the parametric equations for the coordinates of P as the circle rolls toward the origin $(0, 0)$. Suppose that when the circle touches $(0, 0)$, we have $\alpha = 0$, so that P is diametrically opposed to the origin at this instant. Assuming that the circle rolls without slipping, find the coordinates for the center C in terms of R and α. Then find the parametric equations for the coordinates of the point P in terms of R and α. This is the equation of the cycloid. What expressions for R and α yield the equation of the tautochrone in Exercise 1? Be sure to justify your answer.

Exercise 3.4. In this exercise the shape of the metal plates in Huygens's pendulum is determined by studying the evolute of a cycloid. Consider for simplicity the cycloid given by (compare with Exercise 3.2)

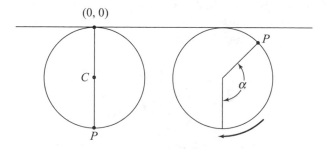

Fig. 3.7. The rolling circle.

$$\varphi(\theta) = \big(\theta + \sin\theta,\ -(1 + \cos\theta)\big), \quad -\pi \le \theta \le \pi,$$

i.e., $x(\theta) = \theta + \sin\theta$, $y(\theta) = -(1 + \cos\theta)$. Recall that the evolute of $\varphi(\theta)$ is given by the locus of the centers of the osculating circles for the graph of $\varphi(\theta)$. First, determine the radius of curvature R of the parametric curve $\varphi(\theta)$ at an arbitrary point P using formula (3.2) from the next section, where $R = 1/k$. The center C of the osculating circle is then located on a line perpendicular to $\varphi(\theta)$ at P, and at a distance R from P. Moreover, C is positioned above $\varphi(\theta)$ (i.e., the osculating circle and $\varphi(\theta)$ have the same concavity at P). Using these geometric principles, find a parametric equation for $E(\theta)$, the evolute of $\varphi(\theta)$. Show that $E(\theta)$ is also a cycloid, congruent to $\varphi(\theta)$, except shifted by an amount π in the x direction and 2 in the y direction.

Exercise 3.5. Of course, the construction of the evolute itself in Exercise 3.4 does not offer proof that the pendulum bob will follow the path of the cycloid $\varphi(\theta)$, simply because the metal plates are themselves cycloids. This requires a separate exercise. Suppose then that the metal plates are shaped according to $E(\theta)$ found above. Sketch a graph of $E(\theta)$ for $-\pi \le \theta \le \pi$, and verify that

$$E(0) = (0,\ 2), \quad E(\pi) = (\pi,\ 0).$$

Suppose that the pendulum thread has a length that is exactly equal to the arc length of the evolute between $\theta = 0$ and $\theta = \pi$. If the thread is completely wound to the right, then the bob is located at $(\pi, 0)$. As the thread unwinds, it remains tangent to $E(\theta)$, and the length of the unwound thread equals the arc length of $E(\theta)$ between the point of tangency and $(\pi, 0)$. Recall that if

$$E(\theta) = \big(x(\theta),\ y(\theta)\big),$$

then the arc length between $E(\alpha)$ and $E(\beta)$ is given by

$$\int_\alpha^\beta \sqrt{\left(\frac{dx}{d\theta}\right)^2 + \left(\frac{dy}{d\theta}\right)^2}\ d\theta.$$

Let $I(\theta)$ be the curve traversed by the bob (at the end of the thread) as this thread unwinds. Use the ideas of tangency and arc length to find parametric equations for $I(\theta)$. Show that $I(\theta)$ is the original cycloid $\varphi(\theta)$.

For any curve $r(t)$, imagine a thread wrapped around $r(t)$ with an endpoint P on the curve. If P is pulled from the curve so that the thread remains taut (and tangent to $r(t)$), then the locus of points traversed by P as the thread unwinds is called the involute I of $r(t)$. What do you conjecture about the involute of an evolute in general? Huygens addressed this very question in his *Horologium oscillatorium* and discovered several striking relations between an evolute and its involute.

Exercise 3.6. In this problem we wish to compare the solution of the differential equation

$$\frac{d^2 s}{dt^2} = -k \sin s, \tag{i}$$

which arises from the simple pendulum, to the solution of

$$\frac{d^2 s}{dt^2} = -ks, \tag{ii}$$

which occurs for the isochronous pendulum. In both cases, suppose that s is a function of t, and that the initial conditions $s(0) = 0$, $s'(0) = \sqrt{k}$ are given. Find the solution to (ii). For (i) start with a series solution

$$s(t) = a_0 + a_1 t + a_2 t^2 + a_3 t^3 + a_4 t^4 + a_5 t^5 + \cdots,$$

and find a_0 and a_1 from the initial conditions. Using the first few terms of the Taylor–Maclaurin series for $\sin(s)$, find a_3, a_4, and a_5 by substituting certain series into both sides of (i). Compare this with (ii) by writing out the terms of the series for (ii) up to and including t^5. Finally compare the value of both solutions when $k = 1$, $t = 0.1$, and comment on the improved accuracy of the isochronous pendulum over the simple pendulum for small values of t.

Exercise 3.7. In this exercise the modern analytic formula for the radius of curvature is derived from Huygens's original geometric description. Since Huygens had neither the concept of derivative nor that of limit, infinitesimals are used in the sequel. Consider a curve in the xy-plane such as ABF in Figure 3.5. Let $B(x_1, y_1)$ and $F(x_2, y_2)$ be two points on the curve that are infinitesimally close, and let

$$dy = y_2 - y_1, \quad dx = x_2 - x_1.$$

Then the derivative of the curve at B or F is given by $\frac{dy}{dx}$. Moreover, the length of the line segment joining $B(x_1, y_1)$ and $F(x_2, y_2)$ is

$$ds = \sqrt{(dx)^2 + (dy)^2},$$

and this segment may be considered tangent to the curve at either of the two points.

Turn now to Huygens's construction of the evolute of the curve ABF (see Figure 3.5). Recall that BG represents the radius of curvature of ABF at the point B, and that the conclusion of his geometric argument may be summarized as

$$\frac{BG}{MG} = \frac{HN}{HL} \frac{KL}{MN}.$$

Notice that the reference line KL in Figure 3.5 serves as the y-axis, while FL can be considered as the x-axis, with the point L being the modern equivalent of the origin. In this interpretation, however, increasing values of x point to the left, while increasing values of y point downward, so that as curve ABF is drawn, point B is reached before F. From the equation

$$\frac{HN}{HL} = \frac{HN}{FH}\frac{FH}{HL},$$

conclude that

$$\frac{HN}{HL} = \left(\frac{ds}{dy}\right)^2$$

by arguing that

$$\frac{HN}{FH} = \frac{ds}{dy} \quad \text{and} \quad \frac{FH}{HL} = \frac{ds}{dy}.$$

Next show that

$$\frac{MN}{KL} = 1 + \left(\frac{LN - KM}{KL}\right).$$

Use geometry to conclude that $LN = x\frac{dx}{dy}$, where x is the horizontal distance between the KL-axis and a point on the curve ABF. For the point F, $x = x_2 = FL$. Explain conceptually why

$$\frac{LN - KM}{KL} = \frac{d}{dy}\left(x\frac{dx}{dy}\right),$$

and use the product rule to compute the latter. Also using the geometry of $\triangle BKM$ and $\triangle BPF$, find an expression for BM and substitute this into the equation

$$BG = BM + MG.$$

Find an expression for BG using algebra with infinitesimals, noting that

$$(ds)^2 = (dx)^2 + (dy)^2, \qquad \frac{d^2x}{dy^2} = \frac{d(dx)}{dy \cdot dy},$$

where $d(dx)$ is the second difference of the quantity x. Express BG as the ratio of two terms with the denominator being simply d^2x/dy^2, and compare this to formula (3.1) in the next section, from which the radius of curvature is $R = 1/k$. You may wish to switch the dependent and independent variables (x and y) to reconcile Huygens's construction with modern conventions for graphing a function $y = f(x)$ in the xy-plane. Also, use of the absolute value of the expression for BG may be more appropriate.

Exercise 3.8. Although many introductory physics texts discuss the simple pendulum, very few describe Huygens's contributions to pendular motion. Postulate why his work on the isochronous pendulum is given so little attention today.

3.3 Newton Derives the Radius of Curvature

With the geometric notions of center and radius of curvature established, analytic expressions for these quantities can now be sought. This is the very issue addressed in problem five of Newton's *De methodis serierum et fluxionum* (Methods of Series and Fluxions), written around 1671, but not published until 1736. Isaac Newton (1642–1727), recognized as an intellectual giant of the human race, developed the differential and integral calculus during the years 1665–1667 when the plague closed the universities in England. Further biographical information about Newton can be found in the chapter on numerical solutions to equations. In his *Methods of Series and Fluxions*, Newton viewed variables as flowing quantities, which he called fluents. The derivative of a fluent x with respect to an independent variable, perhaps time, is called a fluxion, and is denoted by $\dot{x}$. During an arbitrarily small time interval o, the value of the variable x changes by $\dot{x} \times o$, which is called the moment of x. In one of history's most controversial events of simultaneous discovery, Leibniz independently developed the calculus around 1675 and introduced the notation d for derivative and $\int$ for integral.

Problem five of the tract on fluxions states "Curvæ alicujus ad datum punctum curvaturam invenire" ("to find the curvature of any curve at a given point"). With this work, a shift toward the abstract is seen, with Newton beginning his treatment of curvature by stating four axioms that describe the properties of this quantity. The utility of curvature, however, is paramount in Newton's mind [178, pp. 150–157]:

∞◌◍◌◍◌◍◌◍◌◍◌∞

Newton, from

Methods of Series and Fluxions

PROBLEM   5

TO FIND THE CURVATURE OF ANY CURVE AT A GIVEN POINT

The problem has the mark of exceptional elegance and of being pre-eminently useful in the science of curves. In preface, however, to its construction it is convenient to set down certain generalities:

1. The same circle has everywhere the same curvature, and the curvatures of unequal circles are inversely proportional to their diameters. If the diameter of one is twice as small as that of a second, the curvature of its circumference will be twice as great; if its diameter is three times as small, its curvature will be three times as great; and so on.

2. If a circle touch some curve on its concave side at a given point and be of such a size that no other tangent circle can be drawn between in the contact angles neighbouring that point, that circle has the same curvature as the curve at that point of contact. For a circle which lies between a curve and another circle in the vicinity of the contact point deviates less from the curve and more approximates its curvature than that second circle, and consequently most approximates its curvature when no other circle can be inserted between it and the curve.

3. Accordingly, the center of curvature at some point of a curve is the center of an equally curved circle; and thus the radius (or semidiameter) of curvature is that portion of the normal which ends at that centre.

4. And the ratio of curvature at its various points is known from the ratio of the curvature of equally curved circles or from the inverse ratio of the radii of curvature.

The problem accordingly reduces to this point, that the radius or centre of curvature is to be found.

Imagine, therefore, (see Figure 3.8, left side) that at the three points δ, D, and d of a curve normals are drawn, and let those at D and δ meet in H, those at D and d in h, the point D being the middle one. If the curvature on the side $D\delta$ be greater than that on Dd, then $\delta H < dh$. But the nearer the normals δH and dh are to the intermediate one, the less will be the distance between the points H and h, and when at length the normals meet they will coincide. Let them coincide in the point C: then will that point C be the centre of curvature at the point D on the curve at which they are normal. This is evident of itself.

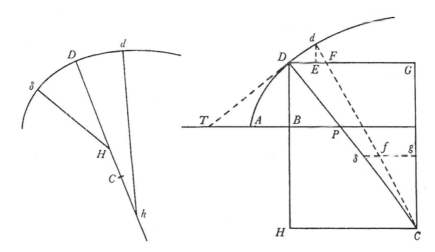

Fig. 3.8. The radius of curvature.

Of this point C, however, there are various defining conditions which can serve to determine it. For instance:

1. It is the meet of normals at indefinitely small distances from DC on its either side.

2. It separates and dichotomizes the intersections of normals in the finitely small neighbourhood on one side and the other, so that those on the more curved portion $D\delta$ meet more rapidly at H, while those on the other, less curved portion Dd do so more remotely at h.

3. If DC while continuing to stand normal to the curve be conceived to move, that point C of it will (if you except its motion towards and away from the point

D at which it stands normal) not move at all but will be in the nature of a centre of motion.[10]

4. If a circle be described with centre C and radius DC, no other circle can be described which shall lie between [it and the curve] in the vicinity of the contact point.

5. If, lastly, the centre, as H or h, of some other tangent circle gradually approach the centre C of this one till at length it coincide with it, then some one of the points in which the former circle cut the curve will simultaneously coincide with the point D of contact.

Each one of these defining properties provides a means of resolving the problem in a different way. We, however, shall choose the first as being the simplest.

At any point D you please of the curve let DT be tangent, DC normal and C the curvature centre, as before (see Figure 3.8, right side). Again, let AB be the base to which DB is applied at right angles and which is met by DC in P. Draw DG parallel to AB and CG perpendicular, taking in it Cg of any given size, and raise the perpendicular $g\delta$ till it meets DC in δ: then will there be $Cg : g\delta$ ($= TB : BD$) the ratio of the fluxion of the base to that of the ordinate. Imagine, furthermore, that the point D advances through the infinitely small distance Dd and draw dE perpendicular to DG and Cd normal to the curve, the latter meeting DG in F and δg in f: then will DE be the moment of the base, dE that of the ordinate and δf the contemporaneous moment of the straight line $g\delta$. Also

$$DF = DE + \frac{dE \times dE}{DE}.$$

Accordingly, when the ratios of these moments—or, what is the same, those of the generating fluxions—are had, there will be obtained the ratio of GC to the given quantity gC (seeing that this is that of DF to δf) and by this the point C will be determined.

Let, therefore, $AB = x$, $BD = y$, $Cg = 1$, and $g\delta = z$ and it will be $1 : z = \dot{x} : \dot{y}$ or $z = \frac{\dot{y}}{\dot{x}}$, calling z's moment $\delta f = \dot{z} \times o$ (the product, namely, of its velocity and an infinitely small quantity): the moment DE will be $\dot{x} \times o$, $dE = \dot{y} \times o$ and thence $DF = \dot{x}o + \frac{\dot{y}\dot{y}o}{\dot{x}}$. Therefore

$$Cg(1) : CG = (\delta f : DF =)\ \dot{z}o : \left(\dot{x}o + \frac{\dot{y}\dot{y}o}{\dot{x}}\right),$$

so that $CG = \frac{\dot{x}\dot{x}+\dot{y}\dot{y}}{\dot{x}\dot{z}}$.

Since, in addition, we are free to assign any velocity at all to the fluxion $\dot{x}$ of the base (to which fluxion, supposed uniform, it is convenient to relate the others), call it unity and there will be $\dot{y} = z$ and $CG = \frac{1+zz}{\dot{z}}$ and thence $DG = \frac{z+z^3}{\dot{z}}$, while

$$DC = \frac{(1+zz)\sqrt{1+zz}}{\dot{z}}.$$

---

[10] The idea of a "center of motion" essentially appears in Huygens's isochronous pendulum (Figure 3.3), where the pendulum bob is instantaneously being forced in a circle with center A and radius AB. The point A is the "center of motion."

∞∞∞∞∞∞∞∞

The above equation for DC is thus an analytic expression for the radius of curvature using Newton's fluxion notation. In modern notation, an equation for the radius of curvature of $y = f(x)$ at the point $(p,\, f(p))$ can be computed from DC by noting that $\dot{x} = 1$, $z = \dot{y} = f'(p)$, and $\dot{z} = f''(p)$. Thus,

$$DC = \frac{\left[1 + f'(p)^2\right]^{3/2}}{f''(p)},$$

and by Newton's observation that curvature itself is given by the reciprocal of the radius of curvature,

$$k = \frac{f''(p)}{\left[1 + f'(p)^2\right]^{3/2}} \tag{3.1}$$

becomes the present-day expression for the curvature of the plane curve $y = f(x)$ when $x = p$. With this equation, curves that are concave upward have positive curvature, and curves that are concave downward have negative curvature, a fact that Euler (as we shall read later) uses tacitly. The reader should verify (Exercise 3.9) that for the parametric curve

$$\alpha(t) = (x(t),\, y(t)),$$

one has

$$k = \frac{\dot{x}\ddot{y} - \dot{y}\ddot{x}}{\left[(\dot{x})^2 + (\dot{y})^2\right]^{3/2}}. \tag{3.2}$$

An equivalent equation for the radius of curvature using Leibniz's dx and dy notation was developed essentially from Figure 3.8 in the first calculus text of 1696, *Analyse des infiniment petits* (Analysis of the infinitely small) [119], written by Guillaume François Antoine Marquis de l'Hospital (1661–1704). Although l'Hospital's name appears as the author of this text, much of its content stems from the lectures and work of Johann Bernoulli (1667–1748) [133, p. 532]. A more modern derivation of the radius and center of curvature can be found in Exercise 3.10, while some computations of curvature appear in Exercises 3.11 through 3.14.

Exercise 3.9. The purpose of this exercise is to develop the modern parametric equation for curvature from Newton's work. Consider the curve α in the xy-plane given parametrically by

$$\alpha(t) = \big(x(t),\, y(t)\big),$$

and suppose that α sweeps through arc AD in Figure 3.8 (right side). Recall that Newton shows that

$$CG = \frac{(\dot{x})^2 + (\dot{y})^2}{\dot{x}\dot{z}}, \quad \text{where} \quad z = \frac{\dot{y}}{\dot{x}}.$$

Explain why each of the following equations holds:

$$\frac{DC}{CG} = \frac{TD}{TB}, \qquad \frac{TD}{TB} = \frac{\sqrt{(\dot{x})^2 + (\dot{y})^2}}{\dot{x}}.$$

Newton uses the simplifying assumption $\dot{x} = 1$ ($\frac{dx}{dt} = 1$), which we wish to avoid for the moment. Use the quotient rule to compute $\dot{z}$ when $\dot{x}$ is not necessarily constant. Calculate the radius of curvature DC, its reciprocal $1/DC$, and compare this with equation (3.2). Finally for the special parameterization

$$x = t, \quad y = f(x) = f(t),$$

show that $\dot{x} = 1$, $\dot{y} = f'(x)$, and the equation for $\frac{1}{DC}$ reduces to the modern equation for curvature given in equation (3.1).

Exercise 3.10. In this exercise we develop equations for the radius and center of curvature by finding the circle that best matches a curve $y = f(x)$ at the point $(p, f(p))$. Suppose that $f''(x)$ is continuous and positive, so that $f(x)$ is concave upward. Sketch a circle

$$(x - a)^2 + (y - b)^2 = r^2$$

with center above the curve $f(x)$ so that the circle and the curve are tangent to each other at $(p, f(p))$. Find an equation that represents the geometric condition that both of these pass through $(p, f(p))$. Equate the first and second derivatives of the curve and the function y of x defined implicitly for the circle when $x = p$ to find a, b, and r. Implicit differentiation may be useful.

Exercise 3.11. Compute the curvature as well as the radius of curvature of $h(x) = 4 - x^2$ at $x = -1$ and compare this with Exercise 3.1 in the previous section.

Exercise 3.12. Find the curvature of $f(x) = x^2$ at the points

$$x = -1, \quad x = 0, \quad x = 1.$$

Exercise 3.13. Find the curvature of $g(x) = x^3$ at the points

$$x = -1, \quad x = 0, \quad x = 1.$$

Exercise 3.14. Graph the curve in the xy-plane given by

$$\sigma(t) = (a \cos t, b \sin t),$$

where a and b are positive constants and $a \geq b$. Compute the curvature k of $\sigma(t)$ at an arbitrary point on the curve. Discuss the value of k in the special case $a = b$ and interpret your answer geometrically. For $a > b$, find those points on the curve at which the curvature attains a maximum value and those points where k attains a minimum value.

3.4 Euler Studies the Curvature of Surfaces

Leonhard Euler (1707–1783) is perhaps history's most prolific mathematician, having published more than 500 books and papers during his lifetime, not to mention those that appeared posthumously. His mathematical tastes were universal, with major works dedicated to analysis, number theory, differential equations, the calculus of variations, and differential geometry. Further biographical information concerning the legendary Euler can be found in the chapter on prime numbers. If any one paper can be said to have inaugurated differential geometry, it would likely be Euler's 1760 publication *Recherches sur la courbure des surfaces* (Researches on the Curvature of Surfaces), presented to the Berlin Academy of Sciences. The Swiss-born Euler spent the years 1741–1766 in the Prussian capital at the invitation of Frederick the Great [18, p. 493].

The study of surfaces began in earnest the analytic investigation of two-dimensional objects in space. The idea of curvature was by 1760 well established as a tool to study one-dimensional curves,[11] but remained unapplied to higher-dimensional objects. Surfaces, such as a sphere, a cylinder, a paraboloid, or a saddle shape, are certainly curved objects and should be subject to study via a precise measurement of curvature at any point on the surface. This is exactly the problem Euler addresses. His approach is to consider the intersection of the surface with a plane, which forms a curve in the plane, thereby reducing the problem of curvature of surfaces to that of such cross-sectional curves. There is, however, an inherent pitfall to this method, since given a point P on the surface, there are infinitely many planes that pass through P. Euler subsequently limits his discussion to planes that are perpendicular to the surface, i.e., planes that pass through what today is called the normal vector. This may seem to be scant progress, since there are still infinitely many planes through P perpendicular to the surface. Here is where Euler makes a decisive contribution to the theory of surfaces: Among all perpendicular cross sections, there is one having maximum curvature (with a value of, say, k_1). At an inclination of 90° to this cross section, the curvature obtains a minimum (call it k_2), and at an inclination of α from the maximum direction, the curvature is

$$k = k_1 \cos^2 \alpha + k_2 \sin^2 \alpha. \qquad (3.3)$$

Thus the problem of determining the infinitely many curvatures of the perpendicular cross sections is reduced to finding only two, the maximum and the minimum. All others are determined by the inclination of the plane forming the cross section.

Although the formula for cross-sectional curvature k given above is implicit in Euler's work, this particular formulation of k is due to Charles Dupin

---

[11] This includes curves in three-space, which had been studied by Alexis Clairaut (1713–1765) in his 1731 book *Recherches sur les courbes à double courbure*.

(1784–1873) [55, p. 109]. Euler instead writes [66, v. 28, p. 22]

$$r = \frac{2fg}{f + g - (f - g)\cos 2\varphi}, \tag{3.4}$$

where we have $r = 1/k$, $g = 1/k_1$, $f = 1/k_2$, $\varphi = 90° - \alpha$ (Exercise 3.15.) A student of Gaspard Monge (1746–1818) at the prestigious École Polytechnique, Dupin published this finding along with several applications of Euler's work in his 1813 text *Développements de Géométrie, avec des Applications à la stabilité des Vaisseaux, aux Déblais et Remblais, au Défilement, à l'Optique, etc.* (Developments of Geometry, with Applications to the Stability of Vessels, to Excavations and Embankments, to Fortifications, to Optics, etc.).

Euler writes using nearly present-day algebraic formulas;[12] the older verbal geometric description of all quantities is by now a style of the past. The facility afforded by algebraic manipulations allows the formulation of complicated expressions that the Swiss master handles adroitly, using a mix of calculus, trigonometry, substitution, and simplification. The modern symbols for partial derivatives, $\frac{\partial z}{\partial x}$, $\frac{\partial z}{\partial y}$, however, are not used, with Euler writing $\frac{dz}{dx}$, $\frac{dz}{dy}$ instead. Moreover, $\sec\varphi \cdot \sec\varphi$ is simply written as $\sec\varphi^2$. It is curious to notice how Euler introduces parametric equations to discuss the planar cross sections of a surface. The idea of a parametric representation of an object is one that reoccurs throughout the study of space. To outline Euler's article, we remark that problem one addresses the curvature of the cross section formed by the intersection of a surface with any plane of the form

$$z = \alpha y - \beta x + \gamma.$$

The solution is in terms of the constants α, β, and the partial derivatives (of the first and second order) of the surface. Problem two is restricted to planes $\mathfrak{P}$ that are perpendicular to the surface, and the solution involves ζ, the angle between the x-axis and EF, where EF is the intersection of $\mathfrak{P}$ with the xy-plane. For problem three Euler identifies a principal plane that is perpendicular to both the given surface and the xy-plane. The equation for cross-sectional curvature is then rewritten in terms of φ, the angle between $\mathfrak{P}$ and the principal plane. This third equation is then simplified in the conclusion, where the extreme values of the radii of curvature are described, and formula (3.4) appears. The excerpt below offers the introduction to Euler's paper and the solution to the first problem. In Euler's own words [64], [66, v. 28, pp. 1–22]:

ⲟⲭⲭⲟⲭⲭⲟⲭⲭⲟⲭⲭⲟ

---

[12] A variable squared, for example x^2, is often written xx.

Euler, from
Researches on the Curvature of Surfaces[13]

In order to know the curvature of curved lines, the determination of the radius of the osculating circle offers the proper method, which for each point on the curve provides us with a circle whose curvature is the same. But when one asks for the curvature of a surface, the question is rather equivocal, and not at all subject to a definitive answer, as in the previous case. It is only spherical surfaces for which one can measure the curvature, considering that the curvature of a sphere is the same as that of its great circles, and its radius can be considered as the proper measure of curvature. But for other surfaces, one would not even know how to compare the curvature of the surface with that of a sphere, as one can always compare the curvature of a curved line with that of a circle. The reason for this is evident, since through each point of a surface, there are infinitely many different curves. One only need consider the surface of a cylinder, where along directions parallel to the axis, there is no curvature, while cross sections perpendicular to the axis, which are circles, have the same curvature, and all other sections taken obliquely to the axis yield particular values of the curvature. Similarly for all other surfaces, where it can even happen that in one direction the curvature might be convex and in another concave, as for surfaces which resemble a saddle.

Thus the question about curvature of surfaces is not amenable to a simple answer, but requires at once infinitely many determinations, because as soon as one is able to draw an infinitude of directions through each point, the curvature must be known along each direction before one is able to form an accurate idea about the curvature of the surface. Now through each point of the surface there are infinitely many cross sections, not only with respect to all the directions on the surface, but also with respect to different inclinations of the sections. But for the matter at hand, of all these infinitely many sections, it suffices to consider only those which are perpendicular to the surface, the number of which is still infinite. To this end one only has to draw a line perpendicular to the surface and all sections which pass through this line are also perpendicular to the surface. Then for each of these sections it remains to find the curvature, or the radius of the osculating circle, and the collection of all these radii will give us an accurate measure of the curvature of the surface at a given point. It must be observed that each of these radii falls along the same perpendicular direction to the surface, and that the elementary arcs of all these sections are part of the shortest curves which can be drawn on the surface.

To render this work more general, I will begin by determining the radius of curvature of an arbitrary planar section which cuts the surface. Then I will apply this solution to sections which are perpendicular to the surface at an arbitrary point, and finally I will compare the radii of curvature for these sections[14] with

---

[13] Translated from the original French, which was the scholarly language of the Berlin *académie des sciences*.

[14] The perpendicular sections.

respect to their mutual inclination, which will allow us to establish a good idea for the curvature of surfaces. All this work reduces then to the following problems.

PROBLEM   1

1. A surface whose nature is known is cut by an arbitrary plane. Determine the curvature of the section which is formed.

SOLUTION

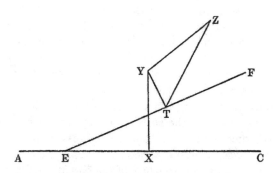

Fig. 3.9. Plane of the cross section.

When one regards (Figure 3.9) the surface with respect to a fixed plane, and from an arbitrary point Z on the surface drops the perpendicular ZY, and from Y drops the perpendicular YX to an axis AC, then the three coordinates $AX = x$, $XY = y$, and $YZ = z$ are given.[15] Since the nature of the surface is known, the quantity z will be equal to a certain function of the two others x and y. Suppose then that by differentiation one obtains $dz = p\,dx + q\,dy$, where

$$p = \left(\frac{dz}{dx}\right) \quad \text{and} \quad q = \left(\frac{dz}{dy}\right).$$

Let the section which cuts the surface pass through the point Z, and let the intersection of the plane of this section and our fixed plane be the line EF. Let

$$z = \alpha y - \beta x + \gamma$$

be the equation which determines the plane of the section, and letting $z = 0$, the equation $y = \frac{\beta x - \gamma}{\alpha}$ will give EF, from which we obtain

$$AE = \frac{\gamma}{\beta} \quad \text{and the tangent of angle } CEF = \frac{\beta}{\alpha}.$$

Thus

$$\text{the sine} = \frac{\beta}{\sqrt{\alpha\alpha + \beta\beta}} \quad \text{and the cosine} = \frac{\alpha}{\sqrt{\alpha\alpha + \beta\beta}}.$$

From this and equating the two values of dz, we will have an equation for the section

---

[15] See Figure 3.10 for a modern sketch of the cross section.

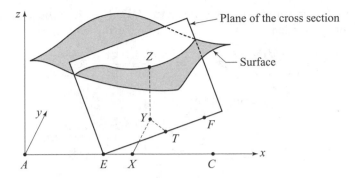

Fig. 3.10. Modern sketch of the cross section.

$$\alpha \, dy - \beta \, dx = p \, dx + q \, dy$$

or just as well

$$\frac{dy}{dx} = \frac{\beta + p}{\alpha - q}.$$

But to reduce this equation to rectangular coordinates, let us draw from Y the perpendicular YT to EF, and the straight line ZT will also be perpendicular to EF. Now, since $EX = x - \frac{\gamma}{\beta}$, we will have

$$ET = -\frac{\alpha x + \beta y}{\sqrt{\alpha\alpha + \beta\beta}} - \frac{\alpha\gamma}{\beta\sqrt{\alpha\alpha + \beta\beta}}$$

and

$$TY = \frac{\alpha y - \beta x}{\sqrt{\alpha\alpha + \beta\beta}} + \frac{\gamma}{\sqrt{\alpha\alpha + \beta\beta}} - \frac{z}{\sqrt{\alpha\alpha + \beta\beta}},$$

and finally

$$TZ = \frac{z\sqrt{1 + \alpha\alpha + \beta\beta}}{\sqrt{\alpha\alpha + \beta\beta}} = \frac{(\alpha y - \beta x + \gamma)\sqrt{1 + \alpha\alpha + \beta\beta}}{\sqrt{\alpha\alpha + \beta\beta}}.$$

Then setting

$$ET = \frac{\alpha x + \beta y}{\sqrt{\alpha\alpha + \beta\beta}} - \frac{\alpha\gamma}{\beta\sqrt{\alpha\alpha + \beta\beta}} = t$$

and

$$TZ = \frac{(\alpha y - \beta x + \gamma)\sqrt{\alpha\alpha + \beta\beta + 1}}{\sqrt{\alpha\alpha + \beta\beta}} = u,$$

we will be able to consider the t and u lines as orthogonal coordinates for the section in question. Thus, if we set $du = s \, dt$, the radius of the osculating circle for the section at the point Z will be

$$= -\frac{dt \, (1 + ss)^{\frac{3}{2}}}{ds}$$

provided that it is turning towards the base EF. Now it is only a matter of reducing this expression to x and y coordinates. To this end, since

$$dt = \frac{\alpha\,dx + \beta\,dy}{\sqrt{\alpha\alpha + \beta\beta}} \quad \text{and} \quad du = \frac{\alpha\,dy - \beta\,dx}{\sqrt{\alpha\alpha + \beta\beta}}\sqrt{1 + \alpha\alpha + \beta\beta},$$

because of $\frac{dy}{dx} = \frac{\beta+p}{\alpha-q}$, we then obtain

$$s = \frac{du}{dt} = \frac{\alpha p + \beta q}{\alpha\alpha + \beta\beta - \alpha q + \beta p}\sqrt{1 + \alpha\alpha + \beta\beta}.$$

Thus

$$1 + ss = \frac{(\alpha\alpha + \beta\beta)(\alpha\alpha + \beta\beta - 2\alpha q + 2\beta p + (\alpha p + \beta q)^2 + pp + qq)}{(\alpha\alpha + \beta\beta - \alpha q + \beta p)^2}.$$

Thus, for the differential of s, we will have

$$ds = \frac{(\alpha\alpha + \beta\beta)(\alpha\,dp + \beta\,dq - q\,dp + p\,dq)\sqrt{1 + \alpha\alpha + \beta\beta}}{(\alpha\alpha + \beta\beta - \alpha q + \beta p)^2}.$$

Let us now notice that

$$dp = dx\left(\frac{dp}{dx}\right) + dy\left(\frac{dp}{dy}\right) \quad \text{and} \quad dq = dx\left(\frac{dq}{dx}\right) + dy\left(\frac{dq}{dy}\right),$$

from which we conclude

$$\frac{dp}{dt} = \frac{(\alpha - q)\left(\frac{dp}{dx}\right) + (\beta + p)\left(\frac{dp}{dy}\right)}{\alpha\alpha + \beta\beta - \alpha q + \beta p}\sqrt{\alpha\alpha + \beta\beta}$$

and

$$\frac{dq}{dt} = \frac{(\alpha - q)\left(\frac{dq}{dx}\right) + (\beta + p)\left(\frac{dq}{dy}\right)}{\alpha\alpha + \beta\beta - \alpha q + \beta p}\sqrt{\alpha\alpha + \beta\beta},$$

and finally[16]

$$\frac{ds}{dt} = \frac{ABC}{D},$$

$$A = (\alpha\alpha + \beta\beta)^{\frac{3}{2}}, \quad C = \sqrt{1 + \alpha\alpha + \beta\beta},$$

$$B = \left[(\alpha - q)^2\left(\frac{dp}{dx}\right) + (\beta + p)^2\left(\frac{dq}{dy}\right) + 2(\alpha - q)(\beta + p)\left(\frac{dp}{dy}\right)\right],$$

$$D = (\alpha\alpha + \beta\beta - \alpha q + \beta p)^3,$$

since $\left(\frac{dq}{dx}\right) = \left(\frac{dp}{dy}\right)$ as is otherwise known. As a consequence, the osculatory radius for the section at the point Z will be expressed in the form[17]

---

[16] The original formula is split for legibility.

[17] Using the simplifying notation above.

$$-\frac{\left(\alpha\alpha + \beta\beta - 2\alpha q + 2\beta p + (\alpha p + \beta q)^2 + pp + qq\right)^{\frac{3}{2}}}{BC}.$$

This is then the veritable expression for the osculatory radius of an arbitrary section which cuts the given surface.

∞∞∞∞∞∞∞∞∞

The reader is invited to follow Euler's study (in English translation) of the above expression for the radius of curvature at [136, Eneström 333], which contains the body of Euler's work on this problem as well as the conclusion concerning perpendicular cross sections. Albeit direct, Euler's proof is rather long and involves a bit of computational endurance. Exercise 3.16 offers a specific example of a surface for which the curvature of the perpendicular cross sections (at a given point) can be readily computed and compared. A hallmark of a great theorem is the work of others to simplify the proof of the result, which is the case with Euler's theorem. Exercise 3.17, which builds from Exercise 3.16, outlines the approach taken by Jean-Baptiste Meusnier (1754–1793) to re-prove the result using what would become the "textbook proof" of Euler's theorem.

Exercise 3.15. Prove Dupin's result (3.3)

$$k = k_1 \cos^2 \alpha + k_2 \sin^2 \alpha$$

from Euler's equation (3.4)

$$r = \frac{2fg}{f + g - (f - g) \cos 2\varphi},$$

where $r - 1/k$, $g = 1/k_1$, $f = 1/k_2$. Here k_1 is the maximum curvature of a perpendicular cross section to a given surface, k_2 is the minimum, and k represents the curvature of an arbitrary perpendicular cross section forming an angle α with the plane that yields k_1. Since Euler is using φ to denote the angle between an arbitrary perpendicular cross section and the plane that yields the maximum radius of curvature (the minimum value of k), we have

$$\varphi = 90° - \alpha.$$

Exercise 3.16. Determine the maximum and minimum values for the curvature of perpendicular cross sections to the surface

$$z = x^2 + 2y^2$$

at the point $(0, 0, 0)$. Begin by considering the line in the xy-plane

$$r(t) = (at, bt),$$

where (a, b) is a fixed point on the unit circle $a^2 + b^2 = 1$. Let $\mathfrak{P}$ be the plane that is perpendicular to $z = x^2 + 2y^2$ at $(0, 0, 0)$, and that passes through the line $r(t)$ in the xy-plane. Find the equation of the curve σ formed by the intersection of $\mathfrak{P}$ with $z = x^2 + 2y^2$, and express the curve as a function of t. Use results from Section three to find the curvature of σ at the point $(0, 0, 0)$. Find k_1, the maximum value of this curvature, and k_2, the minimum value. Verify that if $\mathfrak{P}$ forms an angle α with the maximum direction, then the curvature is given by

$$k = k_1 \cos^2 \alpha + k_2 \sin^2 \alpha$$

(which is Dupin's result).

Exercise 3.17. In 1776, Jean-Baptiste Meusnier (1754–1793) presented to the French *Académie royale des sciences* a concise and highly original proof of Euler's results on curved surfaces [231, pp. 106–107]. In this exercise we retrace the steps taken by Meusnier in his 1785 publication *Mémoire sur la courbure des surfaces* (Memoir on the curvature of surfaces) [167] and arrive at a proof of Euler's result that isolates the key ingredients for determining the curvature of a perpendicular cross section.

In modern language, let $z = f(x, y)$ be a function with continuous second partial derivatives, and let $Q = (x_0, y_0, z_0)$ be a point on the surface $z = f(x, y)$ where the curvature is sought. Suppose further that T is the tangent plane to $z = f(x, y)$ at the point Q. Meusnier's insight is to consider T as the xy-plane, and choose $Q = (0, 0, 0)$, so that the origin is the point of tangency. Since the curvature of any planar section is determined by the first and second derivatives, it suffices to consider the second Taylor–Maclaurin polynomial of $z = f(x, y)$ expanded about $(0, 0)$, namely

$$P(x, y) = f(0, 0) + \left(\frac{\partial f}{\partial x}(0, 0)\right) x + \left(\frac{\partial f}{\partial y}(0, 0)\right) y$$
$$+ \left(\frac{1}{2}\frac{\partial^2 f}{\partial x^2}(0, 0)\right) x^2 + \left(\frac{\partial^2 f}{\partial x\, \partial y}(0, 0)\right) xy + \left(\frac{1}{2}\frac{\partial^2 f}{\partial y^2}(0, 0)\right) y^2.$$

Justify why $f(0, 0) = 0$, $\frac{\partial f}{\partial x}(0, 0) = 0$, $\frac{\partial f}{\partial y}(0, 0) = 0$ when the xy-plane is considered as the tangent plane. Thus

$$P(x, y) = Ax^2 + Bxy + Cy^2$$

for some constants A, B, C. Consider the line in the xy-plane given by

$$r(t) = (t \cos\theta, \ t \sin\theta),$$

where θ is constant, and let $\mathfrak{P}$ be the plane that passes through $r(t)$ and is perpendicular to $z = f(x, y)$ at the point Q. Let σ_θ be the curve formed by the intersection of $\mathfrak{P}$ with $z = f(x, y)$. Find an expression for $k(\theta)$, the curvature of σ_θ at Q. Considering $k(\theta)$ as a function of θ, what conditions relating A, B, C, and θ are necessary for $k(\theta)$ to achieve a maximum or minimum value?

Suppose then that $k(\theta)$ achieves a maximum when $\theta = \theta_0$ with a value of $k_1 = k_{\theta_0}$. Show that a minimum value of $k(\theta)$ occurs when $\theta = \theta_0 + 90°$, and let k_2 denote this minimum. Consider a new set of axes x' and y' in the xy-plane so that x' lies along the line

$$r(t) = (t \cos\theta_0,\ t \sin\theta_0),$$

and y' lies along the line

$$r(t) = \big(t \cos(\theta_0 + 90°),\ t \sin(\theta_0 + 90°)\big).$$

Let θ' be an angle in the $x'y'$-plane, measured counterclockwise from the x'-axis, such that

$$\theta' + \theta_0 = \theta.$$

Show that the curvature at Q of the perpendicular cross section above the line

$$r(t) = (t \cos\theta',\ t \sin\theta')$$

has a general expression of the form

$$k(\theta') - 2\big(A' \cos^2 \theta' + B' \cos\theta' \sin\theta' + C' \sin^2 \theta'\big)$$

for some constants A', B', C'. Find expressions for A', B', and C' in terms of the original constants A, B, C, and the angle θ_0. Using a conceptual argument, algebraic substitution, or both, conclude that

$$k_1 = 2A', \quad k_2 = 2C', \quad B' = 0.$$

Finally, explain why (a conclusion that apparently Meusnier did not reach)

$$k(\theta') = k_1 \cos^2 \theta' + k_2 \sin^2 \theta',$$

where θ' is the angle of the cross section as measured from the direction that yields the maximum curvature. Thus, by an appropriate choice of axes, the curvature at a given point of a surface is determined by a quadratic function

$$q(x,\ y) = ax^2 + by^2$$

for certain constants a and b.

A student of Monge at the school of Mézières, Meusnier re-proved Euler's result on the day of his arrival at Mézières after a preliminary discussion with Monge, but without having read Euler's manuscript [236, p. 234]. Monge writes of Meusnier:

> The next morning in the classroom, he handed me a short paper which contained this proof, but what was remarkable was that all considerations which he had employed were more direct, the path which he followed was more rapid than those of which Euler had made use [236, p. 234].

3.5 Gauss Defines an Independent Notion of Curvature

On October 8, 1827, Carl Friedrich Gauss (1777–1855) presented to the Göttingen Royal Scientific Society his essay *Disquisitiones generales circa superficies curvas* (General Investigations of Curved Surfaces) [84, 86, 87], the fruit of at least 15 years of intense intellectual effort, inspired in part by his research into non-Euclidean geometry, as well as his field work as surveyor of Hanover [51]. One of the most brilliant mathematicians in all history, Gauss published major works in algebra, number theory, complex analysis, error analysis, and, of course, differential geometry. Further biographical information about Gauss can be found in the chapter on prime numbers. The problem of map projection, so eloquently solved in *Disquisitiones superficies*, was the subject of two other papers by the Göttingen professor, the first in 1822: *Allgemeine Auflösung der Aufgabe die Theile einer gegebenen Fläche auf einer andern gegebenen Fläche so abzubilden dass die Abbildung dem Abgebildeten in den kleinsten Theilen ähnlich wird* (General solution to the problem of mapping a given surface onto another given surface so that the image will be similar to the first surface in the smallest details) [84, pp. 189–216]. The second paper, *Neue allgemeine Untersuchungen über die krummen Flächen* (New General Investigations of Curved Surfaces) [85, pp. 408–442], apparently written in 1825, but not published until 1900 in *Gauss Werke*, was a draft of the 1827 *Disquisitiones superficies*. Two years of concentrated effort following the 1825 draft were devoted to finding an equation for curvature in terms of the metric data (E, F, G in Gauss's notation), needed in the proof of the *theorema egregium*. In the master's own words, "If a curved surface is developed upon any other surface whatever, the measure of curvature in each point remains unchanged" [84, p. 237]. The term *development* is used here to describe a function between two surfaces that preserves distances, while modern geometry calls such a function an isometry.

Gauss's own unpublished work on non-Euclidean geometry served in part to motivate the results of *Disquistiones superficies*. In 1817 he had written privately in a letter to H. W. M. Olbers (1758–1840), "I am becoming more and more convinced that the necessity of our [Euclidean] geometry cannot be proved, at least not by human reason nor for human reason. Perhaps in another life we will be able to obtain insight into the nature of space" [101], [225, p. 55]. With his *Disquisitiones superficies* Gauss had sown the seeds for what would become the nature of space, and in the year before his death, 1854, he would hear a lecture by his successor, Bernhard Riemann (1826–1866), entitled *On the Hypotheses That Lie at the Foundations of Geometry*, which would serve as the standard for the classification of space [223]. Before then, however, an axiomatic study of geometry continued with the simultaneous and independent discovery of hyperbolic geometry by János Bolyai (1802–1860) and Nikolai Ivanovich Lobachevsky (1792–1856). Gauss had also arrived at many of the results of hyperbolic geometry, yet to avoid controversy, published almost nothing in this regard. Lobachevsky published his findings in an 1829 memoir entitled *On the Principles of Geometry*, which is an account of his lecture *Exposition succincte des principes de la géométrie avec une démonstration rigoureuse du théorème des parallèles*, read in 1826 to the Physical Mathematical Section of the University of Kazan (Russia) [17, p. 85]. By 1823 Bolyai was in possession of the key ideas behind hyperbolic geometry, and published his work *Appendix scientiam spatii absolute* (The Science of Absolute Space) in 1831 as an appendix to the text *Tentamen* authored by his father, Wolfgang Bolyai (1775–1856) [17, pp. 98–99]. Hyperbolic geometry results from replacing the parallel postulate with the statement, "there is some line L and some point P not on L such that there is more than one line through P parallel to L," and is understood today as the geometry on a surface of constant negative Gaussian curvature.

The selection from *Disquisitiones superficies* that follows develops the idea of Gaussian curvature for a surface, while the reader is referred to the original paper for a complete proof of the *theorema egregium*. Unlike Euler, Gauss does not consider planar cross sections of the surface, but instead begins his inquiry with normal vectors to the surface. For example, given the saddle surface

$$z = y^2 - x^2,$$

then all normal vectors at

$$P_0 = (x_0, y_0, z_0) - (x_0, y_0, y_0^2 - x_0^2)$$

are scalar multiples of the vector $(-2x_0, 2y_0, -1)$. Gauss considers two possible directions for the normals, which we denote by

$$N_1 - (-2x_0, 2y_0, -1) \quad \text{and} \quad N_2 = (2x_0, -2y_0, +1). \tag{3.5}$$

From the notation in *Disquisitiones superficies*, the saddle surface may be described as

$$W(x, y, z) = z + x^2 - y^2 \equiv 0,$$

and the vectors N_1 and N_2 may be classified as an inward-pointing normal and an outward-pointing normal respectively, since

$$W(P_0 + N_1) = W((x_0, y_0, z_0) + (-2x_0, 2y_0, -1)) = -(8y_0^2 + 1) < 0,$$
$$W(P_0 + N_2) = W((x_0, y_0, z_0) + (2x_0, -2y_0, 1)) = (8x_0^2 + 1) > 0.$$

Choosing the outward-pointing normal and dividing by its length, we see that

$$\frac{1}{\sqrt{4x_0^2 + 4y_0^2 + 1}} \left(2x_0, -2y_0, +1\right)$$

is a point on the unit sphere,

$$XX + YY + ZZ = 1,$$

and in this way, "each point on the curved surface is made to correspond to a definite point on the sphere" [86, p. 9]. In modern language, the function

$$\varphi(x, y, z) = \frac{1}{\sqrt{4x^2 + 4y^2 + 1}} \left(2x, -2y, 1\right)$$

from points on the surface $z = y^2 - x^2$ to points on the sphere is called the *Gauss map*. (In the next section, Exercise 3.31 describes a space for which the image of a related "Gauss map" does not depend on the choice of an inward- or outward-pointing normal vector.)

The German geometer then considers an infinitesimally small triangle on the surface that contains the point P_0. Letting T denote the triangle, then the area of $\varphi(T)$ is called the integral (total) curvature of T, while the measure of curvature at P_0 is the limiting value of

$$\frac{\text{area of } \varphi(T)}{\text{area of } T}$$

as T shrinks to P_0. This particular ratio had been studied by Olinde Rodrigues[18] (1794–1851) in an 1815 publication [197], although he did not discover the *theorema egregium*. The measure of curvature is assigned a positive or negative value, depending on whether φ maps T to a "similar" region on the sphere or to an "opposite" (inverse) region. Today, the terms orientation-preserving and orientation-reversing are used to describe the effect of mapping a region to a "similar" or "opposite" one respectively. In Exercise 3.18 the reader is asked to verify, using the criteria set forth by Gauss, that φ is orientation-reversing for the saddle surface. Thus $z = y^2 - x^2$ is a surface of negative Gaussian curvature.

In an elegant and unifying theorem, Gauss then relates his measure of curvature to results proven by Euler: "The measure of curvature at any point whatever of the surface is equal to a fraction whose numerator is unity, and whose denominator is the product of the extreme radii of curvature of the sections by normal planes" [86, p. 15]. The extreme radii of curvature R_1 and R_2 refer to the largest and smallest osculatory radii found by Euler. Thus the Gaussian curvature k may be simply written as

$$k = \frac{1}{R_1 R_2}.$$

In memoirs of 1821 and 1826 [88, 89], Sophie Germain (1776–1831) had proposed that the measure of curvature be given essentially by the mean (average) of $1/R_1$ and $1/R_2$, a quantity that proved to be crucial in her work on elasticity [24, p. 113]. She introduced the notion of a referent sphere to a surface, which is the sphere whose radius is the reciprocal of

$$\frac{1}{2}\left(\frac{1}{R_1} + \frac{1}{R_2}\right).$$

---

[18] Rodrigues was a student of Monge at the École Polytechnique.

After reading Gauss's treatise on curved surfaces, Mademoiselle Germain wrote to Gauss comparing her notion of curvature to his and lamented the lack of recognition her own work had received [24, p. 114]:

> In chatting with Monsieur Bader[19] about the current subject of my study, I provided him with the occasion to speak to me, and subsequently, to show me, the learned memoir in which you compare the curvature of surfaces to that of the sphere. ...
>
> I cannot tell you, Monsieur [Gauss], how astonished, and at the same time, how satisfied I was in learning that a renowned mathematician, almost simultaneously, had the idea of an analogy that seems to me so rational that I neither understood how no one had thought of it sooner, nor how no one has wished to give any attention to date to what I have already published in this regard.

Germain's *mean curvature* today plays a vital role in differential geometry as a tool for finding what are called minimal surfaces, i.e., those surfaces of minimum surface area with a prescribed boundary [224].

For the proof of the *theorema egregium* Gauss does not describe the surface in terms of the equation

$$W(x, y, z) = 0.$$

Instead, he introduces a "second method [that] expresses the coordinates in the form of functions of two variables p, q " [86, p. 7], a crucial idea, going back to Euler's *De solidis quorum superficiem in planum explicare licet* (On Solids Whose Surfaces Can Be Developed in the Plane) [66, v. 28, pp. 161–186]. Consider, for example, the circle in the yz-plane with equation

$$(y - 2)^2 + z^2 = 1,$$

and suppose that the circle is rotated around the z-axis to generate a surface in three-space (see Figure 3.11). The resulting figure is called a torus and consists only of points on the shell of the surface, and not inside the surface. Such a point (x, y, z) can be written in terms of the parameters p and q (see Exercise 3.19):

$$x = (2 + \cos p) \sin q,$$
$$y = (2 + \cos p) \cos q,$$
$$z = \sin p.$$

The parameterization itself, written as a function

$$\varphi : (pq\text{-plane}) \to (xyz\text{-space})$$
$$\varphi(p, q) = \big((2 + \cos p) \sin q, \ (2 + \cos p) \cos q, \ \sin p \big),$$

is today called a coordinate chart, which provides "two systems of curved lines on the curved surface, one system for which p is variable, q constant; the other for which q is variable, p constant" [86, p. 9].

---

[19] A student of Gauss who was visiting Paris in 1829.

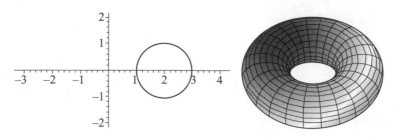

Fig. 3.11. Torus.

Gauss continues with an incisive analysis of the partial derivatives (first and second order) of the coordinates x, y, z with respect to p, q for a general surface. The *theorema egregium*, which is a statement about distance-preserving map projections, requires for its proof a method of computing distances on the surface in terms of the parameters p and q. If γ represents a curve on a surface with

$$\gamma(t) = \big(x(t),\ y(t),\ z(t)\big),$$

then the arc length of γ between $t = a$ and $t = b$ is given by

$$\int_a^b \sqrt{\left(\frac{dx}{dt}\right)^2 + \left(\frac{dy}{dt}\right)^2 + \left(\frac{dz}{dt}\right)^2}\ dt. \tag{3.6}$$

Equation (3.6) is in terms of the coordinates x, y, z for three-space and not in terms of p and q. In the specific example of the torus, we have

$$\frac{dx}{dt} = -\frac{dp}{dt} \sin p \sin q + \left(2 + \cos p\right)\left(\frac{dq}{dt} \cos q\right),$$
$$\frac{dy}{dt} = -\frac{dp}{dt} \sin p \cos q - \left(2 + \cos p\right)\left(\frac{dq}{dt} \sin q\right),$$
$$\frac{dz}{dt} = \frac{dp}{dt} \cos p,$$

and from (3.6) the arc length between $t = a$ and $t = b$ reduces to

$$\int_a^b \sqrt{\left(\frac{dp}{dt}\right)^2 + 0\left(\frac{dp}{dt}\right)\left(\frac{dq}{dt}\right) + \left(2 + \cos p\right)^2 \left(\frac{dq}{dt}\right)^2}\ dt.$$

In Gauss's notation we have the coefficients $E = 1$, $F = 0$, $G = (2 + \cos p)^2$, which become the "metric data" for the torus.

On a general surface S_1, arc length (in a certain coordinate chart φ) is computed according to the formula

$$\int_a^b \sqrt{E\left(\frac{dp}{dt}\right)^2 + 2F\left(\frac{dp}{dt}\right)\left(\frac{dq}{dt}\right) + G\left(\frac{dq}{dt}\right)^2}\ dt$$

for certain functions E, F, and G (the "metric data"), which depend on p and q. If f is a distance-preserving map projection from part of S_1 to part of another surface S_2, then the composition $f \circ \varphi$ would be a coordinate chart for S_2. On the image of $f \circ \varphi$, which is now on the second surface S_2, arc length is computed according to the formula

$$\int_a^b \sqrt{E' \left(\frac{dp}{dt}\right)^2 + 2F' \left(\frac{dp}{dt}\right)\left(\frac{dq}{dt}\right) + G' \left(\frac{dq}{dt}\right)^2}\, dt\,,$$

where E', F', and G' are new functions of p and q computed from the composition $f \circ \varphi$. A necessary condition for f to be a development of S_1 on S_2 is that

$$E = E'\,, \quad F = F'\,, \quad G = G'\,,$$

i.e., the formulas for computing distance must be the same on the two surfaces. Gauss proves that the curvature of the surface depends only on E, F, G, and the first and second partial derivatives of these quantities with respect to p and q. Realizing the implications of this when $E = E'$, $F = F'$, $G = G'$, we have, "If a curved surface is developed upon any other surface whatever, the measure of curvature in each point remains unchanged" [86, p. 20].

The reading selection begins with Gauss's own abstract to *Disquisitiones superficies*, which speaks to the motivation and origin of some of the ideas behind his treatment of curvature. The body of the text can be read with a knowledge of calculus of functions of two variables, noting that a partial derivative $\frac{\partial z}{\partial x}$ is still written as $\frac{dz}{dx}$. Also, early in the paper, the points $(1, 0, 0)$, $(0, 1, 0)$, and $(0, 0, 1)$ are referred to as (1), (2), and (3) respectively. Moreover, $\cos(1)L$ denotes the cosine of the angle formed by the x-axis and the ray **OL**, where O is the origin and L is a point in three-space. In the words of the master [86]:

Gauss, from
General Investigations of Curved Surfaces

Gauss's Abstract

Although geometers have given much attention to general investigations of curved surfaces and their results cover a significant portion of the domain of higher geometry, this subject is still so far from being exhausted, that it can well be said that, up to this time, but a small portion of an exceedingly fruitful field has been cultivated. Through the solution of the problem, to find all representations of a given surface upon another in which the smallest elements remain unchanged, the author sought some years ago to give a new phase to this study. The purpose of the present discussion is further to open up other new points of view and to develop some of the new truths which thus become accessible. We shall here give an account of those things which can be made intelligible in a few words. But we wish to remark at the outset that the new theorems as well as the presentations of new ideas, if the greatest generality is to be attained, are still partly in need of some limitations or closer determinations, which must be omitted here.

In researches in which an infinity of directions of straight lines in space is concerned, it is advantageous to represent these directions by means of those points upon a fixed sphere, which are the end points of radii drawn parallel to the lines. The centre and the radius of this auxiliary sphere are here quite arbitrary. The radius may be taken equal to unity. This procedure agrees fundamentally with that which is constantly employed in astronomy, where all directions are referred to a fictitious celestial sphere of infinite radius. Spherical trigonometry and certain other theorems, to which the author has added a new one of frequent application, then serve for the solution of the problems which the comparison of the various directions involved can present.

If we represent the direction of the normal at each point of the curved surface by the corresponding point of the sphere, determined as above indicated, namely, in this way, to every point on the surface, let a point on the sphere correspond; then, generally speaking, to every line on the curved surface will correspond a line on the sphere, and to every part of the former surface will correspond a part of the latter. The less this part differs from a plane, the smaller will be the corresponding part on the sphere. It is, therefore, a very natural idea to use as the measure of the total curvature, which is to be assigned to a part of the curved surface, the area of the corresponding part of the sphere. For this reason the author calls this area the *integral curvature* of the corresponding part of the curved surface. Besides the magnitude of the part, there is also at the same time its position to be considered. And this position may be in the two parts similar or inverse, quite independently of the relation of their magnitudes. The two cases can be distinguished by the positive or negative sign of the total curvature. This distinction has, however, a definite meaning only when the figures are regarded as upon definite sides of the two surfaces. The author regards the figure in the case of the sphere on the outside, and in the case of the curved surface on that side upon which we consider the normals erected. It follows then that the positive sign is taken in the case of convexo-convex or concavo-concave surfaces (which are not essentially different), and the negative in the case of concavo-convex surfaces. If the part of the curved surface in question consists of parts of these different sorts, still closer definition is necessary, which must be omitted here.

The comparison of the areas of two corresponding parts of the curved surface and of the sphere leads now (in the same manner as, e.g., from the comparison of volume and mass springs the idea of density) to a new idea. The author designates as *measure of curvature* at a point of the curved surface the value of the fraction whose denominator is the area of the infinitely small part of the curved surface at this point and whose numerator is the area of the corresponding part of the surface of the auxiliary sphere, or the integral curvature of that element. It is clear that, according to the idea of the author, integral curvature and measure of curvature in the case of curved surfaces are analogous to what, in the case of curved lines, are called respectively amplitude and curvature simply. He hesitates to apply to curved surfaces the latter expressions, which have been accepted more from custom than on account of fitness. Moreover, less depends upon the choice of words than upon this, that their introduction shall be justified by pregnant theorems.

The solution of the problem, to find the measure of curvature at any point of a curved surface, appears in different forms according to the manner in which the nature of the curved surface is given. When the points in space, in general, are distinguished by three rectangular coordinates, the simplest method is to express one coordinate as a function of the other two. In this way we obtain the simplest expression for the measure of curvature. But, at the same time, there arises a remarkable relation between this measure of curvature and the curvature of the curves formed by the intersections of the curved surface with planes normal to it. EULER, as is well known, first showed that two of these cutting planes which intersect each other at right angles have this property, that in one is found the greatest and in the other the smallest radius of curvature; or, more correctly, that in them the two extreme curvatures are found. It will follow then from the above mentioned expression for the measure of curvature that this will be equal to a fraction whose numerator is unity and whose denominator is the product of the extreme radii of curvature. The expression for the measure of curvature will be less simple, if the nature of the curved surface is determined by an equation in x, y, z. And it will become still more complex, if the nature of the curved surface is given so that x, y, z are expressed in the form of functions of two new variables p, q. In this last case the expression involves fifteen elements, namely the partial differential coefficients of the first and second orders of x, y, z with respect to p and q. But it is less important in itself than for the reason that it facilitates the transition to another expression, which must be classed with the most remarkable theorems of this study. If the nature of the curved surface be expressed by this method, the general expression for any linear element upon it, or for $\sqrt{dx^2 + dy^2 + dz^2}$, has the form

$$\sqrt{E\,dp^2 + 2F\,dp.dq + G\,dq^2},$$

where E, F, G are again functions of p and q. The new expression for the measure of curvature mentioned above contains merely these magnitudes and their partial differential coefficients of the first and second order. Therefore we notice that, in order to determine the measure of curvature, it is necessary to know only the general expression for a linear element; the expressions for the coordinates x, y, z are not required. ...

GENERAL INVESTIGATIONS
of
CURVED SURFACES

1.

Investigations, in which the directions of various straight lines in space are to be considered, attain a high degree of clearness and simplicity if we employ, as an auxiliary, a sphere of unit radius described about an arbitrary centre, and suppose the different points of the sphere to represent the directions of straight lines parallel to the radii ending at these points. As the position of every point in space is determined by three coordinates, that is to say, the distances of the point from three mutually perpendicular fixed planes, it is necessary to consider, first of all, the directions of the axes perpendicular to these planes. The points on the sphere, which represent these directions, we shall denote by (1), (2), (3). The distance of any one of these points from either of the other two will be a quadrant; and we shall suppose that the directions of the axes are those in which the corresponding coordinates increase.

2.

. . . VII. Let L, L', L'' be the three points on the sphere and set, for brevity,

$$\cos(1)L = x, \qquad \cos(2)L = y, \qquad \cos(3)L = z,$$
$$\cos(1)L' = x', \qquad \cos(2)L' = y', \qquad \cos(3)L' = z',$$
$$\cos(1)L'' = x'', \qquad \cos(2)L'' = y'', \qquad \cos(3)L'' = z'',$$

and also

$$xy'z'' + x'y''z + x''yz' - xy''z' - x'yz'' - x''y'z = \Delta$$

. . . Whence, finally, it is clear that the expression $\pm\frac{1}{6}\Delta$ expresses generally the volume of any pyramid contained between the origin of coordinates and the three points whose coordinates are[20] x, y, z; x', y', z'; x'', y'', z''.

3.

A curved surface is said to possess continuous curvature at one of its points A, if the directions of all the straight lines drawn from A to points of the surface at an infinitely small distance from A are deflected infinitely little from one and the same plane passing through A. This plane is said to *touch* the surface at the point A. If this condition is not satisfied for any point, the continuity of the curvature is here interrupted, as happens, for example, at the vertex of a cone. The following investigations will be restricted to such surfaces, or to such parts of surfaces, as have the continuity of their curvature nowhere interrupted. We shall only observe now that the methods used to determine the position of the tangent plane lose their meaning at singular points, in which the continuity of the curvature is interrupted, and must lead to indeterminate solutions.

4.

---

[20] In modern language, the (signed) volume of a parallelepiped with edges along three (linearly independent) vectors is given by the determinant of the vector components, as developed in Exercise 3.21. In Section five, the sign of this determinant is used to determine whether the three vectors have the same orientation as the standard unit vectors $(1, 0, 0)$, $(0, 1, 0)$, $(0, 0, 1)$.

The orientation[21] of the tangent plane is most conveniently studied by means of the direction of the straight line normal to the plane at the point A, which is also called the normal to the curved surface at the point A. We shall represent the direction of this normal by the point L on the auxiliary sphere, and we shall set

$$\cos(1)L = X, \quad \cos(2)L = Y, \quad \cos(3)L = Z \,;$$

and denote the coordinates of the point A by x, y, z. Also let $x + dx$, $y + dy$, $z + dz$ be the coordinates of another point A' on the curved surface; ds its distance from A, which is infinitely small; and finally, let λ be the point on the sphere representing the direction of the element AA'. Then we shall have

$$dx = ds.\cos(1)\lambda, \quad dy = ds.\cos(2)\lambda, \quad dz = ds.\cos(3)\lambda$$

and, since λL must be equal to $90°$,

$$X\,\cos(1)\lambda + Y\,\cos(2)\lambda + Z\,\cos(3)\lambda = 0.$$

By combining these equations we obtain

$$X\,dx + Y\,dy + Z\,dz = 0.$$

There are two general methods for defining the nature of a curved surface. The *first* uses the equation between the coordinates x, y, z, which we may suppose reduced to the form $W - 0$, where W will be a function of the indeterminates x, y, z. Let the complete differential of the function W be

$$dW = P\,dx + Q\,dy + R\,dz,$$

and on the curved surface we shall have

$$P\,dx + Q\,dy + R\,dz = 0$$

and consequently,

$$P\,\cos(1)\lambda + Q\,\cos(2)\lambda + R\,\cos(3)\lambda = 0.$$

Since this equation, as well as the one we have established above, must be true for the directions of all elements ds on the curved surface, we easily see that X, Y, Z must be proportional to P, Q, R respectively, and consequently, since

$$XX + YY + ZZ = 1,$$

we shall have either

$$X = \frac{P}{\sqrt{PP+QQ+RR}}, \quad Y = \frac{Q}{\sqrt{PP+QQ+RR}}, \quad Z = \frac{R}{\sqrt{PP+QQ+RR}},$$

---

[21] The Latin "situs" could have also been translated as "position" instead of "orientation."

or

$$X = \frac{-P}{\sqrt{PP+QQ+RR}}, \quad Y = \frac{-Q}{\sqrt{PP+QQ+RR}}, \quad Z = \frac{-R}{\sqrt{PP+QQ+RR}}.$$

The *second* method expresses the coordinates in the form of functions of two variables p, q. Suppose that differentiation of these functions gives

$$dx = a\,dp + a'\,dq,$$
$$dy = b\,dp + b'\,dq,$$
$$dz = c\,dp + c'\,dq.$$

Substituting these values in the formula given above, we obtain

$$(aX + bY + cZ)\,dp + (a'X + b'Y + c'Z)\,dq = 0.$$

Since this equation must hold independently of the values of the differentials dp, dq, we evidently shall have

$$aX + bY + cZ = 0, \quad a'X + b'Y + c'Z = 0.$$

From this we see that the X, Y, Z will be proportioned to the quantities

$$bc' - cb', \quad ca' - ac', \quad ab' - ba'.$$

Hence, on setting, for brevity,

$$\sqrt{(bc' - cb')^2 + (ca' - ac')^2 + (ab' - ba')^2} = \Delta$$

we shall have either

$$X = \frac{bc' - cb'}{\Delta}, \quad Y = \frac{ca' - ac'}{\Delta}, \quad Z = \frac{ab' - ba'}{\Delta},$$

or

$$X = \frac{cb' - bc'}{\Delta}, \quad Y = \frac{ac' - ca'}{\Delta}, \quad Z = \frac{ba' - ab'}{\Delta}.$$

With these two general methods is associated a *third*, in which one of the coordinates, z, say, is expressed in the form of a function of the other two, x, y. This method is evidently only a particular case of the first method, or of the second. If we set

$$dz = t\,dx + u\,dy$$

we shall have either

$$X = \frac{-t}{\sqrt{1 + tt + uu}}, \quad Y = \frac{-u}{\sqrt{1 + tt + uu}}, \quad Z = \frac{1}{\sqrt{1 + tt + uu}},$$

or

$$X = \frac{t}{\sqrt{1 + tt + uu}}, \quad Y = \frac{u}{\sqrt{1 + tt + uu}}, \quad Z = \frac{-1}{\sqrt{1 + tt + uu}}.$$

5.

The two solutions found in the preceding article evidently refer to opposite points of the sphere, or to opposite directions, as one would expect, since the normal may be drawn toward either of the two sides of the curved surface. If we wish to distinguish between the two regions bordering upon the surface, and call one the exterior region and the other the interior region, we can then assign to each of the two normals its appropriate solution by aid of the theorem derived in Art. 2 (VII), and at the same time establish a criterion for distinguishing the one region from the other.

In the first method, such a criterion is to be drawn from the sign of the quantity W. Indeed, generally speaking, the curved surface divides those regions of space in which W keeps a positive value from those in which the value of W becomes negative. In fact, it is easily seen from this theorem that, if W takes a positive value toward the exterior region, and if the normal is supposed to be drawn outwardly, the first solution is to be taken. Moreover, it will be easy to decide in any case whether the same rule for the sign of W is to hold throughout the entire surface, or whether for different parts there will be different rules. As long as the coefficients P, Q, R have finite values and do not all vanish at the same time, the law of continuity will prevent any change.

If we follow the second method, we can imagine two systems of curved lines on the curved surface, one system for which p is variable, q constant; the other for which q is variable, p constant. The respective positions of these lines with reference to the exterior region will decide which of the two solutions must be taken. In fact, whenever the three lines, namely, the branch of the line of the former system going out from the point A as p increases, the branch of the line of the latter system going out from the point A as q increases, and the normal drawn toward the exterior region, are *similarly* placed as the x, y, z axes respectively from the origin of abscissas (e.g., if, both for the former three lines and for the latter three, we can conceive the first directed to the left, the second to the right, and the third upward), the first solution is to be taken. But whenever the relative position of the three lines is opposite to the relative position of the x, y, z axes, the second solution will hold.

In the third method, it is to be seen whether, when z receives a positive increment, x and y remaining constant, the point crosses toward the exterior or the interior region. In the former case, for the normal drawn outward, the first solution holds; in the latter case, the second.

6.

Just as each point on the curved surface is made to correspond to a definite point on the sphere, by the direction of the normal to the curved surface which is transferred to the surface of the sphere, so also any line whatever, or any figure whatever, on the latter will be represented by a corresponding line or figure on the former. In the comparison of two figures corresponding to one another in this way, one of which will be as the map of the other, two important points are to be considered, one when quantity alone is considered, the other when, disregarding quantitative relations, position alone is considered.

The first of these important points will be the basis of some ideas which it seems judicious to introduce into the theory of curved surfaces. Thus, to each point of a curved surface inclosed within definite limits we assign a *total* or *integral curvature*, which is represented by the area of the figure on the sphere corresponding to it. From this integral curvature must be distinguished the somewhat more specific curvature which we shall call the *measure of curvature*. The latter refers to a *point* of the surface, and shall denote the quotient obtained when the integral curvature of the surface element about a point is divided by the area of the element itself; and hence it denotes the ratio of the infinitely small areas which correspond to one another on the curved surface and on the sphere. The use of these innovations will be abundantly justified, as we hope, by what we shall explain below. As for the terminology, we have thought it especially desirable that all ambiguity be avoided. For this reason we have not thought it advantageous to follow strictly the analogy of the terminology commonly adopted (though not approved by all) in the theory of plane curves, according to which the measure of curvature should be called simply curvature, but the total curvature, the amplitude. But why not be free in the choice of words, provided they are not meaningless and not liable to a misleading interpretation?

The position of a figure on the sphere can be either similar to the position of the corresponding figure on the curved surface, or opposite (inverse). The former is the case when two lines going out on the curved surface from the same point in different, but not opposite directions, are represented on the sphere by lines similarly placed, that is, when the map of the line to the right is also to the right; the latter is the case when the contrary holds. We shall distinguish these two cases by the positive or negative *sign* of the measure of curvature. But evidently this distinction can hold only when on each surface we choose a definite face on which we suppose the figure to lie. On the auxiliary sphere we shall use always the exterior face, that is, that turned away from the centre; on the curved surface also there may be taken for the exterior face the one already considered, or rather that face from which the normal is supposed to be drawn. For, evidently, there is no change in regard to the similitude of the figures, if on the curved surface both the figure and the normal be transferred to the opposite side, so long as the image itself is represented on the same side of the sphere.

The positive or negative sign, which we assign to the *measure* of curvature according to the position of the infinitely small figure, we extend also to the integral curvature of a finite figure on the curved surface. However, if we wish to discuss the general case, some explanations will be necessary, which we can only touch here briefly. So long as the figure on the curved surface is such that to *distinct* points on itself there correspond distinct points on the sphere, the definition needs no further explanation. But whenever this condition is not satisfied, it will be necessary to take into account twice or several times certain parts of the figure on the sphere. Whence for a similar, or inverse position, may arise an accumulation of areas, or the areas may partially or wholly destroy each other. In such a case, the simplest way is to suppose the curved surface divided into parts, such that each part, considered separately, satisfies the above condition; to assign to each of the parts its integral curvature, determining this magnitude by the area of the corresponding figure on the sphere, and the sign by the position of this figure; and, finally, to assign to the total figure the integral curvature arising from the addition of the integral curvatures which correspond to the single parts. So, generally, the integral curvature of a figure is equal to $\int k \, d\sigma$, $d\sigma$ denoting the element of area of the figure, and k the measure of curvature at any point. The principal points concerning the geometric representation of this integral reduce to the following. To the perimeter of the figure on the curved surface (under the restriction of Art. 3) will correspond always a closed line on the sphere. If the latter nowhere intersect itself, it will divide the whole surface of the sphere into two parts, one of which will correspond to the figure on the curved surface; and its area (taken as positive or negative according as, with respect to its perimeter, its position is similar, or inverse, to the position of the figure on the curved surface) will represent the integral curvature of the figure on the curved surface. But whenever this line intersects itself once or several times, it will give a complicated figure, to which, however, it is possible to assign a definite area as legitimately as in the case of a figure without nodes; and this area, properly interpreted, will give always an exact value for the integral curvature. However, we must reserve for another occasion the more extended exposition of the theory of these figures viewed from this very general standpoint.

7.

We shall now find a formula which will express the measure of curvature for any point of a curved surface. Let $d\sigma$ denote the area of an element of this surface; then[22] $Z \, d\sigma$ will be the area of the projection of this element on the plane of the coordinates x, y; and consequently, if $d\Sigma$ is the area of the corresponding element on the sphere, $Z \, d\Sigma$ will be the area of its projection on the same plane. The positive or negative sign of Z will, in fact, indicate that the position of the projection is similar or inverse to that of the projected element. Evidently these projections have the same ratio as to quantity and the same relation as to position as the elements themselves. Let us consider now a triangular element on the curved surface, and let us suppose that the coordinates of the three points which form its projection are

$$x, \qquad y,$$
$$x + dx, \quad y + dy,$$
$$x + \delta x, \quad y + \delta y.$$

The double area of this triangle will be expressed by the formula[23]

---

[22] Given an arbitrary point P on the surface, let (X, Y, Z) be the corresponding point on the unit sphere $X^2 + Y^2 + Z^2 = 1$. See Exercise 3.22 for a discussion of the area of the projected surface elements.

[23] The (signed) area of a parallelogram with edges along two vectors is given by the determinant of the vector components.

$$dx.\delta y - dy.\delta x$$

and this will be in a positive or negative form according as the position of the side from the first point to the third, with respect to the side from the first point to the second, is similar or opposite to the position of the y-axis of coordinates with respect to the x-axis of coordinates.

In a like manner, if the coordinates of the three points which form the projection of the corresponding element on the sphere, from the centre of the sphere as origin, are

$$X, \qquad Y,$$
$$X + dX, \ Y + dY,$$
$$X + \delta X, \ Y + \delta Y,$$

the double area of this projection will be expressed by

$$dX.\delta Y - dY.\delta X$$

and the sign of this expression is determined in the same manner as above. Wherefore the measure of curvature at this point of the curved surface will be

$$k = \frac{dX.\delta Y - dY.\delta X}{dx.\delta y - dy.\delta x}.$$

If now we suppose the nature of the curved surface to be defined according to the third method considered in Art. 4, X and Y will be in the form of functions of the quantities x, y. We shall have, therefore,

$$dX = \left(\frac{dX}{dx}\right)dx + \left(\frac{dX}{dy}\right)dy,$$
$$\delta X = \left(\frac{dX}{dx}\right)\delta x + \left(\frac{dX}{dy}\right)\delta y,$$
$$dY = \left(\frac{dY}{dx}\right)dx + \left(\frac{dY}{dy}\right)dy,$$
$$\delta Y = \left(\frac{dY}{dx}\right)\delta x + \left(\frac{dY}{dy}\right)\delta y.$$

When these values have been substituted, the above expression becomes

$$k = \left(\frac{dX}{dx}\right)\left(\frac{dY}{dy}\right) - \left(\frac{dX}{dy}\right)\left(\frac{dY}{dx}\right).$$

Setting, as above,

$$\frac{dz}{dx} = t, \quad \frac{dz}{dy} = u$$

and also

$$\frac{ddz}{dx^2} = T, \quad \frac{ddz}{dx.dy} = U, \quad \frac{ddz}{dy^2} = V$$

or

$$dt = T\,dx + U\,dy, \quad du = U\,dx + V\,dy,$$

we have from the formulae given above

$$X = -tZ, \quad Y = -uZ, \quad (1 + tt + uu)ZZ = 1$$

and hence

$$dX = -Z\,dt - t\,dZ,$$
$$dY = -Z\,du - u\,dZ,$$
$$(1 + tt + uu)dZ + Z(t\,dt + u\,du) = 0,$$

or

$$dZ = -Z^3(t\,dt + u\,du),$$
$$dX = -Z^3(1 + uu)dt + Z^3 tu\,du,$$
$$dY = +Z^3 tu\,dt - Z^3(1 + tt)du,$$

and so

$$\frac{dX}{dx} = Z^3\big(-(1 + uu)T + tuU\big),$$
$$\frac{dX}{dy} = Z^3\big(-(1 + uu)U + tuV\big),$$
$$\frac{dY}{dx} = Z^3\big(tuT - (1 + tt)U\big),$$
$$\frac{dY}{dy} = Z^3\big(tuU - (1 + tt)V\big).$$

Substituting these values in the above expression, it becomes

$$k = Z^6(TV - UU)(1 + tt + uu) = Z^4(TV - UU)$$
$$= \frac{TV - UU}{(1 + tt + uu)^2}.$$

∞∞∞∞∞∞∞∞∞∞

The reader is encouraged to consult Gauss's original essay [87] for reformulations of the above equation for curvature k, as well as a proof of the *theorema egregium*, both of which are somewhat lengthy for this chapter. Calculations of curvature can be found in Exercises 3.20 and 3.23 through 3.26, with an application of the *theorema egregium* in Exercise 3.24, and a model for hyperbolic geometry in Exercise 3.26.

Exercise 3.18. Recall that for the surface $z = y^2 - x^2$, the Gauss map is given by

$$\varphi(x, y, z) = \frac{1}{\sqrt{4x^2 + 4y^2 + 1}} (2x, -2y, 1).$$

We wish to investigate the curvature of the saddle surface at the point

$$P_0 = (x_0, y_0, z_0).$$

Consider the curves on this surface given by

$$\alpha_1(t) = \left(x_0 + t, \ y_0, \ y_0^2 - (x_0 + t)^2\right), \quad t \in \mathbf{R},$$
$$\alpha_2(t) = \left(x_0, \ y_0 + t, \ (y_0 + t)^2 - x_0^2\right), \quad t \in \mathbf{R},$$

and compare the relative position of $\alpha_1(t)$ and $\alpha_2(t)$ to the relative position of $\varphi(\alpha_1(t))$ and $\varphi(\alpha_2(t))$. Using Gauss's description of a figure on the sphere corresponding to the opposite figure on the surface (Article 6 of *Disquisitiones superficies*), explain why φ is orientation-reversing. Verify that if φ had been defined using the inward-pointing normal (3.5), then φ would still be orientation-reversing. Finally, compute the Gaussian curvature of the saddle surface at P_0. (You may wish to read Exercise 3.22, which relates the tangent plane of a surface at a point to the tangent plane of the sphere at a corresponding point of the Gauss map.)

Exercise 3.19. Consider the circle in the yz-plane with equation

$$(y - 2)^2 + z^2 = 1,$$

and suppose that the circle is revolved around the z-axis to generate a torus. Show that any point (x, y, z) on the torus may be described by the parametric equations

$$x = (2 + \cos p) \sin q,$$
$$y = (2 + \cos p) \cos q,$$
$$z = \sin p,$$

for some values of p and q. (Of course, there are other parameterizations of the torus. For example, in the above, $\sin q$ and $\cos q$ may be switched, yet the same surface results from plotting values of p and q that run through a period of 2π.)

Exercise 3.20. Look up Gauss's formula for the curvature of a surface given in terms of two parameters, p, q [87, p. 18], and use this to compute the curvature of the torus in Exercise 3.19 at an arbitrary point.

Exercise 3.21. Let $\mathbf{a}$, $\mathbf{b}$, $\mathbf{c}$ be three (linearly independent) vectors in $\mathbf{R}^3$, all of which begin at the origin. Consider the parallelepiped P with three adjacent sides formed by the vectors $\mathbf{a}$, $\mathbf{b}$, and $\mathbf{c}$. Many modern calculus texts show that the (signed) volume of P is given by the "scalar triple product"

$$\mathbf{a} \cdot (\mathbf{b} \times \mathbf{c}),$$

where "$\cdot$" denotes the dot product of vectors, and "$\times$" denotes the cross product of vectors. Compute the (signed) volume of P in the case

$$\mathbf{a} = (x,\, y,\, z), \quad \mathbf{b} = (x',\, y',\, z'), \quad \mathbf{c} = (x'',\, y'',\, z''),$$

and compare the result with what Gauss calls Δ in article 2 of *Disquisitiones superficies*.

Exercise 3.22. Consider now an arbitrary surface S given by $z = f(x,\, y)$ with continuous partial derivatives. Suppose that P_0 is a point on S, and P_0 is sent to the point $(X_0,\, Y_0,\, Z_0)$ on the unit sphere via the Gauss map. Using normal vectors, justify why the tangent plane to S at P_0 is parallel to the tangent plane to the unit sphere at $(X_0,\, Y_0,\, Z_0)$. Let $d\sigma$ be the area of a triangle on the tangent plane to S at P_0 (with one vertex of the triangle being P_0). Explain why the area of the projected triangle on the xy-plane is given by $Z_0(d\sigma)$.

Exercise 3.23. Compute the Gaussian curvature of the ellipsoid

$$\left(\frac{x}{a}\right)^2 + \left(\frac{y}{b}\right)^2 + \left(\frac{z}{c}\right)^2 = 1$$

at the point $(x_0,\, y_0,\, z_0)$. Be sure to simplify your answer.

Exercise 3.24. Sketch the graph of $x^2 + z^2 = 1$ in xyz-space, and compute the Gaussian curvature of this surface at the point $(x_0,\, y_0,\, z_0)$. Is it possible to find a development of the surface

$$S_1 = \{\, (x,\, y,\, z) \in \mathbf{R}^3 \mid -1 \leq x \leq 1,\ z = \sqrt{1 - x^2}\,\}$$

onto the surface

$$S_2 = \{\, (x,\, y,\, z) \in \mathbf{R}^3 \mid -\pi/2 \leq x \leq \pi/2,\ z = 0\,\}\, ?$$

If so, find such a development $\varphi : S_1 \to S_2$. If not, prove that such a development cannot exist.

Exercise 3.25. Compute the Gaussian curvature k of the surface

$$z = x^2 + 2y^2$$

at the origin $x = 0$, $y = 0$. From Exercise 3.16 determine R_1 and R_2, the maximum and minimum radii of curvature of cross sections perpendicular to the surface at $(0,\, 0,\, 0)$. Verify, as Gauss claims, that

$$k = \frac{1}{R_1 R_2}.$$

Exercise 3.26. Consider the surface in $\mathbf{R}^3$ given by

$$f(x, y) = \sqrt{1 - x^2 - y^2} - \ln \left(\frac{1 + \sqrt{1 - x^2 - y^2}}{\sqrt{x^2 + y^2}} \right),$$

where $0 < x^2 + y^2 \leq 1$.
(a) Graph this surface in $\mathbf{R}^3$ by first considering the curve

$$g(x) = \sqrt{1 - x^2} - \ln \left(\frac{1 + \sqrt{1 - x^2}}{x} \right), \quad 0 < x \leq 1,$$

in the xy-plane, and then realize that $z = f(x, y)$ is a surface of revolution.
(b) Compute the Gaussian curvature of $z = f(x, y)$.
(c) Report on the historical significance of the curve known as the tractrix (which is given by $g(x)$). Why might a description of $f(x, y)$ in terms of hyperbolic trigonometric functions

$$f(x, y) = \sqrt{1 - x^2 - y^2} - \cosh^{-1} \left(\frac{1}{\sqrt{x^2 + y^2}} \right)$$

be more appropriate?

3.6 Riemann Explores Higher-Dimensional Space

Born the son a Lutheran pastor in the northern German kingdom of Hanover, Georg Friedrich Bernhard Riemann (1826–1866) showed a striking aptitude for arithmetic calculations. The young Riemann read Legendre's lengthy treatise on number theory in less than one week, and later used some of these ideas in his own work [173, p. 5]. In 1846 he began a study of theology at the University of Göttingen, only to transfer to mathematics. At this time, Riemann apparently had very little contact with Gauss, who, although the preeminent faculty member at Göttingen, remained distant from beginning students [173, p. 5]. After one year there, Riemann moved to Berlin, where he continued his studies at the prestigious Humboldt University, and met the prominent mathematicians Carl Gustav Jacob Jacobi (1804–1851) and Peter Gustav Lejeune Dirichlet (1805–1859), from whom he learned a great deal.

   Students within the German university system enjoyed a particular freedom to move from one institution to another, and two years later Riemann returned to Göttingen, where he authored his dissertation *Grundlagen für eine allgemeine Theorie der Functionen einer veränderlichen complexen Grösse* (On the General Theory of Functions of a Complex Variable), published in 1851. In words of rare praise, Gauss wrote, "The dissertation submitted by Herr Riemann offers convincing evidence of the author's thorough and penetrating investigations . . . of a creative, active, truly mathematical mind, of a gloriously fertile originality" [212, p. 844].

Photo 3.2. Riemann.

In 1854 Riemann submitted his inaugural paper to obtain the position of an unsalaried lecturer (discussed in the introduction), which bore the title *Über die Darstellbarkeit einer Function durch eine trigonometrische Reihe* (On the Representability of a Function by a Trigonometric Series). Among other things, this paper set forth necessary and sufficient conditions for an integral to exist, and these ideas are today known as the "Riemann integral." Later in life he collaborated with Dirichlet, who had become a full professor at Göttingen, and in 1857 Riemann published the memoir *Theorie der Abelschen Functionen* (Theory of Abelian Functions), which further developed his ideas on complex analysis. In 1859 appeared a short yet profound paper of Riemann, *Über die Anzahl der Primzahlen unter einer gegebenen Grösse* (On the Number of Primes Less Than a Given Value), in which he studied the now famous "Riemann zeta function" (for a complex number s)

$$\zeta(s) = \sum_{n=1}^{\infty} \frac{1}{n^s}$$

and its relation to number theory (see the chapter on the bridge between the continuous and the discrete). We now turn to the visionary's work on manifold theory.

On June 10, 1854, Georg Friedrich Bernhard Riemann delivered his inaugural lecture *On the Hypotheses That Lie at the Foundations of Geometry* to the philosophical faculty of Göttingen. Considering geometry as an empirical science, Riemann boldly regards the parallel postulate not as an axiom, but as a hypothesis, whose truth depends on the nature of space. "These data [the data of Euclidean geometry] are—like all data—not necessary, but only of empirical certainty; they are hypotheses . . . " [166, p. 269]. In the Euclidean plane the shortest path between two points is, of course, a straight line, while on a curved surface, Gauss and others had shown that a geodesic (a curve of shortest distance between two given points) depends on the curvature of the surface. With the introduction of a paradigm-breaking concept, Riemann then generalizes the ideas of space and distance (metric relations) from two-dimensional surfaces to objects of arbitrary dimension. His new concept of n-fold extended quantities, also called n-dimensional manifolds, remains a vibrant topic of current research [168].

Riemann's lecture was meant to be understood by an audience of scholars from several disciplines, not just mathematics, and as such, his lecture takes the form of a philosophical essay that contains virtually no formulas. The mathematician of the time most likely to appreciate the work of the young genius was Gauss, who attended Riemann's lecture, and whose reaction is described as having[24] "surpassed all his expectations in the greatest surprise, and on the return from the faculty meeting he expressed to Wilhelm Weber his highest approval for the depth of Riemann's ideas with excitement rare for him" [194, p. 549]. The text of the lecture did not appear in print until 1868 [193], two years after the author's death, in part since Riemann made no particular effort to publish it.

Shortly thereafter, Riemann's ideas attracted the attention of Hermann von Helmholtz (1821–1894) and William Clifford (1845–1879), whose further work on space solidified Riemann's legacy [133, p. 781]. In an 1868 paper entitled *Über die Tatsachen die der Geometrie zu Grunde liegen* (On the Facts That Lie at the Foundations of Geometry) [114], Helmholtz lists a set of hypotheses that, in his view, would serve as the basis for the study of geometry. In particular he makes the observation that the non-Euclidean geometry of Bolyai and Lobachevsky fits into the context of Riemann's work. Clifford also pursued the nature of space, making this the topic of a lecture series in England in the early 1870s, and attempted to devise strategies to distinguish between Euclidean and non-Euclidean spaces [133, p. 782].

Riemann's mathematical interests were broad, with major works devoted to the theory of functions of a complex variable, trigonometric series, the theory of the integral, electricity, number theory, and what would become the

---

[24] The quotation is by R. Dedekind.

theory of manifolds. Poverty in his early career [212, p. 844], however, took a toll on his health and may have contributed to his death from tuberculosis at the age of 39. During the latter years of his life his work garnered the recognization it merited with an appointment as full professor at Göttingen (1859), membership in the Berlin Academy of Sciences (1859), the Paris Academy, and the London Royal Society (both in 1866) [173]. The following passage [194, p. 272–277] from *On the Hypotheses That Lie at the Foundations of Geometry* presents Riemann's discussion of manifolds and only the beginnings of his treatment of metric relations (the length of line segments). An inquiry into the foundations of the parallel postulate would require further analysis of lines in curved space, as Riemann states, "The problem then is to set up a mathematical expression for the length of a line" [166, p. 272], a problem that the German geometer solves in terms of the curvature of space. With further commentary suppressed until after Riemann's essay, the reader is encouraged to formulate a precise notion of what a manifold ought to be as we read in the master's own words [194, p. 272–277]:

∞∞∞∞∞∞∞∞∞

Riemann, from
On the Hypotheses That Lie at the Foundations of Geometry[25]

Plan of Inquiry

As is well known, geometry presupposes both the concept of space and the basic notions for construction within space. Geometry provides only nominal definitions for these, with their essential expressions appearing in the form of axioms. The relation of the presuppositions to one another remains in the dark. We see neither how, nor to what extent their relation is necessary, nor a priori possible.

From Euclid to Legendre, to name the most famous of the new geometers, this darkness has been dispelled neither by the mathematicians, nor by the philosophers who have occupied themselves with this. The reason for this is that the general concept of a multiply extended quantity, including spatial quantities, has been completely undeveloped. I have thus posed myself the task of constructing the notion of a multiply extended quantity from general notions of quantity. It will follow that a multiply extended quantity is susceptible to different metric relations and that Space is simply a special case of a three-fold extended quantity. A necessary consequence of this is that the theorems of geometry cannot be derived from general notions of quantity. Rather the characteristics of Space that distinguish it from other conceivable three-fold extended quantities can only be deduced from experience. Thus arises the task of finding the simplest facts from which the metric relations of Space can be determined. This task, by its very nature, is not completely prescribed, since several systems of simple data can be given that are sufficient for determining the metric relations of Space.

---

[25] Translated from the original German.

For the present purpose the most important system is that established by Euclid. These data, like all data, are not necessary, but of empirical certainty; they are hypotheses. We can thus investigate their likelihood, which within the limits of observation may be quite high. After which we may judge the admissibility of their extension beyond the limits of observation, both for the immeasurably large, and the immeasurably small.

I. Concept of an n-fold Extended Quantity

As to the solution of the first of these problems, namely the development of the concept of a multiply extended quantity, I request an indulgence of leniency, since I am little practiced in arguments of a philosophical nature, where the difficulties lie more in the concepts than in the constructions. Moreover, I was unable to use any previous works, other than a few philosophical results of Herbart[26] and a short reference given by Privy Councillor Gauss in his second memoir on biquadratic residues, appearing in the Göttingen Learned Notices and the Göttingen Jubilee Book.

1

Notions of quantity are only possible when a general concept exists that admits particular instances. According to whether or not a continuous transition from one instance to another occurs, the instances form either a continuous or discrete manifold. In the first case the individual instances are called points, in the second case, elements of the manifold. Concepts whose particular instance form a discrete manifold are so numerous, that a general concept can always be found to describe an arbitrary collection of objects, at least in the more developed languages. (In the theory of discrete quantities, mathematicians could thus proceed unhesitatingly from the claim that given objects can be considered as all of the same kind.) On the other hand, occasions for developing concepts whose instances form a continuous manifold occur so seldom in day-to-day life, that position and color are perhaps the only simple concepts whose instances form a multiply extended manifold. More frequent occasions for the generation and development of these concepts first appear in higher mathematics.

Certain parts of a manifold, distinguished by a mark or a boundary, are called quanta. Their quantitative comparison occurs in the case of discrete quantities by counting, in the case of continuous quantities by measurement. Measuring consists of the superposition of the quantities to be compared, and requires a means of transporting a standard quantity to be used to measure others. Lacking this, we can compare two quantities only when one is part of the other, and then decide only as to "more" or "less," not as to "how much." The investigations that

---

[26] Johann Friedrich Herbart (1776–1841) was a professor at Göttingen whose interests were philosophy, psychology, and mathematics. "In line with the spirit of Herbart, Riemann felt that the task of science was to comprehend and explain nature logically by means of precise concepts" [173, p. 9].

can be employed here form a general part of the theory of of quantity, independent of measurement, where quantities are regarded not as existing independent of position, nor as expressible in terms of a unit, but as regions in a manifold. ... From this part of the theory of extended quantity, which proceeds without any further assumptions, it suffices for the present purposes to emphasize two points, the first of which concerns the development of the concept of a multiply extended manifold, the second concerns position fixing in a given manifold in terms of numerical determination, and clarifies the essential character of an n-fold extension.

2

In a concept whose instances form a continuous manifold, when passage from one instance to another occurs in a well-determined way, then the instances of passage form a simply extended manifold, whose essential characteristic is that a continuous movement from a point is possible in only two directions, forwards and backwards. Suppose that this [simply extended] manifold transitions into a completely different one so that each point of the first passes into a well-determined point of the second. Then the totality of the instances so formed become a two-fold extended manifold. Similarly we obtain a three-fold extended quantity if we imagine that a two-fold extended quantity transitions along a completely different [simply extended] quantity in a well-determined way. It is easily seen how we can continue this construction. If we view the beginning object of this concept as changeable instead of fixed, then this construction can be recognized as the formation of an $(n + 1)$ dimensional variability from an n-dimensional variability and a one-dimensional variability.

3

Conversely, I will now show how a variability whose region is given can be decomposed into a variability of fewer dimensions and a variability of one dimension. To this end, we consider a piece of the manifold as a variability of one dimension, beginning from a fixed point so that the value of other points along it can be compared with one another. This establishes a continuously changing value for the points. In other words, we suppose a continuous function of position within the given manifold, so that the function is not constant along a piece of the manifold. Each system of points where the function has a constant value then forms a continuous manifold of fewer dimensions than the given manifold. By changing the value of the function, these manifolds continuously transition into one another. We thus suppose that from one of them all others emerge, occurring in general so that each point in one passes to a specific point of the other. The exceptional cases, whose study is important, need not be considered here. In this way the determination of position in the given manifold is reduced to a numerical specification and to a determination of position in a manifold of fewer dimensions. It is easily shown that this manifold has $n - 1$ dimensions, if the given manifold has n dimensions. By repetition of this procedure n times, the determination of position in an n-fold extended manifold is thus reduced to n numerical specifications. Hence, determining position in a given manifold, when possible, is reduced to determining a finite number of quantities. There are also manifolds for which a finite number of numerical specifications does not suffice, but instead require either an infinite sequence or a continuous manifold of numerical specifications. Such manifolds form, for example, all possible functions on a given region, or all possible forms of a spatial figure, etc.

II. Metric relations of which a manifold of n dimensions is susceptible, under the condition that lines have length independent of position, and thus each line is measurable by any other

Now that the concept of an n-fold extended manifold has been constructed, and its essential character found in that position fixing in the manifold can be reduced to n numerical specifications, there follows the second of the above tasks, namely a study of the metric relations of which such a manifold is susceptible, and of the conditions that suffice to determine these metric relations. The metric relations can only be studied in abstract terms and in a given context represented only by formulas. Under certain conditions, we can decompose them in terms of relations that are individually capable of a geometric representation, allowing calculations to be expressed geometrically. Although an abstract study via formulas cannot be avoided, to gain solid ground, the results are expressed in a geometric guise. The foundations of both questions are contained in the famous memoir of Privy Councillor Gauss on curved surfaces.

1

Measurement requires an independence of quantity from position, which can occur in more than one way. The next assumption that I wish to pursue is that indeed the length of a line is independent of position, so that any line is measurable by any other. If determining position is reduced to numerical specifications, the location of a point in a given n-fold extended manifold can be expressed in terms of n variable quantities x_1, x_2, x_3, and so on to x_n. Thus, the specification of a line amounts to writing the quantity x as a function of one variable. The task then becomes to establish a mathematical expression for the length of a line, for which purposes the quantity x must be expressible in terms of units.

∞∞∞∞∞∞∞∞∞

Early in his essay, Riemann distinguishes between discrete and continuous manifolds, with the latter being the focus of study. In either case a manifold is constructed by a transition, replication, or repetition of an instance with the transition required to be a continuous motion for a continuous manifold, and a simple replication of an instance for a discrete manifold. For example, if a single (square) tile is considered as an instance, then duplicating that tile to cover a floor would provide a pattern that qualifies as a discrete manifold. As a further example, modern artist Andy Warhol's (1929–1987) *32 Soup Cans* [91, p. 24] could also be considered a discrete manifold, since one instance of a Campbell's soup can is repeatedly juxtaposed to itself 31 times. As Riemann observes, "concepts whose instances form a discrete manifold are so numerous" [166, p. 270] that attention is restricted to continuous manifolds whose instances "first occur in higher mathematics" [166, p. 270]. In the sequel, manifold is synonymous with continuous manifold, which, to paraphrase Riemann, is defined as *a continuous transition of an instance*.

The most elementary of instances is a point, a continuous transition of which could form a straight line or a circle, both being simply extended manifolds, where "movement is possible in only two directions, forwards and backwards" [166, p. 270]. Riemann then proceeds by induction to describe multiply extended quantities (i.e., higher-dimensional manifolds). If a circle in the xy-plane is chosen as an example of an instance, then translation of the circle parallel to the z-axis would form a cylinder, while revolution of the circle (around a fixed axis) in three-space would form a torus. Both a cylinder and a torus provide examples of doubly extended manifolds (two-dimensional manifolds). Can the reader visualize how the unit sphere

$$x^2 + y^2 + z^2 = 1$$

can be realized by a continuous transition of the circle $x^2 + y^2 = 1$? By revolving the circle $x^2 + y^2 = 1$ about the y-axis, a complete sphere is generated, except both the north and south poles remain fixed during this generation, while other points on the sphere are reached via a continuous motion of the circle. Today certain technical restrictions are placed on how points within a manifold are described as a transition of an instance, with the above description valid only for a portion of the sphere, and not valid at either pole. Exercise 3.29 outlines a different approach to showing that the sphere is a manifold. Visualization of manifolds, however, is rather restricted once we encounter the third dimension [244], where, in Riemann's words, "[xyz-]Space constitutes only a special case of a three-fold extended quantity." A graphic example of a three-dimensional manifold is M. C. Escher's (1898–1972) "High and Low" (see Figure 3.12), where any two parts of space are joined by a continuous transition, yet the entire sketch cannot be constructed in Euclidean three-space. Escher's drawing is that of a three-dimensional manifold with nonzero curvature. Euclidean three-space, of course, has zero curvature.

For manifolds of dimension four and beyond, visualization becomes exceedingly difficult, and an alternative description is sought so that "the determination of position in an n-fold extended manifold is reduced to n numerical determinations" [166, p. 271]. Thus, the issue arises whether an analytic expression (or equation) for the manifold or at least part of the manifold can be found. The visionary Riemann describes his solution to this problem by first identifying "an arbitrary piece of a manifold of one dimension," i.e., a curve on the manifold is chosen to play the role of the x_1-axis. This choice then establishes "within the given manifold a continuous function of position," namely the function that yields the x_1-component of a point within the manifold. Then "Every system of points where the function has a constant value

Fig. 3.12. Escher's "High and Low."

forms a continuous manifold of fewer dimensions than the given one" [166, p. 271], i.e., points within the manifold that have a constant x_1-component,

$$x_1 = c, \quad c \text{ constant,}$$

form a submanifold of dimension $n - 1$. Then proceeding by induction, an x_2-component, x_3-component, ... and an x_n-component can be identified in the submanifold. Thus, a point p in the original manifold is determined by the n numerical values

$$(x_1, x_2, x_3, \ldots, x_n).$$

Exercise 3.28 describes a method for choosing an x_1-component and an x_2-component on the torus.

Riemann's description of position fixing within a manifold is today achieved by the use of *coordinate charts*, the idea for which can be traced to the work of Gauss and Euler. Given any point p in an n-dimensional manifold M, then all other points near p can be referenced by a coordinate map

$$\varphi_p : U \to \mathbf{R}^n,$$

where U contains the point p and all points of M close to p, and $\mathbf{R}^n$ denotes n-dimensional Euclidean space. The function φ_p maps certain curves in U to the actual x_1-, x_2-, ..., x_n-axes of $\mathbf{R}^n$, and these curves play the role of coordinate axes within U. Modern mathematics places a list of technical conditions on M, U, and the φ_p's before M itself qualifies as a manifold [222, 243], with the proper definition of a differentiable manifold due to Hermann Weyl (1885–1955). Some practice with coordinate charts can be had in Exercises 3.29 through 3.31.

Riemann's essay continues with an investigation of metric relations and curvature in manifolds. Although virtually no formulas for measuring distances are introduced in the paper, Riemann clearly conceived of a metric as given by a generalization of the functions E, F, and G that Gauss had identified for computing arc length on surfaces. In a subsequent paper (1861) [194, pp. 391–404], submitted to the Paris Academy in response to a contest to study heat conduction, Riemann does introduce specific formulas for distance and curvature. All claims in Riemann's inaugural lecture can be substantiated, including his reduction of metric relations to curvature within a manifold: "If the curvature is given in $\frac{1}{2} n(n - 1)$ surface directions at every point, then the metric relations of the manifold may be determined" [166, p. 274]. (For a modern proof, see [223].) The surfaces referred to are certain two-dimensional surfaces within the n-dimensional manifold, and curvature refers to the Gaussian curvature of these surfaces. Thus, curvature determines the metric, which in turn determines the geodesics on the manifold, and these provide us with the version of the parallel postulate that holds in a given manifold (as discussed in the introduction). In this sense, curvature determines the nature of space.

Even after the 1868 publication of Riemann's lecture, differential geometry continued to develop along several lines. Important problems continued to be the study of distance-preserving mappings of one surface onto another as well as the search for surfaces and higher-dimensional manifolds of constant curvature. Eugenio Beltrami (1835–1900) investigated these problems, and combining ideas of Gauss, Lobachevsky, and Riemann, found a surface that serves as a model for hyperbolic geometry [228]. Another major figure in the Italian school is Luigi Bianchi (1856–1928), who studied many of these problems as well as non-Euclidean geometry, and published the influential textbook *Lezioni di geometria differenziale* (1893), which emphasizes the higher-dimensional aspects of differential geometry [231, p. 186]. Important new directions were forged by Gaston Darboux (1824–1917) in Paris and Sophus Lie (1842–1899) in Norway. Darboux studied mutually perpendicular surfaces and authored the four-volume text *Leçons sur la théorie générale des surfaces* (Lessons on the General Theory of Surfaces) with installments in 1887, 1889, 1894, and 1896. Lie studied the geometry of transformation groups, which are special cases of higher-dimensional manifolds that carry the additional structure of a group, which allows points of the manifold M to be multiplied together in a continuous way to yield a new point of M. See [115] for the definition of a group, [222, 243] for the definition of a Lie group, and Exercise 3.30 for a specific example of a Lie group.

A particular impetus for the study of manifolds arose in the work of Albert Einstein (1879–1955) and Hermann Minkowski (1864–1909), who viewed space-time as a four-dimensional manifold with a certain choice for its metric. In an exposition on relativity, Einstein uses the notion of a continuum to describe space-time:

> The surface of a marble table is spread out in front of me. I can get from any one point on this table to any other point by passing continuously from one point to a "neighbouring" one, and repeating this process a (large) number of times, or, in other words, by going from point to point without executing "jumps." ... We express this property of the surface by describing the latter as a continuum [58, p. 98].

A continuum is Riemann's notion of manifold. The space-time continuum is a four-dimensional manifold in which the fourth dimension is time, and represents the possibility of a continuous transition between space and time. The reader is invited to consider further examples of manifolds in Exercise 3.32, and to reflect on the progress toward the formulation of the notion of a manifold in Exercise 3.27.

Since Riemann and Einstein, the classification of manifolds has continued in all dimensions, with an outstanding problem being the three-dimensional Poincaré conjecture, mentioned in the introduction. The French mathematician Henri Poincaré (1854–1912) had an interest in virtually every field of mathematics. He developed a unit disk model for hyperbolic geometry [150], and is the founder of what today is called algebraic topology, a subject that has as its goal the classification of space via combinatorial and algebraic methods.

Exercise 3.27. Describe the intellectual progress achieved toward defining the notion of a manifold by comparing the following quotations from Euler, Gauss, and Riemann respectively.

Then setting

$$ET = \frac{\alpha x + \beta y}{\sqrt{\alpha\alpha + \beta\beta}} - \frac{\alpha\gamma}{\beta\sqrt{\alpha\alpha + \beta\beta}} = t$$

and

$$TZ = \frac{(\alpha y - \beta x + \gamma)\sqrt{\alpha\alpha + \beta\beta + 1}}{\sqrt{\alpha\alpha + \beta\beta}} = u,$$

we will be able to consider the t and u lines as orthogonal coordinates for the section in question [64], [66, v. 28, p. 3].

We can imagine two systems of curved lines on the surface, one system for which p is variable, q constant; the other for which q is variable, p constant [86, p. 9].

The determination of position in an n-fold extended manifold is reduced to n numerical determinations, and therefore the determination of position in a given manifold is reduced, whenever this is possible, to a finite number of numerical determinations [166, p. 271].

Exercise 3.28. Recall that the torus T in Exercise 3.19 is described via the "coordinate chart"

$$\varphi : (pq\text{-plane}) \to (xyz\text{-space}),$$
$$\varphi(p, q) = \big((2 + \cos p) \sin q, \ (2 + \cos p) \cos q, \ \sin p \big).$$

Plot the "x_1-axis" on T that results by considering the points

$$\varphi(p, 0), \quad p \in \mathbf{R}.$$

Similarly plot the "x_2-axis" on the torus. How might the values of p be restricted so that the x_1-axis does not intersect itself? How should the values of q be restricted so that the x_2-axis is not self-intersecting?

Exercise 3.29. In this exercise we borrow an idea from Ptolemy's work (100–178 C.E.) on cartography [18, p. 171] to construct a coordinate chart for the unit 2-sphere

$$S^2 = \{ (a, b, c) \in \mathbf{R}^3 \mid a^2 + b^2 + c^2 = 1 \}.$$

Consider the line L through the points $(0, 0, 1)$ and (a, b, c), where (a, b, c) is a point on S^2 and

$$(a, b, c) \neq (0, 0, 1).$$

Show that L may be written parametrically as

$$L(t) = \big(at,\ bt,\ 1 + (c-1)t\big), \quad t \in \mathbf{R}.$$

Letting $(x_0,\ y_0,\ 0)$ be the point where L intersects the xy-plane, find equations for x_0 and y_0 in terms of a, b, c. The function

$$\alpha : S^2 - \{(0,\ 0,\ 1)\} \to \mathbf{R}^2,$$
$$\alpha(a,\ b,\ c) = (x_0,\ y_0)$$

is called the stereographic projection. Show that α is one-to-one, i.e., if

$$\alpha(a,\ b,\ c) = \alpha(d,\ e,\ f),$$

then $a = d$, $b = e$, $c = f$, where both $(a,\ b,\ c)$ and $(d,\ e,\ f)$ are points on

$$S^2 - \{(0,\ 0,\ 1)\}.$$

Show that α is onto, i.e., given any $(x,\ y) \in \mathbf{R}^2$, then there is some

$$(a,\ b,\ c) \in S^2 - \{(0,\ 0,\ 1)\}$$

with $\alpha(a,\ b,\ c) = (x,\ y)$. Setting

$$\alpha^{-1} : \mathbf{R}^2 \to S^2 - \{(0,\ 0,\ 1)\},$$
$$\alpha^{-1}(x,\ y) = (a,\ b,\ c),$$

then α^{-1} is a coordinate chart for S^2 (see the discussion following the Riemann source). The above construction could also be applied to lines through the points $(0,\ 0,\ -1)$ and $(a,\ b,\ c)$, where $(a,\ b,\ c)$ is a point on

$$S^2 - \{(0,\ 0,\ -1)\},$$

thus covering all of S^2 with two coordinate charts. Finally, discuss whether S^2 satisfies Riemann's definition of a manifold.

Exercise 3.30. Generalize the construction for the coordinate chart in Exercise 3.29 to find a similar chart for S^3, where

$$S^3 = \{\,(a,\ b,\ c,\ d) \in \mathbf{R}^4 \mid a^2 + b^2 + c^2 + d^2 = 1\,\}.$$

Using the quaternions [115, p. 124] $\mathbf{i}$, $\mathbf{j}$, $\mathbf{k}$, note that every point in S^3 may be written as a "unit quaternion," i.e.,

$$(a,\ b,\ c,\ d) = a + b\mathbf{i} + c\mathbf{j} + d\mathbf{k}.$$

Using the multiplication of quaternions, explain how S^3 can be given the structure of a group [115, p. 28], which thus becomes both a group and a manifold. Such "group manifolds" were studied intensely by the Norwegian mathematician Sophus Lie (1842–1899), and are today called Lie groups [158], [159].

Exercise 3.31. Recall that the Gauss map associates normal vectors of a given surface to points on the unit sphere

$$S^2 = \{\, (a,\, b,\, c) \in \mathbf{R}^3 \mid a^2 + b^2 + c^2 = 1 \,\}.$$

Simply put, if $\mathbf{n} = (x,\, y,\, z)$ is normal to the surface at some point P, then

$$\frac{\mathbf{n}}{\|\mathbf{n}\|} = \frac{1}{\sqrt{x^2 + y^2 + z^2}}\,(x,\, y,\, z)$$

can be interpreted as a point on the unit sphere. Of course,

$$-\mathbf{n} = (-x,\, -y,\, -z)$$

is also normal to the surface at P and

$$\frac{1}{\sqrt{x^2 + y^2 + z^2}}\,(-x,\, -y,\, -z) = \frac{-1}{\sqrt{x^2 + y^2 + z^2}}\,(x,\, y,\, z)$$

could equally be chosen as the corresponding point on the sphere. Gauss takes great care to distinguish between inward-pointing normals and outward-pointing normals so that a consistent choice can be made for the Gauss map. In this exercise we analyze a construction that permits us to define a "Gauss map" regardless of which of the two normals is chosen. Suppose that every point $Q = (a,\, b,\, c)$ of S^2 is identified with $-Q = (-a,\, -b,\, -c)$, which is also on S^2. Then define

$$\mathbf{R}P^2 = \{\, \{Q,\, -Q\} \mid Q \in S^2 \,\},$$

i.e., every element in $\mathbf{R}.P^2$ consists of a pair of diametrically opposed points on the sphere. If the target of the Gauss map ν is taken to be $\mathbf{R}P^2$ (instead of S^2), explain why the choice of normal does not affect the image of ν, i.e., argue why

$$\nu(\mathbf{n}) = \nu(-\mathbf{n}).$$

Is $\mathbf{R}P^2$ a manifold in the sense of Riemann? Justify your answer either in philosophical terms or by using coordinate charts. Today $\mathbf{R}P^2$ is called real projective space and is an important example in the study of topology and geometry. Extra credit: Can $\mathbf{R}P^2$ be constructed in Euclidean three-space? Why or why not?

Exercise 3.32. Using the criteria set forth in Riemann's philosophical discussion of space, determine whether each of the following is a manifold:

1. the surface $x^2 + y^2 - z^2 = 1$ in $\mathbf{R}^3$;
2. a figure eight;
3. a knot formed by your shoelace;
4. the virtual space of a three-dimensional computer game;
5. a doughnut with two holes;
6. a Möbius band, which is formed from a thin strip of paper by giving it a half twist, and then attaching both ends together;
7. the crystalline structure of a diamond;
8. the universe.

4

Patterns in Prime Numbers:
The Quadratic Reciprocity Law

4.1 Introduction

The ancient Greek philosopher Empedocles (c. 495–c. 435 B.C.E.) postulated that all known substances are composed of four basic elements: air, earth, fire, and water. Leucippus (fifth century B.C.E.) thought that these four were indecomposable. And Aristotle (384–322 B.C.E.) introduced four properties that characterize, in various combinations, these four elements: for example, fire possessed dryness and heat. The properties of compound substances were aggregates of these. This classical Greek concept of an element was upheld for almost two thousand years. But by the end of the nineteenth century, 83 chemical elements were known to exist, and these formed the basic building blocks of more complex substances. The European chemists Dmitry I. Mendeleyev (1834–1907) and Julius L. Meyer (1830–1895) arranged the elements approximately in the order of increasing atomic weight (now known to be the order of increasing atomic numbers), which exhibited a periodic recurrence of their chemical properties [205, 248]. This pattern in properties became known as the *periodic law* of chemical elements, and the arrangement as the *periodic table*. The periodic table is now at the center of every introductory chemistry course, and was a major breakthrough into the laws governing elements, the basic building blocks of all chemical compounds in the universe (Exercise 4.1).

The prime numbers can be considered numerical analogues of the chemical elements. Recall that these are the numbers that are multiplicatively indecomposable, i.e., divisible only by 1 and by themselves. The world in all its aspects is governed by whole numbers and their relationships, according to the Pythagoreans, a very influential group of philosophers and mathematicians gathered around Pythagoras during the sixth century B.C.E., after whom the Pythagorean theorem in geometry is named. One of the results about positive whole numbers known in antiquity is that every number can be written in essentially only one way as a product of prime numbers. This fact is now known as the *fundamental theorem of arithmetic* (see the Appendix). For instance, a consequence of the theorem is not only that 42 is a product of prime numbers, namely 2, 3, and 7, but moreover that this is unique, in that it is the only way that 42 can be decomposed as a product of primes (Exercise 4.2). The uniqueness of the prime decomposition is incredibly important, underlying virtually all aspects of number theory.[1] Thus the prime numbers can be viewed as basic building blocks of whole numbers.[2] Like chemists, mathematicians too have striven to discover the laws that govern prime numbers, and to answer some of the most fundamental questions about them: How many prime numbers are there? Can we find them all?

The first question was already answered by the mathematicians of ancient Greece. The great Greek mathematician Euclid of Alexandria, who lived around 300 B.C.E., published a collection of results in geometry and number theory, *The Elements* [61], which went on to become one of the all-time best-sellers. One of the results, Proposition 20 in Book IX, states that "Prime numbers are more than any assigned multitude of prime numbers." Today we would say that there are infinitely many prime numbers. Euclid's very clever proof in fact provides a bit more information than that. A few hours of calculation will easily demonstrate to the reader that prime numbers seem to appear quite irregularly among all numbers. So one might ask the question whether, beginning with a given prime number, there is an estimate as to when we will encounter the next prime number. Euclid's proof contains one such estimate, even though it is not a very good one.

Euclid proves the result by showing that if we take the first so many prime numbers, there has to be another one that is not among them. He concludes this from the observation that, if $p_1, \ldots, p_n$ are the first n prime numbers, then the number $x = (p_1 \cdot p_2 \cdots p_n) + 1$ can be written uniquely as a product of prime numbers, by the fundamental theorem of arithmetic. But none of the primes p_i divides this number, as the reader is invited to verify, so that there must be other primes. Furthermore, these other primes must appear between p_n and x (Exercise 4.4).

As to the second question, since the answer to the first question is "an infinite number," no one can hope to write out all prime numbers any more than one can write out all positive integers. However, there are ways to find all prime numbers up to a certain size. The sieve of Eratosthenes is an ancient one (Exercise 4.5). One may also find the nth prime, for any n, by simply extending the sieve far enough. No efficient closed formula to accomplish this

---

[1] For instance, we will often use without mention the fact that if a prime divides a product, then it divides one of its factors. This is a consequence of the uniqueness of prime factorization (Exercise 4.3), or can be proven separately and used to deduce the fundamental theorem of arithmetic. It is hard to overemphasize how often we use the fundamental theorem of arithmetic without even thinking about it.

[2] Note that the fundamental theorem of arithmetic builds numbers "multiplicatively" from primes. For an attempt at an "additive" analogy for numbers, see the later footnote about Lagrange's "four squares" theorem, when we discuss his work under the heading Divisor plus Descent.

is known that invokes only elementary functions. However, Juri Matiya-sevič, as a by-product of showing that Hilbert's tenth problem[3] is unsolvable [42, 43, 44, 130], discovered a hefty polynomial (with integral coefficients and many variables) that, with positive integral inputs, outputs only primes and eventually any given prime (see also [250] for other formulas). Since primes seem at first sight to occur totally irregularly among the natural numbers, it would be wonderfully surprising to find patterns in their appearance. This chapter is about exactly that: the discovery, proof, and applications of the first big pattern discovered in the occurrence of prime numbers, emerging from the ancient study of which numbers can be expressed as sums of squares.

Much of modern number theory revolves directly or indirectly around prob-lems related to prime numbers. However, despite some ancient Greek interest in the subject, it was not until the work of the Frenchman Pierre de Fermat (1601–1665) that the foundations of the subject began to be laid. Most famous for his *last theorem*,[4] Fermat worked on many problems, some of which had ancient origins, most notably in the *Arithmetica* of Diophantus of Alexandria, who lived during the third century, one of the last great mathematicians of Greek antiquity. The *Arithmetica* is a collection of 189 problems on the solu-tion, using fractions, of equations in one or more variables, originally divided into thirteen "books," of which only ten are preserved (four were rediscovered only in 1972) [12, 189, 206]. The solutions are presented in terms of specific numerical examples. Studying a Latin edition of the *Arithmetica* published in 1621 [49] (see also [109]), Fermat was inspired to begin his own number-theoretic researches. For instance, Problem 9 in Book V asks for an odd num-ber to be expressed as a sum of two squares, with several side conditions. In the course of presenting a solution to this problem, Diophantus seems to assume that every integer can be written as a sum of at most four squares (Exercise 4.6). Another problem (Problem 19 in Book III) asserts that "It is in the nature of 65 that it can be written in two different ways as a sum of two squares, viz., as $16 + 49$ and as $64 + 1$; this happens because it is the product of 13 and 5, each of which is a sum of two squares" (Exercise 4.7).

---

[3] David Hilbert (1862–1943), one of the most renowned mathematicians at the beginning of the twentieth century, proposed a list of 23 unsolved problems for consideration in the coming century, saying, "As long as a branch of science offers an abundance of problems, so long is it alive" [18, p. 657]. Hilbert's problems have served as major inspiration and guideposts to mathematics now for more than a hundred years [253].

Hilbert's tenth problem was, "Does there exist a universal algorithm for solving Diophantine equations?" A Diophantine equation is a polynomial equation with integer coefficients for which only integer solutions are allowed, such as the Fermat equation in the next footnote.

[4] See the chapter on Fermat's last theorem in [150]. Fermat claimed that the equa-tion $x^n + y^n = z^n$ has no solution in positive whole numbers x, y, z when $n > 2$. One of the greatest triumphs of twentieth-century mathematics was the proof of his famous long-standing claim.

DIOPHANTI

ALEXANDRINI

ARITHMETICORVM

LIBRI SEX,

ET DE NVMERIS MVLTANGVLIS

LIBER VNVS.

CVM COMMENTARIIS C. G. BACHETI V. C.
& obſeruationibus D. P. de FERMAT *Senatoris Toloſani.*

Acceſſit Doctrinæ Analyticæ inuentum nouum, collectum
ex varijs eiuſdem D. de FERMAT Epiſtolis.

TOLOSÆ,

Excudebat B E R N A R D V S B O S C , è Regione Collegij Societatis Ieſu.

M. DC. LXX.

Photo 4.1. Diophantus's *Arithmetica.*

One of the well-known successors of Diophantus who also worked on sums of squares was Leonardo of Pisa (1180–1250), better known as Fibonacci, after whom the Fibonacci numbers are named [156].

Fermat, in a letter to Sir Kenelm Digby (an English adventurer and double agent) from June 1658, proudly presents some of his discoveries about sums of squares:

> In most of the questions of volumes IV and V, Diophantus supposes that every whole number is either a square, or a sum of two, three, or four squares. In his commentary to Problem IV.31, Bachet admits that he was not able to completely prove this proposition. René Descartes himself, in a letter which will soon be published, and whose content I have learnt, ingeniously admits that he is ignoring the proof and declares that the road to obtaining it seems to him to be one of the most difficult and most obstructed. So I don't see how one could doubt the importance of this proposition. Well, I am announcing to your distinguished correspondents that I have found a complete demonstration. I can add a number of very celebrated propositions for which I also possess irrefutable proof. For example:
> *Every prime number of the form $4n + 1$ is the sum of two squares, such as 5, 13, 17, 29, 37, 41, etc.*
> *Every prime number of the form $3n + 1$ is the sum of a square and the triple of another square, for instance, 7, 13, 19, 31, 37, 43, etc.*
> *Every prime number of the form $8n + 1$ or $8n + 3$ is the sum of a square and double another square, such as 3, 11, 17, 19, 41, 43, etc.*
> [73, pp. 314–315].

No proofs of these results came forth to back up Fermat's amazing claims, and it was not until over a hundred years later that all of them were shown to be true.

Fermat had a law degree and spent most of his life as a government official in Toulouse. There are many indications that he did mathematics partly as a diversion from his professional duties, solely for personal gratification. That was not unusual in his day, since a mathematical profession comparable to today's did not exist. Very few scholars in Europe made a living through their research accomplishments. Fermat had one especially unusual trait: characteristically he did not divulge proofs for the discoveries he wrote of to others; rather, he challenged them to find proofs of their own. While he enjoyed the attention and esteem he received from his correspondents, he never showed interest in publishing a book with his results. He never traveled to the centers of mathematical activity, not even Paris, preferring to communicate with the scientific community through an exchange of letters, facilitated by the Parisian theologian Marin Mersenne (1588–1648), who served as a clearinghouse for scientific correspondence from all over Europe, in the absence as yet of scientific research journals. While Fermat made very important contributions to the development of the differential and integral calculus and to analytic geometry

Photo 4.2. Fermat.

[163, Chapters III, IV], his life-long passion belonged to the study of properties of the integers, now known as number theory, and it is here that Fermat has had the most lasting influence on the subsequent course of mathematics.

In hindsight, Fermat was one of the great mathematical pioneers, who built a whole new paradigm for number theory on the accomplishments of his predecessors, and laid the foundations for a mathematical theory that would later be referred to as the "queen of mathematics." But, as is the fate of some scientific pioneers, during his lifetime he tried in vain to kindle serious interest among the larger scientific community in pursuing his number-theoretic researches. Christiaan Huygens (1629–1695), the object of Fermat's final effort to arouse interest in his work, commented in a 1658 letter [245, p. 119] to John

Wallis (1616–1703), "There is no lack of better things for us to do." Whether it was the sentiment of the times, or Fermat's secretiveness about his methods of discovery and his lack of proofs, he was singularly unsuccessful in enticing the great minds among his contemporaries to follow his path. It was to be a hundred years before another mathematician of Fermat's stature took the bait and carried on his work.

Leonhard Euler (1707–1783) was without doubt one of the greatest mathematicians the world has ever known. A native of Switzerland, Euler spent his working life at the Academies of Sciences in St. Petersburg and Berlin. His mathematical interests were wide-ranging, and included number theory, which he pursued almost as a diversion, in contrast to the more mainstream areas of mathematics to which he contributed [136]. A large part of Euler's number-theoretic work consisted of a systematic program to provide proofs for all the assertions of Fermat, including Fermat's last theorem. An excellent detailed description can be found in [245].

One of the first things that caught Euler's attention in Fermat's work, which he became aware of in 1730 through his correspondence with Christian Goldbach (1690–1764), was Fermat's claim that every whole number was a square or a sum of two, three, or four squares. For the rest of his life he was to search for a proof of it, in vain. All he succeeded in showing is that every whole number is a sum of at most four *rational*[5] squares. (An English translation of Euler's proof can be found in [220, pp. 91–94].) After discovering proofs for many of Fermat's claims about sums of squares, he turned his attention to the following more general question:

Representation Problem. For a given nonzero integer a, which prime numbers can be *represented* in the form $x^2 + ay^2$ for a suitable choice of positive integers x and y?

As Euler certainly realized, this representability question is multiplicative in nature, in the sense that if m and n are of the form $x^2 + ay^2$, then so is mn (Exercise 4.8). So it makes sense first to pose and solve the problem of representing prime numbers in the form $x^2 + ay^2$. For the cases $a = 1, 2, 3$, a solution was essentially claimed by Fermat, as the letter to Digby, quoted above, shows. This is because, first, Fermat claims in each of these cases that every prime number of a certain "linear" form, i.e., lying in certain arithmetic progressions, will also be of the desired "quadratic" form $x^2 + ay^2$. For instance, for $a = 2$, Fermat asserts that every prime lying in the arithmetic progressions $8n + 1$ or $8n + 3$ (where n may be any nonnegative integer) can also be represented in the quadratic form $x^2 + 2y^2$. Second, the converse is easy to show in each case, namely that any (odd) prime represented in the desired quadratic form must also have the specified linear form (Exercise 4.9). Fermat's assertions thus suggest that in general one look for certain linear

---

[5] A *rational* number is one that can be expressed as a fraction.

forms (arithmetic progressions) whose primes are exactly those represented by certain quadratic forms.

Another amazing claim of Fermat's, that a sum $x^2 + y^2$, with x and y relatively prime positive integers, i.e., having no common prime divisors, can never have a divisor of the form $4n - 1$ no matter how x and y are chosen, suggested a related problem to Euler:

Divisor Problem. For a given nonzero integer a, find all nontrivial prime divisors, e.g., those in specified arithmetic progressions, of numbers of the form $x^2 + ay^2$, again with x and y running through all positive integers.

Let us begin by analyzing this second problem, which became a major focus for Euler. We begin by deciding what we mean by nontrivial divisors. First, any common divisor of x and y will trivially be a divisor of a number of the form $x^2 + ay^2$, so we may as well assume that x and y are relatively prime. Second, any divisor of a will clearly also be a divisor of some number of this form. Third, 2 is always a divisor of some number of the form $x^2 + ay^2$. So to summarize, the *nontrivial* prime divisors p of numbers of the form $x^2 + ay^2$ will be those for which p is odd, x and y are relatively prime, and p does not divide a. As a consequence we see that the divisor p will also not divide x, nor therefore y.

Having set the stage, our nontrivial prime divisor satisfies

$$pm = x^2 + ay^2$$

for some integer m. From this point on we will freely use basic notation and properties of modern congruence arithmetic, outlined in the Appendix to this chapter, even though this did not come into use until around the beginning of the nineteenth century; it is amazing how helpful this notation and way of thinking is. Since y is relatively prime to p, we can find an integer z such that $yz \equiv 1 \pmod{p}$ (see the Appendix). Multiplying $x^2 + ay^2$ by z^2, we obtain

$$-a \equiv (xz)^2 \pmod{p},$$

that is, $-a$ is a square or *quadratic residue* modulo p (we reserve the term quadratic residue for nonzero squares modulo p).

Conversely, suppose that $-a \equiv n^2 \pmod{p}$ with n not divisible by p. Then $-a = n^2 + mp$ for some integer m, and

$$(-m)p = n^2 + a \cdot 1^2.$$

Thus we have the following statement:

Divisors and Quadratic Residues. The nontrivial prime divisors p of numbers of the form $x^2 + ay^2$ are precisely the odd primes p for which $-a$ is a nonzero quadratic residue modulo p.

Therefore to solve the problem of finding nontrivial prime divisors of numbers of the form $x^2 + ay^2$ it is enough to find those odd primes for which $-a$ is a nonzero quadratic residue. But of course this seems like an infinite task, to be calculated one prime at a time, with no pattern in sight! On the bright side, Fermat's claims, both positive and negative, enticingly suggest that there may be an undiscovered pattern to the nontrivial prime divisors of such quadratic forms, namely that for each quadratic form its prime divisors might be precisely those in certain arithmetic progressions. For instance, if we put together Fermat's two claims about the quadratic form $x^2 + y^2$, that odd primes of the form $4n + 1$ are always of the form $x^2 + y^2$, and that no number of the form $4n - 1$ can ever be a divisor of a number of the form $x^2 + y^2$, we see that the nontrivial prime divisors of numbers of the form $x^2 + y^2$ are precisely the primes in the arithmetic progression $4n + 1$, solving the divisor problem for $a = 1$. Rephrased in the language of congruences and quadratic residues, we can say that -1 is a quadratic residue modulo primes of the form $4n + 1$, and a quadratic nonresidue modulo primes of the form $4n + 3$.

Euler sought precisely such patterns, and amassed vast calculational evidence, enough that he was able to discover and state general patterns for the nontrivial prime divisors of all quadratic forms $x^2 + ay^2$, for arbitrary positive and negative values of a. Already in 1744, in the earlier part of his career, Euler published the paper *Theoremata circa divisores numerorum in hac forma contentorum paa ± qbb* (Theorems about the divisors of numbers expressed in the form *paa ± qbb*), in which he presents the results of his extensive experimental calculations, and displays and states what patterns he has observed for the prime divisors of such quadratic forms. In the next section, excerpts from this paper will form our first primary source, and in hindsight we can see in the general patterns he asserted in 1744 the essence of a fundamental law governing prime numbers [140]. One of the delightful aspects of reading Euler's work is how transparently and expansively he shows us his train of investigation and exploration leading to the patterns that he conjectures to hold in general. Despite efforts spanning much of his life, though, he was able to prove these assertions in only a very few cases, essentially those of Fermat's claims. He eventually managed to find proofs for the nontrivial prime divisors of numbers of the form $x^2 + ay^2$ for $a = 1, \pm 2, 3$, and in these cases he could even prove Fermat's assertions above that prime numbers in certain arithmetic progressions are always actually represented by these forms, not merely divisors of them.

After a lifelong search for ways to settle the question of prime divisors of numbers of the form $x^2 + ay^2$, Euler published his final formulation of the still generally unproven magical property of primes he had discovered that provides the solution. We will also read excerpts from this later paper, *Observationes circa divisionem quadratorum per numeros primos* (Observations on the Division of Square Numbers by Primes), published in the year of his death. The property Euler discovered is a precursor of the *quadratic reciprocity law* (QRL), the cornerstone of our chapter. It enables one to answer the divisor problem. In more modern form and terminology, we shall see that it allows the determination of the quadratic character (quadratic residue or not) of $-a$ modulo p in terms of the quadratic character of p modulo primes dividing $-a$, i.e., with the roles of $-a$ and p reversed! Note how this helps solve the original problem of finding all nontrivial prime divisors of a fixed form $x^2 + ay^2$, which we already translated into the problem of finding the primes p for which $-a$ is a quadratic residue modulo p. Since the QRL will convert this into the question of which primes p are quadratic residues modulo each of the prime divisors of $-a$, we are now dealing with finitely many fixed moduli, for which such calculations are highly tractable. Prior to this advance we needed to consider infinitely many prime moduli p. We will illustrate this in detail as soon as we have a proper statement of the QRL in hand.

Let us now return to the representability problem, namely, which numbers are actually of the form $x^2 + ay^2$, not merely divisors of such a form? It is of course still true, as above, that if a prime p is actually of the form $x^2 + ay^2$, then $-a$ must be a quadratic residue modulo p. But this is only a necessary condition, and not always a sufficient one. For instance, 3 divides a number of the form $x^2 + 5y^2$, since $1^2 + 5 \cdot 2^2 = 21$; thus -5 is a quadratic residue modulo 3. But clearly 3 is not of the form $x^2 + 5y^2$. So while 3 is a nontrivial divisor of (a number of) the form $x^2 + 5y^2$, it fails to be actually represented by it. This simple example shows that the representability problem is a related but in general much harder problem than the divisor problem, and it took a fresh paradigm to begin any real headway on it, established by the most distinguished of Euler's young contemporaries, Joseph Louis Lagrange (1736–1813).

Divisor Plus Descent Can Produce Representability. A solution to the divisor problem can often be combined with a method called *descent*, pioneered by Fermat, used by Euler, and vastly extended by Lagrange, to solve the representability problem.

We will sketch an initial illustration here, combining a divisor problem solution with a descent to prove Fermat's most famous representation claim, that any prime p of the form $4n+1$ can be written as a sum of two squares. The first step is to have in hand a solution to the divisor problem for $x^2 + y^2$, which we will obtain from our extract of Euler's second paper: -1 is a quadratic residue modulo the prime $p = 4n + 1$, i.e., any prime p of the form $4n + 1$ is a nontrivial divisor of some number of the form $x^2 + y^2$. We also need a descent result, which will be provided by our extract from Lagrange's work: If $x^2 + y^2$ is a number with x and y relatively prime, then any divisor of this number is likewise a sum of two relatively prime squares. This is called descent because we have descended from one sum of squares to a smaller number of the same quadratic form. Now we combine these, following the divisor solution by the descent: Given a prime $4n + 1$, by our divisor solution it must nontrivially divide a sum of two squares. But the descent solution says that any divisor of this sum of two squares is again a sum of two squares. Voilà, our original prime $4n + 1$ is a sum of two squares.

To apply this two-step "divisor plus descent" technique to solve a representability problem, one would in general need a solution to the divisor problem (to be provided by the quadratic reciprocity law), and a descent result of some kind. But we caution that from the example above, in which $x^2 + 5y^2$ nontrivially represents 21, but not its divisor 3, we see that descent does not always work as simply as one would wish. Lagrange's analysis is what was needed next.

There was only a small number of scholars during the second half of the eighteenth century interested in pure mathematics. Fortunately, one of them, Lagrange, devoted part of his career to the pursuit of number theory. In 1766 Lagrange became the successor of Euler at the Academy of Sciences in Berlin, after Euler returned to St. Petersburg. Inspired by Euler's work on number theory, Lagrange produced a string of publications on the subject during the following decade. Going beyond the scattered results of Fermat and Euler on sums of squares, Lagrange proposed a powerful abstraction, to make the representational forms themselves the object of study, rather than merely the integers represented by them. And he realized that in order to get a coherent theory, he needed to consider more general quadratic expressions than $x^2 + ay^2$. In other words, Lagrange proposed to study formally general *quadratic forms*, expressions of the form

$$ax^2 + bxy + cy^2,$$

as well as their properties and relationships, where a, b, c are integers. In particular, he studied what the possible quadratic and linear forms could be for nontrivial divisors of a given quadratic form; this provides a basis for general descent results, as we shall see. Lagrange went on to lay the foundations of the theory of quadratic forms, which would be deepened and extended later by the great Carl Friedrich Gauss (1777–1855).

Amazingly, the general descent results Lagrange obtained, cleverly combined with just a few divisor problem solutions, produced a fountain of theorems about representability of primes by forms $x^2 + ay^2$, far beyond what Fermat and Euler had been able to show. To top it all off, Lagrange was able to find a proof for the long-standing claim that every positive integer is a sum of four integer squares.[6] Our next original source in the chapter will be an excerpt from Lagrange's work on quadratic forms in *Recherches d'Arithmétique* (Researches in Arithmetic), published in 1773–1775 as a Memoir of the Berlin Academy of Sciences [141, vol. III, pp. 695–795], showing how his abstract analysis of quadratic forms enabled him to obtain many new representability results. The reader with some knowledge of algebraic number theory can find an extensive treatment of representations of integers as sums of squares in [102]. The problem of representing primes as values of quadratic forms is discussed from a mathematically sophisticated point of view in the excellent exposition [41].

---

[6] This had become part of a much broader claim of Fermat's, that every number is the sum of at most three triangular numbers (see the bridge chapter), or four squares, or five pentagonal numbers, etc. [245], indicating that these particular types of numbers are additive building blocks for all numbers in a certain sense.

The next link in the precarious chain between Fermat and modern number theory was the French mathematician Adrien-Marie Legendre (1752–1833). After receiving an education in mathematics and physics in Paris, Legendre spent some time teaching at the military academy there. In 1782 he attracted the attention of Lagrange by winning the prize of the Berlin Academy of Sciences for a paper on applications of mathematics to ballistics. Legendre went on to a distinguished career at the Paris Academy of Sciences. He made significant contributions to several areas of mathematics, in particular number theory. In the tradition of his number-theoretic predecessors Fermat, Euler, and Lagrange, Legendre too studied the problem of representing prime numbers by quadratic forms, including some results on quadratic forms in more than two variables. In Part IV of the memoir *Recherches d'Analyse Indéterminée* (Researches in Indeterminate Analysis), published in 1788 (and submitted to the Paris Academy of Sciences in 1785), Legendre attempted a proof of what he called later "a law of reciprocity between primes," now known as the quadratic reciprocity law. Curiously enough, there is no mention anywhere of Euler's work on the QRL. Even though Legendre had carefully studied Euler [245, p. 326], he had apparently missed this gem. And Euler would certainly have deserved mention, even though he had not made major progress on providing a proof.

Later on, in his treatise *Théorie des Nombres* (Theory of Numbers) [153], one of the first books on number theory, Legendre reformulated the law using what is now called the *Legendre symbol*, and presented a proof of it. Unfortunately, his proof was not complete. In it he assumed, among other things, that in every arithmetic progression of the form

$$\{an + b \mid n = 0, 1, 2, \dots\},$$

with a and b relatively prime, there exist infinitely many prime numbers. Despite Legendre's certainty of this truth about arithmetic progressions, he had no proof for it. Proof was finally provided in 1837 by Lejeune Dirichlet (1805–1859) [135, p. 829f], and it stands now as one of the deep results about prime numbers. Our final original source on the discovery of the quadratic reciprocity law is excerpted from Legendre's *Theory of Numbers*. In it not only will we see him state the law as a genuine reciprocity principle, but we will see that he uses it to prove many of the results that eluded Euler on divisors of quadratic forms and representation of primes via certain forms.

In order to state the quadratic reciprocity law as presented by Legendre, we need first to mention a result discovered by Euler, which we shall see derived in our Legendre source. Euler derived a criterion, essentially a formula, for whether a given integer a is a quadratic residue modulo a given prime p. Stated in modern congruence language, it says the following:

Euler's Criterion. *Let p be an odd prime and a not divisible by p. Then a is a quadratic residue modulo p if and only if*

$$a^{(p-1)/2} \equiv 1 \pmod{p}.$$

While one could in principle actually use this formula to calculate whether a number is a quadratic residue, the calculations can be long, and Euler's criterion is much more valuable for its theoretical use. In fact, everything from now on will be based on it. When we read Legendre, we shall see that even more is going on than is stated in the criterion above, namely that when p is an odd prime and a not divisible by p, the expression $a^{(p-1)/2}$ is always congruent to either 1 or -1 modulo p, according to whether or not a is a quadratic residue modulo p. Legendre then introduces the symbolism $\left(\frac{a}{p}\right)$ for this resulting value, so we have

$$\left(\frac{a}{p}\right) = \begin{cases} 1 & \text{if } a \text{ is a quadratic residue modulo } p, \\ -1 & \text{if not.} \end{cases}$$

For example, if $p = 11$, then $\left(\frac{a}{11}\right)$ equals 1 for $a = 1, 3, 4, 5, 9$, and equals -1 for $a = 2, 6, 7, 8, 10$. One can confirm this for each a by brute force by calculating and listing the squares of the ten nonzero residues modulo 11, or alternatively, using Euler's criterion, by calculating for each a above whether $a^{(p-1)/2}$ is congruent to 1 or -1 modulo 11 (Exercise 4.10). Since

$$\left(\frac{a}{p}\right) = \left(\frac{b}{p}\right)$$

if a and b have the same remainder modulo p, as the reader should verify, evaluating other Legendre symbols with denominator 11 merely requires first calculating a remainder modulo 11.

We are now ready to state the QRL and actually use it to do a computation. Given two odd primes p and q, it establishes an amazingly simple relationship between the Legendre symbol $\left(\frac{p}{q}\right)$ and its "reciprocal" $\left(\frac{q}{p}\right)$. But before reading further, the reader is encouraged to work Exercise 4.11 and guess the law.

Quadratic Reciprocity Law. *If p, q are odd prime numbers, then*

$$\left(\frac{p}{q}\right)\left(\frac{q}{p}\right) = (-1)^{\frac{p-1}{2} \cdot \frac{q-1}{2}}.$$

The reader is strongly encouraged to make this very compact formula more meaningful by reinterpreting it as a statement about how the two Legendre symbols $\left(\frac{p}{q}\right)$ and $\left(\frac{q}{p}\right)$, each of which is either 1 or -1, compare with each other, depending on whether each of p or q has the form $4n + 1$ or $4n + 3$ (Exercise 4.12).

Two extremely useful supplementary results are commonly proven along with the reciprocity law, the first of which is a straightforward calculation if we allow ourselves to use Euler's criterion (Exercise 4.13). We shall see Euler's proof of it later, too, and we will see how the second part follows from the same argument that proves the main QRL.

Supplementary Theorem. *If p is an odd prime, then*

$$
1. \qquad \left(\tfrac{-1}{p}\right) = (-1)^{\frac{p-1}{2}} = \begin{cases} 1 & \text{if } p \equiv 1 \pmod{4}, \\ -1 & \text{if } p \equiv 3 \pmod{4}. \end{cases}
$$

$$
2. \qquad \left(\tfrac{2}{p}\right) = (-1)^{\frac{p^2-1}{8}} = \begin{cases} 1 & \text{if } p \equiv 1, 7 \pmod{8}, \\ -1 & \text{if } p \equiv 3, 5 \pmod{8}. \end{cases}
$$

Let us illustrate the use of the QRL and the supplementary theorem to make computations with the example $a = -6$, $p = 101$. For this we need just one more tool (Exercise 4.14).

Multiplicativity of the Legendre Symbol. *If a and b are integers relatively prime to p, then*

$$
\left(\frac{ab}{p}\right) = \left(\frac{a}{p}\right)\left(\frac{b}{p}\right).
$$

With all this we then easily calculate

$$
\begin{aligned}
\left(\tfrac{-6}{101}\right) &= \left(\tfrac{-1}{101}\right)\left(\tfrac{2}{101}\right)\left(\tfrac{3}{101}\right) \\
&= (-1)^{50}(-1)^{\frac{1}{8}(101^2-1)}(-1)^{\frac{1}{4}(3-1)(101-1)}\left(\tfrac{101}{3}\right) \\
&= 1 \cdot (-1)\left(\tfrac{101}{3}\right) \\
&= -\left(\tfrac{2}{3}\right) \\
&= -(-1) \\
&= 1.
\end{aligned}
$$

Thus, -6 is a quadratic residue modulo 101. Without the QRL we would have possibly had to compute all quadratic residues modulo 101, or calculate the remainder of $(-6)^{\frac{101-1}{2}}$ mod 101 (Exercise 4.15). Our example suggests that with the tools now at hand, it is easy to calculate Legendre symbols, and this is indeed the case, especially by using the QRL repeatedly as necessary during a calculation (Exercise 4.16).

Near the end of our discussion of the divisor problem, we promised that once we had the quadratic reciprocity law in hand, we would illustrate how it helps solve the original problem of finding all nontrivial prime divisors of a fixed quadratic form $x^2 + ay^2$. Let us take $a = 6$ as our example, so the question translates into asking for which odd primes p not dividing 6 is $\left(\tfrac{-6}{p}\right) = 1$ (recall "Divisors and Quadratic Residues" above). Euler claimed explicitly in his 1744 paper, based on his experimental evidence, that the nontrivial prime divisors of numbers of the form $x^2 + 6y^2$ are exactly those primes of the forms $24n + 1$, $24n + 5$, $24n + 7$, $24n + 11$. So let us see whether we can determine that these are exactly the odd primes not dividing 6 for which $\left(\frac{-6}{p}\right) = 1$. We calculate, as above, that

$$\left(\frac{-6}{p}\right) = \left(\frac{-1}{p}\right)\left(\frac{2}{p}\right)\left(\frac{3}{p}\right)$$
$$= (-1)^{\frac{p-1}{2}}(-1)^{\frac{1}{8}(p^2-1)}(-1)^{\frac{(3-1)}{2}\frac{(p-1)}{2}}\left(\frac{p}{3}\right)$$
$$= (-1)^{\frac{1}{8}(p^2-1)}\left(\frac{p}{3}\right).$$

We can now analyze each of the factors in this final product separately. By analyzing the parity of $\frac{1}{8}(p^2-1)$, we see that the first factor is 1 if $p \equiv 1, 7 \pmod 8$, but is -1 if $p \equiv 3, 5 \pmod 8$ (note that this covers all possibilities, since p is odd). And we see directly that the second factor is 1 if $p \equiv 1 \pmod 3$, but is -1 if $p \equiv 2 \pmod 3$ (this too covers all possibilities, since p does not divide 6, so $p \not\equiv 0 \pmod 3$). Now from this information we should be able to check that $\left(\frac{-6}{p}\right)$, which is the product of these two factors, is 1 precisely for primes not dividing 6 that lie in one of the four arithmetic progressions given above by Euler, thereby proving his claimed identification of all the nontrivial prime divisors of numbers of the form $x^2 + 6y^2$. We leave this final step to the reader, which amounts to melding the relevant congruences (Exercises 4.17 and 4.18).

We have already seen how the QRL was discovered in pursuit of questions about representability and divisibility of quadratic forms. We can see the utility of the QRL from another perspective as well. The algorithm provided by the quadratic formula for finding the solutions of a quadratic equation is almost as old as mathematics itself. A natural generalization of quadratic equations are quadratic congruences

$$ax^2 + bx + c \equiv 0 \pmod n,$$

for various values of n, and integers a, b, c. One might ask whether there is an analogue of the quadratic formula. Let us explore the case in which the modulus n is an odd prime, and we may as well assume that a is not divisible by n, since otherwise the congruence is a linear one. (The case of a more general modulus can be reduced to this one; see [96, Section 9.4].) We can then complete the square and rearrange to obtain the equivalent

$$(2ax + b)^2 \equiv b^2 - 4ac \pmod n.$$

To solve this equation for x modulo n, we need to be able to find a square root for $b^2 - 4ac$ and to divide by $2a$, both modulo n. Division by $2a$ is possible modulo n from our assumption that a is an odd prime not dividing n (see Appendix). Thus the original quadratic congruence is solvable if and only if its "discriminant" $d = b^2 - 4ac$ is a square modulo n, that is, if and only if $\left(\frac{d}{n}\right) = 1$. The situation is similar to our example above. If we vary n, it seems that for every choice of n we need to check whether d is a square, an infinite process to accomplish for all n. However, if d is prime, we can use the QRL to translate the computation of $\left(\frac{d}{n}\right)$ into calculating $\left(\frac{n}{d}\right)$ instead (if d is composite, we simply first split the calculation up multiplicatively, as in the example calculation above, in terms of the prime factors of d). Now since computing $\left(\frac{n}{d}\right)$ requires only that we know $\left(\frac{r}{d}\right)$ for the remainder of n modulo d, we see that we need only compute these d values once and for all, and can then easily determine whether the original congruence is solvable modulo any given n. So the QRL saves the day again (Exercise 4.19).

As to congruences of higher degree, the natural question arises whether there are *higher reciprocity laws* that help us solve these congruences. It was Gauss who first formulated one such higher law, namely a fourth-degree, or so-called biquadratic, reciprocity law (its proof was left to Gotthold Eisenstein (1823–1852)). In his first memoir on biquadratic residues Gauss makes it clear that he believes these higher laws to be a whole new ball game:

> The theory of quadratic residues can be reduced to the most beautiful jewel among the fundamental theorems of higher arithmetic, which, as is known, were first discovered easily by inductive methods and then were proved in so many ways that nothing remains to be desired.
> However, the theory of cubic and biquadratic residues is more difficult by far. In 1805, as we began to investigate these, except for the first results which gave several special theorems that stand out both because of their simplicity and because of the difficulty of their proofs, we soon recognized that the principles of arithmetic which were usable until then were in no way sufficient to build a general theory. Rather such a theory necessarily required an infinite enlargement to some extent of the field of higher arithmetic ... [96, p. 224].

This is still one of the important unsolved problems in modern number theory: the search for further higher reciprocity laws [155]. (See Wyman [252] for a beautiful exposition of general reciprocity laws.) Of course, a similar generalization is suggested by quadratic form theory when one asks what happens if the forms are allowed to have degree higher than two.

For the quadratic reciprocity law itself, it was the genius of Carl Friedrich Gauss that finally provided a complete rigorous proof, in his *Disquisitiones Arithmeticae* (Arithmetical Investigations), published in 1801 [80]. An unbelievable tour de force, this book of the twenty-four-year-old Gauss opened up number theory as a full-fledged mathematical subject, established notation that is still standard today, provided an extensive set of tools and methods, and proved a plethora of astounding results, with the QRL being one of them. Another of Gauss's major achievements was a new theory of quadratic forms, comprising a vast extension of Lagrange's foundational work (for a detailed discussion of the contents of the *Disquisitiones* see [26, Chapter 3]).

In our first section on Gauss's work, we shall read excerpts showing how he stated the QRL (which he calls the "fundamental theorem"), as well as his assessment of the work of his predecessors. Interestingly, Gauss considers Euler's work as falling short of an actual discovery of the QRL, and Legendre complained bitterly about not receiving enough credit from Gauss (see Kronecker [140] on the history of the QRL for details concerning this priority issue). At that juncture in the chapter we shall summarize by providing a unified mathematical view connecting the various claims of Euler, Legendre, and Gauss that we have read up to that point.

Gauss gave altogether six different proofs of the QRL, the first one in the *Disquisitiones Arithmeticae*. In introducing his third published proof in 1808, he says:

The questions of higher arithmetic often present a remarkable characteristic which seldom appears in more general analysis, and increases the beauty of the former subject. While analytic investigations lead to the discovery of new truths only after the fundamental principles of the subject (which to a certain degree open the way to these truths) have been completely mastered; on the contrary in arithmetic the most elegant theorems frequently arise experimentally as the result of a more or less unexpected stroke of good fortune, while their proofs lie so deeply embedded in the darkness that they elude all attempts and defeat the sharpest inquiries. Further, the connection between arithmetical truths which at first glance seem of widely different nature, is so close that one not infrequently has the good fortune to find a proof (in an entirely unexpected way and by means of quite another inquiry) of a truth which one greatly desired and sought in vain in spite of much effort. These truths are frequently of such a nature that they may be arrived at by many distinct paths and that the first paths to be discovered are not always the shortest. It is therefore a great pleasure after one has fruitlessly pondered over a truth and has later been able to prove it in a round-about way to find at last the simplest and most natural way to its proof.

The theorem which we have called in sec. 4 of the *Disquisitiones Arithmeticae* the *fundamental theorem*, because it contains in itself all the theory of quadratic residues, holds a prominent position among the questions of which we have spoken in the preceding paragraph. We must consider Legendre as the discoverer of this very elegant theorem, although special cases of it had previously been discovered by the celebrated geometers Euler and Lagrange. I will not pause here to enumerate the attempts of these men to furnish a proof; those who are interested may read the above mentioned work. An account of my own trials will suffice to confirm the assertions of the preceding paragraph. I discovered this theorem independently in 1795 at a time when I was totally ignorant of what had been achieved in higher arithmetic, and consequently had not the slightest aid from the literature on the subject. For a whole year this theorem tormented me and absorbed my greatest efforts until at last I obtained a proof given in the fourth section of the above-mentioned work. Later I ran across three other proofs which were built on entirely different principles. One of these I have already given in the fifth section, the others, which do not compare with it in elegance, I have reserved for future publication. Although these proofs leave nothing to be desired as regards rigor, they are derived from sources much too remote, except perhaps the first, which however proceeds with laborious arguments and is overloaded with extended operations. I do not hesitate to say that until now a natural proof has not been produced. I leave it to the authorities to judge whether the following proof which I have recently been fortunate enough to discover deserves this description [82], [220, pp. 112–118].

As mentioned earlier, it was the *Disquisitiones* that finally established number theory, or higher arithmetic as it was called then, as a full-fledged mathematical subject. One of those inspired by the new subject was Gotthold Eisenstein, who gives his view of the subject:

> Already early in my youth I was attracted by the beauty of a subject which differs from other subjects not only in its content but, most importantly, in the nature and variety of its methods. In it, it is not enough to just lay out the consequences of a single idea in a long sequence of deductions; almost each step requires one to conquer new difficulties and apply new principles.
> A little over fifty years ago, number theory consisted only of a collection of isolated facts, unknown to most mathematicians, and practiced only occasionally by a few, even though Euler already found in it leisure from his other activities. It was through Gauss and some of his successors that number theory has reached such heights that now it is not inferior to any other mathematical discipline in depth and breadth, and has had a fruitful influence on many of them. A school has arisen which counts the most eminent mathematical talents among its disciples, and which I too proudly am a part of, if only one of its lowliest [60, pp. 762–763].

Naturally, Eisenstein too worked on the QRL and higher laws. Among the many contributions he made to number theory during his very short life were several new proofs of the QRL, including a version of Gauss's third proof that used a tool from geometry. Writing to a friend, he says:

> I did not rest until I freed my geometric proof, which delighted you so much, and which also, incidentally, particularly pleased Jacobi, from the Lemma [of Gauss] on which it still depended, and it is now so simple that it can be communicated in a couple of lines [60, pp. 879–904].

It is this proof of Eisenstein's that we shall study in our section on the proof of the quadratic reciprocity law, after reading Gauss's own short original statement of his version of the QRL in the *Disquisitiones*, and his discussion of previous work by others. While Eisenstein's proof takes a bit more than "a couple of lines," it is very accessible, quite ingenious, and beautiful in its elegance and economy. Gauss's own proofs often tended to obscure the paths by which he obtained his insights, in stark contrast to Euler and Lagrange wanting to show us their paths. For instance, the opaqueness of motivation and context for many aspects of Gauss's third proof of the QRL even make it hard to see that Eisenstein's geometric proof apparently evolved from it (see [148, 149] for a comparison of the Gauss and Eisenstein proofs).

Since the time of Gauss, many different proofs of the QRL have been given and its role has evolved with the subject itself. (See [11] for a comprehensive review of different proofs of the QRL.) In fact, together with the Pythagorean theorem, the QRL probably qualifies as the theorem with the largest collection of different proofs to its name (and which seems to be growing; see, e.g., [90]). The reader might wonder why mathematicians would bother re-proving over and over again something already known. One could give a number of technical answers to this question, but the likely essence is that, like a complex and challenging mountain peak, it provides many possible routes for an ascent, each with its unique difficulties and rewards. It appears in many different guises, and its modern formulations are hardly recognizable. It is now properly a result that is formulated in abstract algebra terms as part of a subject called class field theory. And the theory of quadratic forms is now intimately connected with the theory of quadratic number fields. In fact we can see in our excerpts from Lagrange and Gauss on quadratic forms that mathematics was already changing significantly, from limited concrete problems to a more global, structurally oriented, and abstract approach. Gauss, in many ways, opened the door from classical to modern mathematics. In our final section, we shall read a bit from Gauss's development in his *Disquisitiones Arithmeticae* of a modern theory of quadratic forms, to see how the subject developed as it entered the nineteenth century, when modern abstract algebra would transform much of mathematics.

Gauss considered number theory the *queen of the mathematical sciences*, and for the last two hundred years it has stood as one of the most pure and abstract disciplines, fundamental to our understanding of the mathematical world. At the same time, number theory seemed to be totally removed from the concerns of everyday life. In his famous *A Mathematician's Apology*, the distinguished British mathematician G. H. Hardy (1877–1947) opines:

> It is undeniable that a good deal of elementary mathematics ... has considerable practical utility. These parts of mathematics are, on the whole, rather dull; they are just the parts which have least aesthetic value. The "real" mathematics of the "real" mathematicians, the mathematics of Fermat and Euler and Gauss and Abel and Riemann, is almost wholly "useless" (and this is as true of "applied" as of "pure" mathematics). It is not possible to justify the life of any

genuine professional mathematician on the ground of the "utility" of his work [107, pp. 119–120].

Poor Hardy would be in for quite a surprise had he lived just a little longer. Ironically, in the last twenty-five years, number theory in general, and the theory of quadratic residues in particular, has found its way into our daily lives in some of the most surprising ways. So-called public key cryptography was invented in 1978 [195] as a means to exchange encrypted messages without having to exchange decryption keys first. The subsequent emergence of the World Wide Web and the ensuing revolution in commerce and information exchange made public key cryptography essential for the protection of information traveling over the Internet. The encryption key in this scheme is based on the product of two prime numbers (and, interestingly, on an application of Fermat's little theorem, which is included in the Appendix). The product is made public and the two primes are kept secret. Anyone can encrypt, but only the person who knows the two primes can decrypt. The security of the method relies on the fact that the factoring of integers into their prime factors is computationally very expensive (at least no one seems to know how to factor cheaply, i.e., quickly). Given the present state of computing, it is essentially impossible to factor a product of two prime numbers of approximately 150 digits each into its factors. Large-scale application of this process and ever faster computers require a constant supply of ever larger prime numbers, or at least numbers that are prime "for all practical purposes." A generalization of the QRL is at the heart of one of the most commonly used probabilistic primality tests [229, Chapter 4.5]. Somewhat more esoteric uses of the theory of quadratic residues, such as for the design of concert hall ceilings, can be found in [203, Chapter 15].

At the end of this chapter, the reader familiar with the topics discussed will surely complain, and justly so, that many important related topics have not been mentioned. Pell's equation does not appear, continued fractions are completely absent, no discussion of a modern field-theoretic presentation of quadratic form theory is given, and many more topics are left unmentioned. The choices of what to include were guided by space considerations, as well as the background and motivation of the intended audience of this book. The reader interested in a more detailed discussion of issues raised by the work of Lagrange and Legendre can consult the excellent source [245]. For a reader with some background in abstract algebra we recommend [25] for a bird's eye view of quadratic form theory, as seen from the vantage point of quadratic number fields, as well as the historically motivated treatment [200]. And, of course, there is [102], mentioned earlier.

Exercise 4.1. Read about the periodic law in [177, vol. 2, pp. 910–932]. In 1829, well before Dmitry Mendeleyev (1834–1907) began arranging the elements to produce a periodic table, Johann W. Döbereiner discovered, among the elements then known, triads of similar elements: for example, lithium (**Li**, atomic weight 6.9), sodium (**Na**, 23.0) and potassium (**K**, 39.1). What is striking is that the atomic weight of sodium is the average of those of lithium and potassium. We now know that these are the elements 3, 11, and 19 in the periodic table. Another triad noticed by him is chlorine (**Cl**), bromine (**Br**), and iodine (**I**). In mathematics, in the same spirit, there is a triad of primes: 3, 5, and 7. We pose the following questions for exploration. Are there any other triads of primes? What about the pattern 11, 13, 17, and 19? When does such a prime pattern occur again? Does either pattern occur infinitely often?

Exercise 4.2. Prove the fundamental theorem of arithmetic, or look up a proof in a book on elementary number theory, e.g., [27].

Exercise 4.3. Show that if a prime divides a product, then it divides one of its factors.

Exercise 4.4. Find the theorem on the infinitude of primes in Euclid's *Elements* [61] and compare his proof with the sketch given in this section.

Exercise 4.5. Look up the sieve of Eratosthenes and find the first 25 primes. Notice the irregular and elusive spacing of these first few primes. Do you see any patterns?

Exercise 4.6. Problem 29 in Book IV of Diophantus's *Arithmetica* states, "To find four square numbers such that their sum added to the sum of their sides makes a given number." Diophantus provides the following solution (by way of a numerical example) [109].

○◌○◌○◌○◌○◌○◌○

Given the number 12. Now $x^2 + x + 1/4 =$ a square. Therefore the sum of four squares + the sum of their sides +1 = the sum of four other squares = 13, by hypothesis. Therefore we have to divide 13 into four squares; then, if we subtract 1/2 from each of their sides, we shall have the sides of the required squares. Now

$$13 = 4 + 9 = \left(\frac{64}{25} + \frac{36}{25} \right) + \left(\frac{144}{25} + \frac{81}{25} \right),$$

and the sides of the required squares are

$$11/10, \quad 7/10, \quad 19/10, \quad 13/10,$$

the squares themselves being

$$121/100, \quad 49/100, \quad 361/100, \quad 169/100.$$

○◌○◌○◌○◌○◌○◌○

Use Diophantus's example to construct a more general solution, one that applies whenever the given number plus one is a sum of two rational squares. Hint: Notice how Diophantus's illustration shows us how to write any rational square as a sum of two rational squares, by paying attention to the role of the 25.

Exercise 4.7. Show that if n is a product of two integers each of which is a sum of two squares, then n can be written as a sum of two squares in two different ways. (Hint: use the hint in Exercise 4.8.)

Exercise 4.8. Show that if m and n are whole numbers, both of the form $x^2 + ay^2$, then mn is also of this form. Hint: Verify and use "Brahmagupta's identity"

$$(x^2 + ay^2)(z^2 + aw^2) = (xz \pm ayw)^2 + a(xw \mp yz)^2.$$

Exercise 4.9. Show the converse of Fermat's claims made to Digby. In other words, show in each case $a = 1, 2, 3$ that any odd prime represented by the given quadratic form $x^2 + ay^2$ must belong to the claimed arithmetic progressions.

Exercise 4.10. Calculate the values of $\left(\frac{a}{11}\right)$ for all $a = 1, \ldots, 10$, by both the methods suggested in the illustration in the text. Hint: To calculate whether $a^{\frac{p-1}{2}}$ is congruent to 1 or -1 in each case, use everything you know about arithmetic modulo p to shorten your calculation (see the Appendix). You should never have to actually work with very large numbers.

Exercise 4.11. Make an array whose rows and columns are labeled by the first ten odd prime numbers 3, 5, 7, 11, ..., 31. In position (p, q) put the value of the Legendre symbol $\left(\frac{p}{q}\right)$. Use the entries of this array to conjecture a relationship between $\left(\frac{p}{q}\right)$ and the "reciprocal" symbol $\left(\frac{q}{p}\right)$. Before you can do all this you will need to find quadratic residues. Make an auxiliary table of squares of integers modulo p. What additional properties of quadratic residues do you discover?

Exercise 4.12. Reinterpret the equality of the quadratic reciprocity law as saying whether $\left(\frac{p}{q}\right)$ and $\left(\frac{q}{p}\right)$ agree or disagree, depending on whether each of p, q has the form $4n + 1$ or $4n + 3$.

Exercise 4.13. Prove the first part of the supplementary theorem.

Exercise 4.14. Let p be an odd prime, and a, b integers that are relatively prime to p. Prove that

$$\left(\frac{ab}{p}\right) = \left(\frac{a}{p}\right)\left(\frac{b}{p}\right).$$

Hint: Use Euler's criterion.

Exercise 4.15. Calculate $\left(\frac{-6}{101}\right)$ using Euler's criterion by doing a calculation modulo 101. How does this compare in difficulty with the calculation in the text using the quadratic reciprocity law?

Exercise 4.16. Calculate various Legendre symbols by repeatedly using only the QRL, the supplementary theorem, and the multiplicativity of the Legendre symbol. You should never have to check a quadratic residue by brute force or Euler's criterion.

Exercise 4.17. Complete the verification in the text that the odd primes not dividing 6 for which -6 is a quadratic residue are precisely those in the four arithmetic progressions given by Euler.

Exercise 4.18. In the next section we shall read Euler's claim in his paper of 1744 that the nontrivial prime divisors of numbers of the form $x^2 - 5y^2$ are precisely those of the form $10m \pm 1$, and that the nontrivial prime divisors of numbers of the form $x^2 - 7y^2$ are precisely those of the form $28m \pm 1$, $28m \pm 3$, $28m \pm 9$. Use the QRL to verify this.

Exercise 4.19. Find all solutions of the congruence

$$x^2 + x + 1 \equiv 0 \ (\text{mod } 31).$$

4.2 Euler Discovers Patterns for Prime Divisors of Quadratic Forms

Without doubt Leonhard Euler was one of the world's mathematical giants, whose work profoundly transformed mathematics. He made extensive contributions to many mathematical subjects, including number theory, and was so prolific that the publication of his collected works, begun in 1911, is still underway, and is expected to fill more than 100 large volumes.

Born in Basel, Switzerland, in 1707, Euler's mathematical career spanned almost the whole eighteenth century, and he was at the heart of all its great accomplishments. His father, a Protestant minister interested in mathematics, was responsible for his son's earliest education. Later, Euler attended the Gymnasium in Basel, a high school that did not provide instruction in mathematics, however. At fourteen, Euler entered the University of Basel, where Johann Bernoulli (1667–1748) had succeeded his brother Jakob (1654–1705) in the chair of mathematics. Though Bernoulli declined to give Euler private lessons (and Bernoulli's public lectures at the university were limited to elementary mathematics), he was willing to help Euler with difficulties in the mathematical texts that Euler studied on his own.

Euler received a degree in philosophy and joined the Department of Theology in 1723, but his studies in theology, Greek, and Hebrew suffered from his devotion to mathematics. Eventually he gave up the idea of becoming a minister. In autumn of 1725, Johann Bernoulli's sons Nikolaus (1687–1759) and Daniel (1700–1782) went to Russia to join the newly organized St. Petersburg Academy of Sciences; at their behest, the following year the academy invited Euler to serve as adjunct of physiology, the only position available at the time. Euler accepted, arriving in St. Petersburg in May of 1727. In spite of having been invited to study physiology, soon after his arrival he was given the opportunity to work in his true field of mathematics. During fourteen years in St. Petersburg, Euler published fifty-five works, making brilliant discoveries in such fields as analysis, number theory, and mechanics.

Photo 4.3. Euler.

In 1740, Euler was invited to join the Berlin Academy of Sciences and accepted, since the political situation in St. Petersburg had deteriorated by that time. During his tenure in Berlin, he remained an active member of the St. Petersburg Academy as well, publishing prolifically in both academies. In 1766, Euler returned to St. Petersburg, where he remained for the rest of his life. Though he went blind shortly after his return, he was able to continue his work with the aid of assistants; indeed, he actually increased his output.

When Euler received a letter from Christian Goldbach (1690–1764) in December 1729, early in his first St. Petersburg period, little did he know that it was going to instill a new passion in him that would last the rest of his life. Continuing an initial exchange of letters earlier that fall, Goldbach mentions in a postscript the assertion of Fermat that every number of the form $2^{2^n}+1$ is a prime, and that nobody seemed to have a proof for it [65, p. 10]. Euler began to read Fermat's works and embarked on a long journey of providing proofs for and generalizing Fermat's number-theoretic insights. In 1735 he found the counterexample $2^{2^5}+1$ to the conjecture that initially aroused his interest. (What are the factors of this number, and how might Euler have discovered this [245]?)

Goldbach was a well-traveled man whose main intellectual interests were languages and mathematics. In 1725 he became professor of mathematics and history at St. Petersburg, and in 1728 went to Moscow as tutor to Tsar Peter II. He knew many of the distinguished mathematicians of his time, including Nicolas and Daniel Bernoulli, who obtained their appointments to the academy in St. Petersburg thanks to his efforts, leading in turn to Euler's appointment there. Goldbach and Euler became lifelong friends and their correspondence provides the most vivid record of Euler's number-theoretic legacy, more so even than his published papers. Today Goldbach is best remembered for the conjecture, emerging from correspondence with Euler in 1742, that every even integer greater than 2 can be represented as the sum of two primes. This is still a famous open problem today.[7]

As discussed in the introduction, one of the many trails on Euler's journeys into questions raised by Fermat led him to look for patterns in the prime divisors of quadratic forms. By 1744 he could state some amazing patterns by "Induction," i.e., extrapolation from experimental evidence (distinct from the proof technique we call the principle of "mathematical induction" today). He published his claims, without proof, in the paper *Theoremata circa divisores numerorum in hac forma contentorum paa ± qbb* [66, v. 2, pp. 194–222] (Theorems about the divisors of numbers expressed in the form *paa ± qbb*). For each of a large variety of quadratic forms, he presented a list of arithmetic progressions whose primes he claimed were precisely all the nontrivial prime divisors of the quadratic form, i.e., solutions of the divisor problem. Most of the quadratic forms he considered were of the type a^2+Nb^2 or a^2-Nb^2, where N is a particular positive integer, usually prime. We shall present just a few of Euler's many examples, followed by excerpts from his extensive comments, in which he describes the patterns he noticed in a way that completely determines all the arithmetic progressions [57]. In his comments we will later recognize the

---

[7] Goldbach's conjecture was mentioned by David Hilbert as part of his famous problem about the distribution of prime numbers (recall the footnote in the introduction about Hilbert's problems).

discovery of the essence of quadratic reciprocity. The two cases $a^2 + Nb^2$ and $a^2 - Nb^2$ (recall $N > 0$) presented quite distinct patterns for Euler to decipher. Of course the claims we have seen by Fermat were all about forms of the type $a^2 + Nb^2$. While for future discussion we present some examples from both cases below, at this point in our story we will focus largely on the case $a^2 - Nb^2$, since it will lead to the simplest expression of patterns, and most directly to the quadratic reciprocity law as our story progresses. We strongly encourage the reader to pretend to be Euler, and try to conjecture exactly which arithmetic progressions are contained in Euler's experimental data for $a^2 - Nb^2$ (at least for N prime), before reading his comments spelling it all out (Exercise 4.20).

∞∞∞∞∞∞∞

Leonhard Euler, from
Theorems
about the divisors of numbers expressed in the form $paa \pm qbb$

THEOREM 10

Numbers of the form $aa + 5bb$ have prime divisors that are always either 2 or 5 or contained in one of the 4 forms $20m + 1$, $20m + 3$, $20m + 7$, $20m + 9$.

THEOREM 11

If a number $20m + 1$, $20m + 3$, $20m + 9$, $20m + 7$ is prime, then it follows that

$$20m + 1 = aa + 5bb, \quad 2\,(20m + 3) = aa + 5bb,$$
$$20m + 9 = aa + 5bb, \quad 2\,(20m + 7) = aa + 5bb.$$

THEOREM 12

No number contained in a sequence of the form $20m - 1$, $20m - 3$, $20m - 9$, $20m - 7$ can be a divisor of a number of the form $aa + 5bb$.

. . .

THEOREM 44

All prime divisors of the form $aa - 5bb$ are either 2 or 5 or contained

in either the formulas   or in the single one
$$20m \pm 1, \quad 20m \pm 9 \qquad 10m \pm 1.$$

Every prime number of this form is also contained in the divisors of the form $aa - 5bb$.

THEOREM 45

All prime divisors of the form $aa - 7bb$ are either 2 or 7 or contained in one of the following formulas:

$$28m \pm 1, \quad 28m \pm 3, \quad 28m \pm 9;$$

all prime numbers contained in these formulas are also contained in the divisors of the form $aa - 7bb$.

THEOREM 46

All prime divisors of the form $aa - 11bb$ are either 2 or 11 or contained in one of the following formulas:

$$44m \pm 1, \quad 44m \pm 5, \quad 44m \pm 7, \quad 44m \pm 9, \quad 44m \pm 19;$$

all prime numbers contained in these formulas are also contained in the divisors of the form $aa-11bb$, and this reciprocation holds also in all succeeding theorems. ...

THEOREM 49

All prime divisors of the form $aa - 19bb$ are either 2 or 19 or contained in the following formulas:

$$76m \pm 1, \quad 76m \pm 3, \quad 76m \pm 9,$$
$$76m \pm 27, \quad 76m \pm 5, \quad 76m \pm 15,$$
$$76m \pm 31, \quad 76m \pm 17, \quad 76m \pm 25.$$

. . .

COMMENT 13

Therefore, all the prime divisors of the numbers expressed in the form $aa-Nbb$ are either 2 or the divisors of the number N or can be expressed in the form $4Nm \pm \alpha$. But for one of the divisors to be in the form of $4Nm + \alpha$, then $4Nm - \alpha$ will also be the form of one of the divisors; and so this is unlike the case of the form $aa + Nbb$; in which if $4Nm + \alpha$ will be a divisor, then $4Nm - \alpha$ can never express a divisor for the same form.

COMMENT 14

Having established therefore $4Nm \pm \alpha$ as the general form of the divisors of the numbers described by the expression $aa - Nbb$, the letters α generally will represent many numbers, always including the number one; truly then, because this conversation is about prime divisors, no α itself will be among the values of the number N nor any of the divisors of N. Then it is also apparent, all these values of α can be arranged to be made less than $2N$. For if $4Nm+2N+b$ is a divisor, then by substituting $m-1$ in place of m, the divisor will be $4Nm-(2N-b)$. Therefore the values of α itself will be odd numbers [relatively] prime to N, less than $2N$, and of all these numbers which are odd and prime to N and less than $2N$, it will be seen that only one half are suitable values for α; the remaining will exhibit a form, in which plainly no divisor may be contained. It is always certain to have just as many forms of divisors, as there are that are not, except for the single case where $N = 1$.

$$\cdots$$

COMMENT 16

But just as the number one is always found among the values of α, so also any squared number which is [relatively] prime to $4N$ will supply a suitable value for α.

∞⊗⊗⊗⊗⊗⊗⊗∞

Let us see what we can glean from these comments on the case $a^2 - Nb^2$. Euler claims from his observations that the nontrivial prime divisors of $a^2 - Nb^2$ ($N > 0$) are all those found in certain arithmetic progressions having period $4N$, and that these progressions always occur in matched pairs of the form $4Nm \pm \alpha$. (He comments that this is quite contrary to the forms $a^2 + Nb^2$ ($N > 0$), for which the relevant progressions never occur in matched pairs.) He then points out that one can always arrange for $0 < \alpha < 2N$. Next Euler points out that since we are seeking prime divisors, only progressions $4Nm \pm \alpha$ in which α is odd and relatively prime to N need be considered. He then claims that among these restricted possibilities, exactly half of them will be "suitable," i.e., will be those containing precisely the prime divisors of $a^2 - Nb^2$. Finally, Euler says that square numbers relatively prime to $4N$ will always supply suitable values.

Let us examine all these claims in the example $N = 7$ from Theorem 45 above. According to Euler, to find the prime divisors of numbers of the form $a^2 - 7b^2$, we should first consider only progressions of the form $28m \pm \alpha$, where $0 < \alpha < 14$ and α is odd and relatively prime to 7. This leads us to the possible values 1, 3, 5, 9, 11, 13 for α. Euler says that exactly half of these will produce the suitable progressions, and moreover that squares relatively prime to 28 will always produce suitable progressions. We wonder, will squares produce all the suitable pairs of progressions?

To examine the progressions arising from squares relatively prime to 28, we begin by listing the squares of odd numbers not divisible by 7, i.e., 1, 9, 25, 81, 121, 169, etc., and then express each of them in the form $28m \pm \alpha$, where $0 < \alpha < 14$, in order to find a suitable α. Thus we have

$$1 = 28 \cdot 0 + 1, \quad 169 = 28 \cdot 6 + 1,$$
$$9 = 28 \cdot 0 + 9, \quad 121 = 28 \cdot 4 + 9,$$
$$25 = 28 \cdot 1 - 3, \quad 81 = 28 \cdot 3 - 3, \quad \text{etc.}$$

So the values for α produced so far are 1, 9, 3. Notice that these are exactly the three pairs of progressions Euler listed in Theorem 45, and according to Euler this list is complete, since he says that only half of the possible list $1, 3, 5, 9, 11, 13$ for α will be suitable. Notice that the three suitable values for α did all actually arise from our list of squares, and in fact occurred right away, from the first three squares, with the values arising thereafter simply repeating with a certain pattern. Although Euler does not say so, he was surely aware of these facts in general. We shall leave it to the reader to verify these latter phenomena in some other examples, and then to prove that it always happens this way: For N an odd prime, the first $\frac{N-1}{2}$ odd squares will produce $\frac{N-1}{2}$ distinct suitable values for α, and will thus, according to Euler's claims, produce all the suitable progressions (Exercises 4.21, 4.22). This means that Euler discovered (without proof) a complete solution to the problem of determining the forms (arithmetic progressions) of nontrivial prime divisors of $a^2 - Nb^2$ ($N > 0$)! Euler also gave an analogous description for nontrivial prime divisors of $a^2 + Nb^2$ ($N > 0$), which he claimed similarly lie in certain arithmetic progressions with periodicity $4N$. We shall not present that description here, but it will emerge in our second Euler source.

Does this mean that Euler discovered the quadratic reciprocity law in 1744? Certainly it is not in the form stated in the introduction. But in hindsight we do see a strong glimmer of reciprocity here. Recall that in looking for prime divisors p of $a^2 - Nb^2$ (still always $N > 0$), we are asking whether N is a square modulo p. From Euler's claims we deduce that for N an odd prime, this happens precisely when p or $-p$, i.e., α or $-\alpha$, is itself a square modulo $4N$. Thus there is a "reciprocity" between N and p here, i.e., they exchange roles, from quadratic residue to modulus and vice versa, except that there is also an introduced $+$ or $-$ sign on p, and N gets multiplied by 4. The significance of all this will be clarified as we examine the evolution of the discovery of the quadratic reciprocity law in the hands of Euler's successors.

For almost another 40 years, Euler strove to prove his claims, and while he succeeded in a few special cases, especially those conjectured by Fermat, the general case eluded him. In the paper *Observationes circa divisionem quadratorum per numeros primos* (Observations on the Division of Squares by Prime Numbers) [66, v. 3, pp. 497–512], [232, pp. 40–46], presented to the St. Petersburg Academy in 1772, but not published until 1783, the year of his death, Euler did two important things. He gave proofs determining the quadratic character of -1 for all primes (this is the special case $a^2 + b^2$ of Euler's claims about divisors of quadratic forms, and was one of the results claimed by Fermat), and then gave a clear statement of his final vision of the role reversal between quadratic residues and moduli. Our second source consists of relevant excerpts from this paper.

Euler begins the paper by developing various basic properties of quadratic residues, some of which we will state here and leave to the reader to prove in exercises. By this time Euler's view had evolved considerably, and in particular he was thinking and writing partly in terms of quadratic residues. But he still lacked the full benefits of thinking and writing in terms of congruences, which we will rely on from our Appendix.

Key properties of quadratic residues. For p an odd prime:

1. There are exactly $\frac{p-1}{2}$ (nonzero) quadratic residues $\bmod\, p$ (by which we mean to count from among the equivalence classes modulo p not containing zero), obtained by squaring the numbers $1, \ldots, \frac{p-1}{2}$. Thus there are also $\frac{p-1}{2}$ (nonzero) nonquadratic residues $\bmod\, p$, since there is a total of $p - 1$ nonzero equivalence classes modulo p (Exercise 4.23).

2. Recall from the Appendix that every number that is nonzero $\bmod p$ has a *reciprocal* $\bmod p$, i.e., we can divide by it $\bmod p$. Then (Exercise 4.24):

   a) A product or quotient of two quadratic residues $\bmod p$ is also a quadratic residue $\bmod p$.

   b) A product or quotient of a quadratic residue with a quadratic nonresidue is a quadratic nonresidue.

   c) A product or quotient of two quadratic nonresidues is a quadratic residue.

Now we are ready to read from Euler's paper, with first a warning on three important matters of notation. Euler will use P for the prime divisor (modulus) in question, utilizing p for something else. He almost always uses the word *residue* to refer to what we call the *remainder*, i.e., a number congruent to P but chosen or restricted to be in the range from 0 to $P-1$. Finally, in this particular paper he always means "quadratic residue" when he says *residue*, i.e., the remainders only of squares $\bmod P$.

∞∞∞∞∞∞∞∞

Leonhard Euler, from
Observations on the Division of Squares by Prime Numbers

23. *Theorem 4. If the divisor P is of the form $4q+3$, then -1 or $P-1$ is certainly a nonresidue.*

Demonstration. When we write $P = 2p+1$, then $p = 2q+1$, an odd number. Hence the number of all [quadratic] residues will be odd. If -1 were to appear in the sequence of residues, then to every residue α would correspond another residue $-\alpha$, and the sequence of residues could be written as follows:

$$+1, \ +\alpha, \ +\beta, \ +\gamma, \ +\delta, \ \text{etc.},$$

$$-1, \ -\alpha, \ -\beta, \ -\gamma, \ -\delta, \ \text{etc.},$$

and the number of residues would be even. But since this number is certainly odd, it is impossible that -1 or $P-1$ should appear in the sequence of residues; hence it belongs to the sequence of nonresidues ...

30. *Theorem 5. If the divisor P is a prime of the form $4q+1$, then the number -1 or $P-1$ is certainly a residue.* ...

Conclusion. These ... theorems,[8] of which the demonstration from now on is desired, can be nicely formulated as follows:

Let s be some prime number, let only the odd squares 1, 9, 25, 49, etc. be divided by the divisor $4s$, and let the residues be noted, which will all be of the form $4q+1$, of which any may be denoted by the letter α, and the other numbers

---

[8] That is, several theorems succeeding number 30.

of the form $4q+1$, which do not appear among the residues, be denoted by some letter $\mathfrak{U}$, then we shall have

divisor a prime number $[P]$ of the form	then $[\text{modulo } P]$
$4ns + \alpha$	$+s$ is a residue, and $-s$ is a residue;
$4ns - \alpha$	$+s$ is a residue, and $-s$ is a nonresidue;
$4ns + \mathfrak{U}$	$+s$ is a nonresidue, and $-s$ is a nonresidue;
$4ns - \mathfrak{U}$	$+s$ is a nonresidue, and $-s$ is a residue.

∞∞∞∞∞∞∞∞∞∞

The text is quite detailed and requires only a little explanation. In Theorem 4 the reader should confirm why all the residues listed and counted are distinct. In proving Theorem 5 Euler matches residues with their reciprocals, rather than their negatives as in Theorem 4, and we leave this interesting proof to the reader (Exercise 4.25). As we explained when discussing the divisor problem in the introduction, Theorem 4 ensures that no sum of two relatively prime squares can have a prime divisor of the form $4q+3$, and Theorem 5 tells us that every prime of the form $4q+1$ is a divisor of a sum of two relatively prime squares. Recall that in the language of quadratic residues, this solution to the divisor problem for $a = 1$ proves the first part of the supplementary theorem to the quadratic reciprocity law stated in the introduction (Exercise 4.26):

The quadratic character of negative one. -1 is a quadratic residue for every prime of form $4q+1$, but not for any prime of form $4q+3$.

This is an extremely important result, and we shall find it useful shortly. In the full text, Euler comments after Theorem 5 that he can use it to solve the harder problem of representation of the quadratic form in this case, the one claimed by Fermat: Every prime of the form $4q+1$ actually is a sum of two squares. As discussed in the introduction, to do this Euler also needed a descent result, that every nontrivial divisor of a sum of two relatively prime squares is again a sum of two squares. We shall have this in hand shortly, from the source by Lagrange that we will read next.

Finally, let us look at the four statements in Euler's *Conclusion*. While the notation is very different from his earlier paper[9] of 1744, what Euler writes here in 1772 is just a crystallized statement of what he already claimed earlier, now phrased partly in the language of (quadratic) residues, a major step toward the congruence viewpoint. This latter paper was Euler's final formulation of the patterns he saw in quadratic residues modulo prime divisors, which will metamorphose into the modern formulation of quadratic reciprocity. The statements about $+s$ above correspond to the forms $a^2 - Nb^2$ $(N > 0)$ we read about in detail in the earlier paper, while the statements about $-s$ correspond

---

[9] From the 1744 paper to the 1772 paper, his notation changes as follows: $N \to s$, $m \to n$. And α has a subtly different meaning: while in the earlier paper he arranges for α always to satisfy $|\pm\alpha| < 2N$, in this later paper he chooses $0 < \alpha < 4s$, i.e., α becomes, as he says, simply the remainder of an odd square upon division by $4s$.

to the forms $a^2 + Nb^2$ ($N > 0$), of which we presented only a single excerpt. We leave it to the reader to confirm detailed agreement between what Euler claims in the papers of 1744 and 1772 (Exercises 4.27, 4.28). Interestingly, using what we know about the quadratic character of -1, and the key property above about products of quadratic residues and nonresidues, it is easy to see that Euler's statements about $+s$ are equivalent to those about $-s$ (Exercise 4.29), so the distinction between the nature of divisors of the forms $a^2 - Nb^2$ and $a^2 + Nb^2$ is now explained.

We make one last comment about Euler's *Conclusion*. At first it appears from his wording that Euler is making claims in only one direction, namely that $+s$ ($-s$) is or is not a quadratic residue modulo prime divisors of certain types, but not necessarily solely of these types. However, the reader may check that the four prime divisor types he lists actually encompass, mutually exclusively, all odd primes. Thus his four claims actually cover all possibilities, and so provide a complete characterization of the relationship between types of prime divisors and quadratic residues modulo those divisors.

Exercise 4.20. Looking just at the lists of arithmetic progressions Euler presents in his Theorems 44, 45, 46, and 49, conjecture a general description of exactly what those arithmetic progressions might be for any quadratic form $a^2 - Nb^2$ where N is an odd prime. Hint: Which values of α appear for all N?

Exercise 4.21. Check Euler's general claims in his Comments 13, 14, 16, and our further observations, against his Theorems 46 and 49 in the same way we did for $N = 7$ against Theorem 45. In other words, carry out his prescription for finding the arithmetic progressions providing all prime divisors of $a^2 - 11b^2$ and $a^2 - 19b^2$, and see whether the suitable progressions are all provided by the first $\frac{N-1}{2}$ odd squares.

Exercise 4.22. Prove that for N an odd prime, the first $\frac{N-1}{2}$ odd squares provide distinct values of α in Euler's analysis. (Be careful: sometimes a square produces α modulo $4N$, sometimes $-\alpha$.) Thus, according to Euler, the odd squares provide all the suitable values.

Exercise 4.23. Prove that for p an odd prime, there are exactly $\frac{p-1}{2}$ nonzero quadratic residues mod p (by which we mean to count among the equivalence classes modulo p not containing zero), obtained by squaring the numbers $1, \ldots, \frac{p-1}{2}$. Thus there are also $\frac{p-1}{2}$ nonzero nonquadratic residues mod p, since there is a total of $p - 1$ nonzero equivalence classes modulo p.

Exercise 4.24. Prove that:

1. A product or quotient of quadratic residues mod p is also a quadratic residue mod p.
2. A product or quotient of a quadratic residue with a quadratic nonresidue is a quadratic nonresidue.
3. A product or quotient of two quadratic nonresidues is a quadratic residue. (Hint: Count the possible products of a nonresidue with all the residues.)

Exercise 4.25. Give a proof of Euler's Theorem 5. (Hint: First show that 1 and -1 are the only remainders that are their own reciprocals mod p. Then use the idea of his proof of Theorem 4, but match numbers with their reciprocals instead of their negatives.)

Exercise 4.26. Prove the first part of the supplementary theorem to the quadratic reciprocity law stated in the introduction.

Exercise 4.27. Compare Euler's claims in his *Conclusion* of 1772 about the quadratic character of $+s$ ($s > 0$) with what we read in Comments 13, 14, 16 of his paper of 1744. Verify that they agree.

Exercise 4.28. Compare Euler's claims in his *Conclusion* of 1772 about the quadratic character of $-s$ with what he claims in his paper of 1744. You will have to find the relevant parts of the earlier paper, and you may need to read some Latin. Verify that they agree.

Exercise 4.29. Show that Euler's claims for $+s$ are equivalent to his claims for $-s$, using the characterization of the quadratic character of -1 that he just proved, along with the key property about multiplicativity of quadratic residues and nonresidues proven in Exercise 4.24.

4.3 Lagrange Develops a Theory of Quadratic Forms and Divisors

Even though the second half of the eighteenth century was not very favorably disposed toward pure mathematics, in the 1770s the torch of studying quadratic reciprocity was being passed to two younger men, Lagrange and Legendre. Since the time of Newton and Leibniz in the late seventeenth century, the geometers, as mathematicians called themselves, were primarily busy working on the development of the calculus, not number theory. Here too Euler's genius and phenomenal output defined the central problems and lines of development. The astonishing practical applications of the new theory left little time to catch one's breath and worry about the somewhat shaky foundations on which people juggled derivatives, integrals, and infinite series. But this shaky foundation was adequate to most eighteenth-century developments, and there was much political and economic gain from solving applied problems, such as accurate navigation at sea (see our chapter on curvature). Thus, there was neither livelihood nor prestige to be found in working primarily on problems such as the nature of patterns in prime numbers.

To be a professional mathematician in the eighteenth century meant to have a wealthy sponsor and be part of the scientific academy of a country, or be independently wealthy. There was no instruction in higher mathematics at universities, leaving only private tutors if one wanted to be led to the edge of mathematical research. That is how Euler earned a living for a while, and so did several of the Bernoullis. The two leading academies during the second half of the eighteenth century were the Academies of Science in Berlin and Paris. The two men who dominated these institutions, and set the mathematical agenda for all of Europe, were Euler in Berlin and Jean-Baptist d'Alembert (1717–1783) in Paris. But both men were nearing the end of their lives, and they died in the same year. Who was going to take their place? (Gauss, who was destined to become the mathematical titan of the nineteenth century, was only six years old in 1783.)

Photo 4.4. Lagrange.

Just in time, a young man from Turin, Italy, Giuseppe Lodovico Lagrangia, later changed to Joseph Louis Lagrange, caught the attention of Euler and d'Alembert. At the early age of 18 he impressed both with his communications on what was later to become the calculus of variations, and applications of it to mechanics. D'Alembert quickly adopted Lagrange as his protégé, and eventually succeeded in 1766 in making him Euler's successor at the Academy of Sciences in Berlin, after Euler left to take a position at the Academy of Sciences in St. Petersburg. Thus, Lagrange became one of the most influential mathematical scientists in Europe, especially after the deaths of d'Alembert and Euler. These men shared a serious concern for the future of mathematics, in light of strong competition from other sciences. This concern is expressed strongly in the correspondence between d'Alembert and Lagrange, with both men being rather pessimistic. As Lagrange says in a letter to d'Alembert in 1781:

> I begin to notice how my inner resistance increases little by little, and I cannot say whether I will still be doing geometry ten years from now. It also seems to me that the mine has maybe already become too deep and unless one finds new veins it might have to be abandoned.
> Physics and chemistry now offer a much more glowing richness and much easier exploitation. Also, the general taste has turned entirely

in this direction, and it is not impossible that the place of Geometry in the Academies will someday become what the role of the Chairs of Arabic at the universities is now [141, vol. 13, p. 368].

(For a more detailed discussion see [202].)

Lagrange's mathematical career can be divided into three periods. The first one was spent in his native Turin, Italy. Then, in 1766, he moved to the Berlin Academy and finally, in 1787, he took a position at the Academy of Sciences in Paris, where he remained until his death. (For a biography of Lagrange see [92].)

Lagrange's number-theoretic investigations extend approximately over a ten-year stretch beginning shortly after his arrival in Berlin in 1766, no doubt inspired by Euler. In 1773 Euler wrote to Lagrange about quadratic forms and their divisors, "I am sure that this will lead to very important discoveries" [245, p. 219], and in 1775, reacting to the results of Lagrange's that we will now read, Euler says, "Thus all the 'theorems' which I formulated long ago in vol. XIV of the old [Petersburg] *Commentarii* [i.e., the paper of 1744] have acquired a much higher degree of certainty ... and there seems to be no doubt that whatever in them is still to be desired will soon receive a perfect proof."

We present here an excerpt from Lagrange's memoir *Researches in Arithmetic* [141, vol. III, pp. 695–705, 707–709, 714] from 1773–1775, in which he takes a whole new approach to quadratic reciprocity by laying the foundations for a theory of quadratic forms and their divisors, including applications to the representability problem.

∞⬤⬤⬤⬤⬤⬤⬤∞

Joseph Louis Lagrange, from
Researches in Arithmetic
FIRST PART

These investigations have as their subject the numbers that can be represented by the formula

$$Bt^2 + Ctu + Du^2,$$

where B, C, D are supposed to be given integers, and t, u also integers, but undetermined. In what follows I will give the method of finding all the different forms that the divisors of these types of numbers can assume. Then I will give a method for reducing these forms to the smallest number possible. I will show how one can make tables in practice, and I will show the use of these tables in researching the divisors of numbers. Finally, I will give proofs of numerous theorems about prime numbers of the same form $Bt^2 + Ctu + Du^2$, of which some are already known, but have not yet been proven, and others which are entirely new ...

2. OBSERVATION. — The formula of first degree $Bt + Cu$, where B and C are arbitrarily given relatively prime numbers, may represent any number; but it

is not the same for the formula of second degree $Bt^2 + Ctu + Du^2$; because we
have proven elsewhere [...] that the equation

$$A = Bt + Cu$$

is always resolvable in whole numbers, no matter what the numbers A, B, C are,
provided that the latter two are relatively prime;[10] but the equation

$$A = Bt^2 + Ctu + Du^2$$

is only resolvable in certain cases, and when certain conditions hold between the
given numbers A, B, C, D. One must say the same thing, with even greater
reason, of formulas of third degree and higher.

3. SCHOLIUM. — Thus there is a great difference between the formulas of
the first degree and those of higher degrees, the former representing all possible
numbers, whereas the latter can only represent certain numbers which are distinguished from all others by a certain special character. Great geometers have
already considered the properties of numbers which can be represented by one of
the formulas of second degree or of higher degrees, such as

$$t^2 + u^2, \quad t^2 + 2u^2, \quad t^2 + 3u^2, \quad t^4 + u^4, \quad t^8 + u^8, \quad \ldots \; .$$

(See the works of Mr. Fermat and the *New Commentaries of Petersburg*, vols.
I, IV, V, VI, VIII). But nobody, to my knowledge, has yet treated this subject in
a direct and general manner, or has given rules for finding *a priori* the principal
properties of numbers which can be produced by the formulas given above.

Since this subject is one of the most curious of Arithmetic, and since it merits
the special attention of geometers due to the great difficulties it harbors, I will
endeavor to treat it more thoroughly than has been done so far. But for the time
being I will limit myself to formulas of the second degree, and I will begin by
examining the form of divisors of numbers that can be expressed by these sorts
of formulas.

THEOREM I.

4. *If the number A is a divisor of a number represented by the formula*

$$Bt^2 + Ctu + Du^2,$$

*assuming that t and u are relatively prime, then I say that this number will
necessarily be of the form*

$$A = Ls^2 + Msx + Nx^2,$$

---

[10] See Exercise 4.57 in the Appendix.

where one will have

$$4LN - M^2 = 4BD - C^2,$$

with s and x also relatively prime to each other.

Let a be the quotient of $Bt^2 + Ctu + Du^2$ divided by A, so that one has

$$Aa = Bt^2 + Ctu + Du^2,$$

and let b be the greatest common divisor of a and u (if a and u are relatively prime, then one will have $b = 1$), so that by letting

$$a = bc, \quad u = bs,$$

c and s are relatively prime. One will then have

$$Abc = Bt^2 + Cbts + Db^2s^2,$$

and consequently Bt^2 will be divisible by b. But t and u being relatively prime (by hypothesis), t will also be prime to b, which is a divisor of u. Hence it follows that B is divisible by b. Since one has $B = Eb$, and dividing the equation by b, it will become

$$Ac = Et^2 + Cts + Dbs^2.$$

Now, since c and s are relatively prime, one may suppose (by the previous observation) that

$$t = \theta s + cx,$$

which, being substituted, gives

$$Ac = (E\theta^2 + C\theta + Db)s^2 + (2E\theta c + Cc)sx + Ec^2x^2,$$

from which it follows that the number $(E\theta^2 + C\theta + Db)s^2$ is divisible by c. And since c and s are relatively prime, it follows that $E\theta^2 + C\theta + Db$ is divisible by c. Hence, dividing the whole equation by c, and setting

$$L = \frac{E\theta^2 + C\theta + Db}{c}, \quad M = 2E\theta + C, \quad N = Ec,$$

one has

$$A = Ls^2 + Msx + Nx^2.$$

Now, $4LN - M^2$ is equal to

$$4E(E\theta^2 + C\theta + Db) - (2E\theta + C)^2 = 4EDb - C^2 = 4BD - C^2,$$

by virtue of $B = Eb$. Hence, etc.

Finally, since t and u are relatively prime (hypothesis), t and s will be also, because $u = bs$. But if x and s were not relatively prime, it is clear that t would have to be divisible by their greatest common divisor, since $t = \theta s + cx$. Since this cannot be, it follows that x and s are necessarily relatively prime, if t and u are.

THEOREM II.

5. *Every formula of degree two, such as*

$$Ls^2 + Msx + Nx^2,$$

in which M is greater than L or N (without regard to the sign of these quantities), can be transformed into another one of the same degree, like

$$L's'^2 + M's'x' + N'x'^2,$$

in which one has

$$4L'N' - M'^2 = 4LN - M^2,$$

and where M' is smaller than M.[11]

Because, if for example $M > L$, one will set

$$s = mx + s',$$

and the proposed formula will become

$$(Lm^2 + Mm + N)x^2 + (2Lm + M)xs' + Ls'^2.$$

Or else, by changing x to x',

$$L's'^2 + M's'x' + N'x'^2,$$

where

$$L' = L,$$
$$M' = 2Lm + M,$$
$$N' = Lm^2 + Mm + N.$$

Whatever the number m may be, one consequently has

$$4L'N' - M'^2 = 4L(Lm^2 + Mm + N) - (2Lm + M)^2 = 4LN - M^2.$$

Now, since L is less than M (hypothesis), it is clear that one can determine the number m from the fact that $2Lm + M$ will become less than M. Therefore, etc.

6. COROLLARY I. — Thus, if one of the numbers L' or N' in the transformed expression

$$L's'^2 + M's'x' + N'x'^2$$

is less than M', one will be able to obtain another transformed expression such as

$$L''s''^2 + M''s''x'' + N''x''^2,$$

---

[11] In magnitude, i.e., $|M'| < |M|$.

in which one will have similarly that

$$4L''N'' - M''^2 = 4L'N' - M'^2 = 4LN - M^2,$$

and where M'' will be smaller than M', and so on. The series of numbers

$$M, \ M', \ M'', \ \ldots$$

cannot go on infinitely, since these numbers are all integers and are decreasing from one to the next. Hence one will necessarily arrive at a transformed expression, which I will represent as

$$Py^2 + Qyz + Rz^2,$$

in which Q will not be greater than P, nor greater than R, and in which one has

$$4PR - Q^2 = 4LN - M^2.$$

7. COROLLARY II. — If the numbers s and x in the proposed formula are relatively prime, it is clear that the numbers s' and x' in the transformed formula are also relatively prime. Because if they are not, then s would necessarily have to be divisible by the greatest common divisor of s' and x, since $x' = x$ and $s = mx + s'$.

Consequently, the numbers s'' and x'' of the second transformed formula are also relatively prime for the same reason, and so on. From this one can conclude that the numbers y and z of the last transformed formula are necessarily relatively prime, if the numbers s and x are.

THEOREM III.

8. *If A is a divisor of a number of the form*

$$Bt^2 + Ctu + Du^2,$$

with t and u relatively prime, then I say that this number A is necessarily of the form

$$Py^2 + Qyz + Rz^2,$$

with y and z relatively prime, and P, Q, R such that

$$4PR - Q^2 = 4BD - C^2.$$

Furthermore, Q is less than or equal to P and to R, disregarding the signs of P, Q, and R.

The proof of this theorem follows naturally from the two preceding theorems and their corollaries.

9. COROLLARY 1. — If $4BD - C^2$ is a positive number, it follows that $4PR$ is also positive. Consequently, since $P \geq Q$ and $R \geq Q$, it is clear that $4PR$ is also greater than or equal to $4Q^2$. And therefore

$$4PR - Q^2 \geq 3Q^2.$$

From this one also has

$$4BD - C^2 \geq 3Q^2$$

and so[12]

$$Q \leq \sqrt{\frac{4BD - C^2}{3}}.$$

. . .

11. COROLLARY III. — Hence, since Q has to be an integer, one can only take positive or negative integers for Q that do not surpass the limits we found; here we include zero among the integers. So one sees that Q can never take on more than a certain number of different values.

Furthermore, it is clear that for the equation

$$4PR - Q^2 = 4BD - C^2$$

to hold in whole numbers, it follows that Q will be even or odd, depending on C being even or odd, which further limits the values of Q.

Knowing Q, one can easily find P and R from the same equation, since from

$$PR = \frac{4BD - C^2 + Q^2}{4}$$

it follows that for P and R one can only take factors of the integer

$$\frac{Q^2 + 4BD - C^2}{4}.$$

And one has to reject those that are smaller than Q.

PROBLEM I.

12. *Find all the possible forms of divisors of numbers that are represented by the formula of degree two*

$$Bt^2 + Ctu + Du^2,$$

with t and u relatively prime.

It is evident, as we have shown above, that each divisor of the proposed formula is reducible to the form

---

[12] Here of course Lagrange means the size (absolute value) of Q to be bounded as shown next.

$$Py^2 + Qyz + Rz^2,$$

with y and z relatively prime. Hence the difficulty reduces to finding the values of the coefficients P, Q, R, when those of B, C, and D are given.

To this effect I distinguish two cases, one where the number $4BD - C^2$ is positive, and the other where this number is negative.

1. Let $4BD - C^2 = K$ (where K denotes a positive number). One then determines Q by these conditions: whether Q is even or odd follows from the parity of K, and Q does not surpass the number $\pm\sqrt{\frac{K}{3}}$. Then one determines P and R by these conditions: that P and R be two factors of the number $\frac{K+Q^2}{4}$, and that each of these factors not be less than Q (9 and 11).

∞∞∞∞∞∞∞

In (10) and the second part of (12) Lagrange makes a similar analysis for K negative, with quite similar results (Exercises 4.30 and 4.31). We note that $K = 4BD - C^2$ (or its negative[13]) is now commonly called the *determinant* (or the *discriminant*) of the form $Bt^2 + Ctu + Du^2$. The reason for this will be discussed near the end of the chapter.

∞∞∞∞∞∞∞

15. REMARK III. It is remarkable that the formulas for the divisors depend only on the value of K, that is, the number $4BD - C^2$. But it is easy to see the reason by remarking that the expression

$$Bt^2 + Ctu + Du^2$$

can be reduced to

$$\frac{(2Bt + Cu)^2 + (4BD - C^2)u^2}{4B},$$

so that the divisors of the expression $Bt^2 + Ctu + Du^2$ can be regarded also as divisors[14] of the much simpler expression[15]

$$x^2 \pm Ku^2.$$

It follows from this that it suffices to consider the formulas of this latter kind, and for this we further add the following Problem, which can be regarded as a special case of the preceding one, but whose essence is of the same generality.

PROBLEM II.

---

[13] There is variation on this in the literature.

[14] Although not necessarily vice versa.

[15] Lagrange writes $x^2 \pm Ku^2$ here because he wants to arrange for K always to be positive in his analysis.

16. *Find all the possible forms of the divisors of numbers of the form*

$$t^2 \pm au^2,$$

where a is any given positive number, and t and u are indeterminate numbers that are relatively prime.

Let us consider the formula

$$t^2 + au^2,$$

and compare it to the general formula of Problem I. One will obtain

$$B = 1, \quad C = 0, \quad D = a.$$

Hence $K = 4a$, so that Q will have to be even, and it cannot be greater than $\pm\sqrt{\frac{4a}{3}}$. Therefore, taking $Q = \pm 2q$, and regarding q as positive, it follows that q cannot be greater than $\sqrt{\frac{a}{3}}$. Hence

$$PR = \frac{4a + 4q^2}{4} = a + q^2.$$

If p and r denote two factors of $a + q^2$, of which neither is smaller than $2q$, then one has

$$py^2 \pm 2qyz + rz^2$$

as the general formula of divisors of $t^2 + au^2$.

It is proper to remark that since $pr = a + q^2$, it follows that p and r have the same sign, and it is clear that it will be necessary to take them to be positive, since the formula

$$py^2 \pm 2qyz + rz^2$$

represents positive numbers. ...

Further, since this formula does not change its form at all if one interchanges p and r, it will not be necessary to take successively for p every factor of $a + q^2$, and for r all the corresponding factors; because of this it suffices in each pair of factors of $a + q^2$ to always take the smaller as p, and the larger as r; and thus we will use this in what follows.

17. COROLLARY. — If one multiplies the formula

$$py^2 \pm 2qyz + rz^2$$

by p, it may be put in the form

$$(py \pm qz)^2 + (pr - q^2) z^2,$$

that is (since $pr = a + q^2$) in the form

$$(py \pm qz)^2 + az^2,$$

which is the same as that of the formula

$$t^2 + au^2.$$

From this it follows that every divisor of a number of the form t^2+au^2 will be either necessarily of the same form, if p has no values other than one, or will become of this form when multiplied by one of the values of p, if there are several. ...

THEOREMS ABOUT THE DIVISORS
OF NUMBERS OF THE FORM $t^2 + au^2$, t AND u RELATIVELY PRIME.

18. I. Let $a = 1$, then q will be no greater than $\sqrt{\frac{1}{3}}$; hence $q = 0$, $pr = 1$. Therefore

$$p = 1, \quad r = 1.$$

Therefore, the divisors of the numbers of the form

$$t^2 + u^2$$

are necessarily contained in the formula

$$y^2 + z^2.$$

That is: *Every divisor of a number equal to the sum of two* [relatively prime] *squares is also a sum of two squares.*

∞∞∞∞∞∞∞∞∞

Note that this is precisely the descent result needed to complete the two-step proof we sketched (under Divisor Plus Descent in the introduction) of Fermat's claim that every prime of form $4n + 1$ is a sum of two squares. After our second Euler source we already remarked that this gave the requisite divisor result, on the quadratic character of -1, and now we have the required descent. While Euler's argument for this descent was elaborate, and we do not present it in this book, Lagrange's falls immediately out of his general theory of quadratic forms.

∞∞∞∞∞∞∞∞∞

II. Let $a = 2$; hence q is no greater than $\sqrt{\frac{2}{3}}$, so that $q = 0$, $pr = 2$. Therefore

$$p = 1, \quad r = 2.$$

Hence the divisors of numbers of the form

$$t^2 + 2u^2$$

are contained in the formula

$$y^2 + 2z^2;$$

that is to say: *Every divisor of a number equal to a [relatively prime] sum of a square and a double square is also the sum of a square and a double square.*

. . .

V. Let $a = 5$; hence q is no greater than $\sqrt{\frac{5}{3}}$, so that $q = 0$ or 1. Taking $q = 0$, one will have $pr = 5$, hence

$$p = 1, \quad r = 5.$$

Taking $q = 1$, one has $pr = 6$, thus[16]

$$p = 2, \quad r = 3.$$

Therefore the divisors of numbers of the form

$$t^2 + 5u^2$$

are necessarily of one or the other of the forms

$$y^2 + 5z^2, \quad 2y^2 \pm 2yz + 3z^2,$$

so that the divisors themselves or their doubles are always (17) of the form $t^2 + 5u^2$.

∝∝∝∝∝∝∝∝∝

Lagrange continues his analysis for all cases up to $a = 12$.

∝∝∝∝∝∝∝∝∝

19. REMARK. — The first three theorems have been known to geometers for a long time, and are due, I believe, to Mr. Fermat. But Mr. Euler is the first who has proven them. One can see the proofs in volumes IV, VI, and VIII of the *New Petersburg Commentaries*. His method is totally different from ours, and is moreover not applicable to the case where the number a is greater than 3. Perhaps it is this that prevented this great geometer from pushing his researches in this subject further.

∝∝∝∝∝∝∝∝∝

This first part of Lagrange's memoir, published in 1773, goes on to finish his analysis of quadratic forms of potential divisors by investigating when two forms

$$py^2 \pm 2qyz + rz^2$$

representing potential divisors of a form $t^2 \pm au^2$, even after the reduction to a finite list as described above, can be transformed into each other in such a way that they represent the same numbers. In other words, he seeks to make

---

[16] To see why $p = 1$, $r = 6$ is not a possibility, the reader should recall Section (6).

his finite list of possible quadratic forms dividing a given form nonredundant. It turns out there are many subtleties here, especially when $K < 0$. In the Lagrange excerpts we presented above in (16), he is led to focus on forms for which $K = 4pr - 4q^2$ is positive, and p, r are also positive. Such forms are called *positive definite*, since the form represents only positive numbers (Exercise 4.32).

In the end Lagrange has a method of obtaining, for each form $t^2 \pm au^2$, a finite list, often very short, of quadratic forms for all potential nontrivial divisors. He produces tables with a complete list for a up to 31. The summary effect of this work was a tremendous advance in overcoming the descent challenge for understanding the possible quadratic forms of divisors of quadratic forms.

In the second part of the memoir, published in 1775, Lagrange tackles the representation problem explained in the introduction. Recall from the introduction that to solve a representation problem, such as Fermat's claim that every prime of the form $8n+1$ or $8n+3$ is necessarily of the form x^2+2y^2, one seems to need a solution to the divisor problem as well as a viable descent. The idea, mimicking the other example we gave earlier, would be to show first that any prime of the given linear form actually is a nontrivial divisor of some number of the form $x^2 + 2y^2$. Then, by Lagrange's descent result (18.II) immediately above, we know that it is also of the form $x^2 + 2y^2$. While Lagrange was aware of Euler's conjectures about the solution to the divisor problem, he knew that these had been proven in only a few cases. Neither was Lagrange himself able to prove them in general, although he was able to prove certain additional cases, and he also developed a very clever method for handling the divisor problem for primes of the form $4n + 3$, which produced numerous representation results. We shall describe Lagrange's ingenious idea for primes of the form $8n + 3$, as above in Fermat's claim, our aim being to confirm Fermat's claim that any such prime is of the form $x^2 + 2y^2$. And we shall see a strong connection to the Euler sources.

First of all, Lagrange developed an algebraic restriction on the potential linear forms of any number actually represented by a quadratic form, and added this information to the tables he already had of potential nontrivial quadratic divisors of numbers of the form $x^2 \pm ay^2$. To do this, Lagrange writes the quadratic form in a certain linear form $4an + b$ (as suggested by Euler's conjectures), where $\pm 4a$ is both the discriminant of the original form and (from his writings above) of the divisor form as well, and where $|b| \le 2a$ (of course, any number can be so written). At this stage, a is determined by the original form, but b may vary among many possibilities. Then he finds algebraic restrictions on possible values of b (we shall illustrate this shortly in an example and exercise). So each value of the quadratic form is found among the values of a finite number of linear forms. He develops extensive tables with results encompassing all possible odd values of b (only odd b are needed, since we are interested in odd primes $4an + b$), for all a up to 30. Figures 4.1 and 4.2 illustrate his results, and the reader should compare the information in

TABLE III.

Formule des nombres proposés................... $t^2 + au^2$.
Formule de leurs diviseurs impairs, et premiers à a.. $py^2 \pm 2qyz + rz^2 = 4an + b$.

VALEURS DE		VALEURS CORRESPONDANTES DE
a	p	b
1	1	1
2	1	1, 3
3	1	1, — 5
5	1	1, 9
	2	3, 7
6	1	1, 7
	2	5, 11
7	1	1, 9, 11, — 3, — 5, — 13
10	1	1, 9, 11, 19
	2	7, 13, — 3, — 17
11	1, 3	1, 3, 5, 9, 15, — 7, — 13, — 17, — 19, — 21
13	1	1, 9, 17, 25, — 3, — 23
	2	7, 11, 15, 19, — 5, — 21
14	1, 2	1, 9, 15, 23, 25, — 17
	3	3, 5, 13, 19, 27, — 11
15	1	1, 19, — 11, — 29
	3	17, 23, — 7, — 13
17	1, 2	1, 9, 13, 21, 25, 33, — 15, — 19
	3	3, 7, 11, 23, 27, 31, — 5, — 29
19	1, 4	1, 5, 7, 9, 11, 17, 23, 25, 35, — 3, — 13, — 15, — 21, — 27, — 29, — 31, — 33, — 37
21	1	1, 25, 37
	2	11, 23, — 13
	3	19, 31, — 29
	5	5, 17, 41
22	1	1, 9, 15, 23, 25, 31, — 7, — 17, — 39, — 41
	2	13, 19, 21, 29, 35, 43, — 3, — 5, — 27, — 37
23	1, 3	1, 3, 9, 13, 25, 27, 29, 31, 35, 39, 41, — 5, — 7, — 11 — 15, — 17, — 19, — 21, — 33, — 37, — 43, — 45
26	1, 3	1, 3, 9, 17, 25, 27, 35, 43, 49, 51, — 23, — 29
	2, 5	5, 7, 15, 21, 31, 37, 45, 47, — 11, — 19, — 33, — 41
29	1, 5	1, 5, 9, 13, 25, 33, 45, 49, 53, 57, — 7, — 23, — 35, — 51
	2, 3	3, 11, 15, 19, 27, 31, 39, 43, 47, 55, — 17, — 21, — 37, — 41
30	1	1, 31, 49, — 41
	2	17, 23, 47, — 7
	3	13, 37, 43, — 53
	5	11, 29, 59, — 19

Fig. 4.1. Lagrange's possible divisors of numbers of the form $t^2 + au^2$.

TABLE IV.

Formule des nombres proposés.................. $t^2 - au^2$.
Formule de leurs diviseurs impairs, et premiers à a.. $py^2 \pm 2qyz + rz^2 = 4an + b$.

VALEURS DE a	p	VALEURS CORRESPONDANTES DE b
1	1	± 1
2	± 1	± 1
3	1	1
	-1	-1
5	± 1	$\pm 1, \pm 9$
6	1	$1, -5$
	-1	$-1, 5$
7	1	$1, 9, -3$
	-1	$-1, -9, 3$
10	± 1	$\pm 1, \pm 9$
	± 2	$\pm 3, \pm 13$
11	1	$1, 5, 9, -7, -19$
	-1	$-1, -5, -9, 7, 19$
13	± 1	$\pm 1, \pm 3, \pm 9, \pm 17, \pm 23, \pm 25$
14	1	$1, 9, 11, 25, -5, -13$
	-1	$-1, -9, -11, -25, 5, 13$
15	1	$1, -11$
	-1	$-1, 11$
	3	$7, -17$
	-3	$-7, 17,$
17	± 1	$\pm 1, \pm 9, \pm 13, \pm 15, \pm 19, +21, \pm 25, \pm 33$
19	1	$1, 5, 9, 17, 25, -3, -15, -27, -31$
	-1	$-1, -5, -9, -17, -25, 3, 15, 27, 31$
21	1	$1, 25, 37, -5, -17, -41$
	-1	$-1, -25, -37, 5, 17, 41$
22	1	$1, 3, 9, 25, 27, -7, -13, -21, -29, -39$
	-1	$-1, -3, -9, -25, -27, 7, 13, 21, 29, 39$
23	1	$1, 9, 13, 25, 29, 41, -7, -11, -15, -19, -43$
	-1	$-1, -9, -13, -25, -29, -41, 7, 11, 15, 19, 43$
26	± 1	$\pm 1, \pm 9, \pm 17, \pm 23, \pm 25, \pm 49$
	± 2	$\pm 5, \pm 11, \pm 19, \pm 21, \pm 37, \pm 45$
29	± 1	$\pm 1, \pm 5, \pm 7, \pm 9, \pm 13, \pm 23, \pm 25, \pm 33, \pm 35,$ $\pm 45, \pm 49, \pm 51, \pm 53, \pm 57$
30	1	$1, 19, 49, -29$
	-1	$-1, -19, -49, 29$
	2	$17, -7, -13, -37$
	-2	$-17, 7, 13, 37$

Fig. 4.2. Lagrange's possible divisors of numbers of the form $t^2 - au^2$.

these tables with the details of all the examples and related exercises that we now discuss. We cannot emphasize strongly enough that at this stage these are just potential linear forms of potential quadratic forms, hopefully giving the nontrivial divisors of the original form. In particular, Lagrange does not know that these numbers are actual divisors of such a form, which is what the divisor problem requires one to show. So this work does not by itself solve the divisor problem. But Lagrange has a trick that sometimes enables him to go from potential divisors to definitive divisors!

Lagrange's insight is to switch the sign in $x^2 + ay^2$ (or in $x^2 - ay^2$) and first examine the potential nontrivial divisors of $x^2 - ay^2$ (or respectively of $x^2 + ay^2$). In our particular example of Fermat's form $x^2 + 2y^2$, his analysis of quadratic divisors and their potential linear forms already showed him that for $x^2 - 2y^2$, its only potential nontrivial quadratic divisors are the forms $x^2 - 2y^2$ and $-x^2 + 2y^2$ (since Lagrange is seeking positive divisors, these two forms, although negatives of each other, represent different possibilities for divisors of $x^2 - 2y^2$), and his algebraic restrictions on linear forms for these show that only ± 1 are potential odd values for the number b in the potential linear divisor forms $8n + b$, as mentioned above (Exercise 4.33). So in particular, no number of the linear form $8n + 3$ can be a nontrivial divisor of a number of the form $x^2 - 2y^2$, so 2 is not a quadratic residue modulo any prime $p = 8n + 3$. Now recall that we know, from Euler's determination above of the quadratic character of -1, that -1 is not a quadratic residue modulo p, since our p has the form $4q + 3$. And from one of the key properties of quadratic residues, also from Euler above, we know that a product of two quadratic nonresidues must be a quadratic residue. Thus $-2 = (-1)(2)$ must be a quadratic residue modulo $p = 8n + 3$, i.e., p must be a nontrivial divisor of some number of the form $x^2 + 2y^2$. Thus Lagrange has very cleverly turned information about the limitations on potential divisors of one form into definitive divisors of another! This is the solution to the divisor problem that we need, since in (18.II) above we already saw Lagrange provide the needed descent. So we conclude that p can itself be represented in the form $x^2 + 2y^2$, which proves Fermat's claim! Notice of course that Fermat also claimed the same for primes of the form $8n + 1$, but Lagrange's method will not work there; he is unable to determine whether a given prime $8n + 1$ is a nontrivial divisor of a number of the form $x^2 + 2y^2$, because he does not yet have complete quadratic reciprocity in hand.

We leave it to the reader to follow a similar route for Fermat's remaining assertion in his letter to Digby. Fermat claimed that every prime of the form $3n + 1$ is the sum of a square and the triple of another square, i.e., is represented by $x^2 + 3y^2$. Lagrange's theory easily provides the descent needed, and Lagrange's trick may be used to solve the divisor problem, at least when $3n + 1$ is also of the form $4q + 3$, which occurs precisely for primes of the form $12n + 7$ (Exercise 4.34).

Lagrange's tables of potential quadratic divisors and their potential linear forms were so successful because they happen to give precisely the actual linear divisors, as one can see by a detailed comparison with the claims in

Euler's papers above, assuming they are all true. However, except in those situations where Lagrange could use his trick above to convert information about nondivisors to actual, not just potential, divisors, Lagrange could not really know for sure that the algebraically potential divisors he had found really always occurred as actual divisors. This is of course what the divisor problem requires, determining that certain primes really are always divisors of certain quadratic forms. So although Lagrange's progress was tremendous, beginning the theory of quadratic forms and their reduction for descent arguments, it still left the divisor problem largely unresolved. Both are needed to solve the representability problem. Thus we are left to pursue the divisor problem further, which we shall do in the rest of the chapter.

Caveat. Lest the reader think that Lagrange's general theory at least makes descent trivial, and that the divisor problem is the only thing needing pursuing, let us first see the incredible light Lagrange's memoir shines on our example from the introduction of a difference in the answers to the divisor and representability problems. Recall that 3 is a solution to the divisor problem for the form $x^2 + 5y^2$, since 3 divides $21 = 1^2 + 5 \cdot 2^2$, i.e., -5 is a quadratic residue modulo 3. But clearly 3 is not actually represented by the form $x^2 + 5y^2$. This means that something must go wrong with the descent part of the two-step "divisor plus descent" technique. It must not be true that any nontrivial divisor of a number of the form $x^2 + 5y^2$ necessarily has the same form. While the first two descent analyses in our Lagrange excerpt above worked out simply, in that nontrivial divisors of each of $t^2 + u^2$ and $t^2 + 2u^2$ always had the same form, notice that for $t^2 + 5u^2$ in (18.V) above, things are not that simple. Lagrange deduces that nontrivial divisors will have the form $y^2 + 5z^2$ or $2y^2 \pm 2yz + 3z^2$. And in (17) Lagrange showed how multiplying the latter form by its first coefficient 2 would convert it to a number of the form $t^2 + 5u^2$. This is our descent information about quadratic forms. So what kind of representability result can we get in this case?

The solution to the divisor problem is that the nontrivial prime divisors of $x^2 + 5y^2$ are precisely those in the arithmetic progressions $20n + 1$, $20n + 3$, $20n + 7$, $20n + 9$ (Exercise 4.35). Moreover, Lagrange's analysis of potential linear forms of quadratic forms yields that primes nontrivially representing the form $y^2 + 5z^2$ could be only $20n + 1$, $20n + 9$, and those of the form $2y^2 \pm 2yz + 3z^2$ could be only $20n + 3$, $20n + 7$ (Exercise 4.36). So combining the divisor and descent information, we learn that primes $20n + 1$, $20n + 9$ are precisely those nontrivially represented in the form $x^2 + 5y^2$, and primes $20n + 3$, $20n + 7$ are precisely those that are of that form after multiplication by 2 (Exercise 4.37). Amazingly, we can count on Euler already to have observed all this explicitly in his paper of 1744, as we already read in his Theorems 10, 11, 12 in our excerpts! The forms $y^2 + 5z^2$ and $2y^2 \pm 2yz + 3z^2$ have the same nontrivial prime divisors, but represent different progressions of primes! In bringing this phenomenon to light in his analysis, Lagrange was noticing the beginnings of a beautiful and important phenomenon called genus theory, to which [41] provides a wonderful introduction.

Exercise 4.30. In his section (9), Lagrange shows that if $4BD - C^2$ is a positive number, then the coefficient Q in the given form has to satisfy the inequality

$$Q \leq \sqrt{\frac{4BD - C^2}{3}}.$$

Show that if $4BD - C^2$ is negative, then we obtain the inequality

$$Q \leq \sqrt{\frac{C^2 - 4BD}{5}}.$$

Exercise 4.31. In his section (12), Lagrange derives criteria for finding Q if K is positive. Find similar conditions if K is negative. (Hint: Use Exercise 4.30.) What happens if $K = 0$?

Exercise 4.32. Show that positive definite forms represent only positive numbers. Hint: One can use Lagrange's section (15).

Exercise 4.33. Show that the only possible odd primes represented by $x^2 - 2y^2$ or $-x^2 + 2y^2$ are those of the form $8n \pm 1$. Hint: Argue, as did Lagrange, that one can write $x = 4m \pm \rho$ where $0 \leq \rho \leq 2$, and $y = 2m' \pm \omega$, where $0 \leq \omega \leq 1$. Then expand and look at the odd possibilities modulo 8.

Exercise 4.34. Show that every prime of form $12n + 7$ is the sum of a square and the triple of another square. To imitate what Lagrange did you will need to use the fact that his tables show that the only nontrivial prime quadratic divisors of $x^2 - 3y^2$ are $x^2 - 3y^2$ and $-x^2 + 3y^2$.

Then determine for yourself that the only possible linear forms for these are $12n \pm 1$. Do this, as in fact did Lagrange, and analogously to the previous exercise, by confirming that one can write $x = 6m \pm \rho$ where $0 \leq \rho \leq 3$, and $y = 2m' \pm \omega$, where $0 \leq \omega \leq 1$, and expand and look at the odd possibilities modulo 12. Then proceed as did Lagrange, using the quadratic character of -1 to obtain your solution to the divisor problem.

Finally, carry out the descent needed for $x^2 + 3y^2$ by imitating Lagrange's examples above in (18), noting that the additional quadratic form that arises as a possible divisor is not relevant here, and explain why.

Exercise 4.35. Use Euler's claims in his *Conclusion* of 1772 about the quadratic character of $-s$ to find the linear forms of nontrivial prime divisors of numbers of the quadratic form $x^2 + 5y^2$.

Exercise 4.36. Expand on the method of Exercises 4.33 and 4.34 to show that the only possible primes nontrivially represented by $x^2 + 5y^2$ are of the form $20n + 1$, $20n + 9$. (Hint: Write $x = 10m \pm \rho$ with $0 \leq \rho \leq 5$, and $y = 2m' \pm \omega$ with $0 \leq \omega \leq 1$, expand, and look at the possibilities.) Then go one step further to determine, using the information from Lagrange's section

(17), that the only possible primes represented by $2y^2 \pm 2yz + 3z^2$ are of the form $20n + 3$, $20n + 7$.

Exercise 4.37. Combine the descent information in the text from Lagrange's sections (17) and (18.V) with the divisor and linear representability information from the previous two exercises to show that primes in the progressions $20n + 1$, $20n + 9$ are precisely those of the form $x^2 + 5y^2$, and primes $20n + 3$, $20n + 7$ are precisely those that are of that form after multiplication by 2.

4.4 Legendre Asserts the Quadratic Reciprocity Law

Photo 4.5. Legendre.

The next protagonist in our story is Adrien-Marie Legendre. Born to a well-to-do family, he received an excellent education in Paris. His family's modest fortune allowed him to devote himself entirely to research, although he did teach mathematics for a time at the Military Academy in Paris. In 1782 Legendre won a prize from the Berlin Academy for his essay on the subject "Determine the curve described by cannonballs and bombs, taking into consideration the resistance of the air; give rules for obtaining the ranges corresponding to different initial velocities and to different angles of projection." Then, as today, political and military considerations had great influence on the directions of mathematical research. This prize, along with work on celestial mechanics, gained Legendre election to the French Academy of Sciences in 1783, and his scientific output continued to grow. Legendre's favorite areas of research were celestial mechanics, number theory, and the theory of elliptic functions, of which he should be considered the founder. From 1787 Legendre was heavily involved in the academy's work with the geodetic research of the Paris and Greenwich observatories, such as the linking of their meridians, and this stimulated him to work on spherical geometry, where he is known for his theorem on spherical triangles.

During the ravages caused by the French Revolution, Legendre lost his small fortune, but nonetheless his career flourished. In 1791 he became one of the commissioners for the astronomical operations and triangulations necessary for determining the meter and establishing our worldwide metric system of measurements. Soon thereafter he became head of part of the National Executive Commission of Public Instruction. At this time Legendre also supervised part of an enormous project to prepare new tables of functions needed for the surveying work emerging from the adoption of the metric system. For instance, the project calculated sines of angles in ten-thousandths of a right angle, correct to twenty-two decimal places. It is hard for us today, in the computer age, to imagine the scope of such an enterprise. Legendre wrote "These ... tables, constructed by means of new techniques based principally on the calculus of differences, are one of the most beautiful monuments ever erected to science." Legendre arranged to prepare the tables by two completely different methods, one based on new formulas of his for successive differences of sines. The other method involved only additions, and could be done by people with few mathematical skills. Then the two independently obtained tables were compared to verify their correctness. In 1813 Legendre succeeded Lagrange at the *Bureau des Longitudes* (Office of Longitudes), a post he retained for the rest of his life [92].

Number theory plays a prominent role in Legendre's mathematical accomplishments, exemplified by his book *Théorie des Nombres* (Theory of Numbers) [153], published in 1798, which went through several editions. Conceived in part as a textbook, it also contained many of his own research accomplishments, including his work on the theory of quadratic forms and their divisors, based on the work of Fermat, Euler, and Lagrange. We will begin with excerpts from Volume I of Legendre's book leading to his proof of Euler's important theoretical criterion for quadratic residues. Then we will see Legendre's assertion of the quadratic reciprocity law as we know it today, and hear him explain that with it he can solve general divisor problems and representation problems.

§ VI. *Théorème contenant une loi de réciprocité qui existe entre deux nombres premiers quelconques.*

(166) Nous avons vu (n° 135) que si m et n sont deux nombres premiers quelconques impairs et inégaux, les expressions abrégées $\left(\frac{m}{n}\right)$, $\left(\frac{n}{m}\right)$ représentent l'une le reste de $m^{\frac{n-1}{2}}$ divisé par n, l'autre le reste de $n^{\frac{m-1}{2}}$ divisé par m; on a prouvé en même temps que l'un et l'autre restes ne peuvent jamais être que $+1$ ou -1. Cela posé, il existe une telle relation entre les deux restes $\left(\frac{m}{n}\right)$, $\left(\frac{n}{m}\right)$, que l'un étant connu, l'autre est immédiatement déterminé. Voici le théorème général qui contient cette relation.

« Quels que soient les nombres premiers m et n, s'ils ne sont pas « tous deux de la forme $4x+3$, on aura toujours $\left(\frac{n}{m}\right)=\left(\frac{m}{n}\right)$, et « s'ils sont tous deux de la forme $4x+3$, on aura $\left(\frac{n}{m}\right)=-\left(\frac{m}{n}\right)$.

« Ces deux cas généraux sont compris dans la formule

$$\left(\frac{n}{m}\right)=(-1)^{\frac{m-1}{2}\cdot\frac{n-1}{2}}\cdot\left(\frac{m}{n}\right). »$$

Photo 4.6. *Théorie des Nombres.*

∞∞∞∞∞∞∞∞∞∞

Adrien-Marie Legendre, from
Theory of Numbers

PART TWO .

...

§I. THEOREMS ON PRIME NUMBERS

(132) LEMMA. *Let c be a prime number, and P a polynomial of degree m whose coefficients are integers, namely $P = \alpha x^m + \beta x^{m-1} + \gamma x^{m-2} + \cdots + \omega$; I say that there cannot be more than m values of x, contained between $+\frac{1}{2}c$ and $-\frac{1}{2}c$, which make the polynomial divisible by c.*

Because supposing k is a first value of x which makes P divisible by c, one will be able to put[17] $P = (x - k) P' + Ac$, and one will have for P' a polynomial in x of degree $m - 1$. Supposing k' is a second value of x which makes P divisible by c, this value must make $(x - k) P'$ divisible by c. But the factor $x - k$, which becomes $k' - k$, cannot be divisible by c, because k and k' were each assumed smaller[18] than $\frac{1}{2}c$; thus P cannot be divisible a second time by c, unless P' is. The polynomial of degree m consequently admits only one more solution than the polynomial P' of degree $m - 1$; thus there cannot be more than m different values of x, contained between $+\frac{1}{2}c$ and $-\frac{1}{2}c$, which make P divisible by c. We regard as a *solution* or *root* of the equation $\frac{P}{c} = e$, any value of x, contained between $+\frac{1}{2}c$ and $-\frac{1}{2}c$, which makes the first member equal to an integer. The number of these solutions, which one could also take to be between 0 and c, can never exceed the exponent m, as has just been shown; but following a solution such as $x = k$, one can set more generally $x = k + cz$, z being a positive or negative integer, and all the values of x contained in this formula satisfy the equation $\frac{P}{c} = e$.

(133) THEOREM. *Let c always denote a prime number, and P a polynomial of degree m, which is a divisor of the binomial $x^{c-1} - 1$. I say that there will always be m values of x, between $+\frac{1}{2}c$ and $-\frac{1}{2}c$, which make this polynomial divisible by c.*

Because let $x^{c-1} - 1 = PQ$, with Q another polynomial of degree $c - 1 - m$. Since there are $c - 1$ values of x, namely $\pm 1, \pm 2, \pm 3, \ldots, \pm \frac{c-1}{2}$, which make the first member divisible[19] by c, each of these values must make P or Q divisible by c.[20] Amongst these $c - 1$ values, there cannot be more than m that make P divisible by c, because P has only degree m; neither can there less than m, since then there would be more than $c - 1 - m$ values of x making Q divisible by c; which is impossible, because Q is only of degree $c - 1 - m$. Thus the number of values of x which make P divisible by c, and contained between $+\frac{1}{2}c$ and $-\frac{1}{2}c$, is precisely m.

Remark. The same proposition will hold, if P is a divisor of $x^{c-1} - 1 + cR$, R being a polynomial of arbitrary degree. (Exercise 4.38)

(134) THEOREM. *If the prime number c is a divisor of $x^2 + N$, where N is a given positive or negative number, I say that the quantity $(-N)^{\frac{c-1}{2}} - 1$ has to be divisible by c. And conversely, if that condition is satisfied, then there exists a number x (less than $\frac{1}{2}c$) such that $x^2 + N$ is divisible by c. (The cases $c = 2$ and when N is divisible by c are excluded.)*

Because, 1. if c is a divisor of $x^2 + N$, then one has, omitting multiples of c, $x^2 = -N$, hence $x^{c-1} - 1 = (-N)^{\frac{c-1}{2}} - 1$. The first member is divisible[21] by c, hence the second one is also.

---

[17] Here P' is the quotient polynomial upon long division of P by $(x - k)$. Since the remainder is the value of P at k, it is a multiple of c.

[18] In absolute value.

[19] Here Legendre is using Fermat's little theorem. See the Appendix.

[20] Here Legendre is using a key property of primes that we call Euclid's lemma (see Theorem 4.2 in the Appendix).

[21] Here Legendre is again using Fermat's little theorem.

2. If one supposes that $(-N)^{\frac{c-1}{2}} - 1$ is divisible by c, which I set equal to cr, then this gives [for all x]

$$x^{c-1} - 1 - cr = x^{c-1} - (-N)^{\frac{c-1}{2}}.$$

But if one sets $c - 1 = 2b$ and $-N = M$ for a moment, then the second member becomes $x^{2b} - M^b$, which is divisible by $x^2 - M$ or $x^2 + N$. Therefore, $x^2 + N$ also divides the first member $x^{c-1} - 1 - cr$. Therefore (Nr. 133), there are necessarily two values of x, less than $\frac{1}{2}c$, which make $x^2 + N$ divisible by c. These two values can only produce one result, so they differ only by their sign.

Remark. We have shown that, if N is a given number, and c is a prime number that does not divide N, the quantity $N^{c-1} - 1$ is always divisible by c. This quantity is the product of the two factors $N^{\frac{c-1}{2}} + 1$ and $N^{\frac{c-1}{2}} - 1$. It follows therefore that one or the other of these factors is divisible by c. From this we conclude that the quantity $N^{\frac{c-1}{2}}$, when divided by c, always leaves the remainder $+1$ or the remainder -1. (Exercise 4.39)

(135) *Since quantities similar to $N^{\frac{c-1}{2}}$ appear frequently in the course of our investigations, we will employ the abbreviation $\left(\frac{N}{c}\right)$ to express the remainder of $N^{\frac{c-1}{2}}$ under division by c, a remainder which, from what we have seen, can only be $+1$ or -1.*

If $\left(\frac{N}{C}\right) = +1$, then one says that N is a *quadratic residue* of c, since if $N^{\frac{c-1}{2}}$ leaves remainder $+1$ under division by c, then this is the condition for c being a divisor of $x^2 - N$. On the other hand, if $\left(\frac{N}{C}\right) = -1$, then one says that N is a *quadratic nonresidue* of c.

∞∞∞∞∞∞

We have thus arrived at Legendre's notation for quadratic residues, called the Legendre symbol, and his proof of Euler's criterion, which we stated and used in the introduction (Exercise 4.40). Together they become

$$\left(\frac{N}{c}\right) \equiv N^{\frac{c-1}{2}} \pmod{c}.$$

∞∞∞∞∞∞

§VI. A THEOREM CONTAINING A LAW OF RECIPROCITY WHICH EXISTS AMONG ANY TWO PRIME NUMBERS.

(166) We have seen (Nr. 135) that if m and n are any two distinct odd prime numbers, the abbreviated expressions $\left(\frac{m}{n}\right)$, $\left(\frac{n}{m}\right)$ represent the remainder of $m^{\frac{n-1}{2}}$ under division by n, respectively $n^{\frac{m-1}{2}}$ under division by m. One has shown at the same time that these remainders can only be equal to $+1$ or -1. This established, there exists a relationship between the two remainders $\left(\frac{m}{n}\right)$, $\left(\frac{n}{m}\right)$ of the

kind that, if one is known, the other can be immediately determined. Here is the general theorem that contains this relationship.

Let m and n be prime numbers, not both of the form $4x+3$, then one always has $\left(\frac{n}{m}\right) = \left(\frac{m}{n}\right)$, and if they are both of the form $4x+3$, then one has

$$\left(\frac{n}{m}\right) = -\left(\frac{m}{n}\right).$$

Both of these general cases are encompassed in the formula

$$\left(\frac{n}{m}\right) = (-1)^{\frac{m-1}{2}\cdot\frac{n-1}{2}} \cdot \left(\frac{m}{n}\right).$$

∝∝∝∝∝∝∝∝

The reader should verify that this is equivalent to the statement of the quadratic reciprocity law in the introduction. Legendre proceeds with a lengthy proof of this result, considering eight different cases deriving from whether each of the primes m and n is congruent to 1 or 3 mod 4, and whether each Legendre symbol has the value $+1$ or -1. As observed in the introduction, Legendre's proof was not quite complete, assuming for instance a result about primes in arithmetic progressions proven only considerably later by Dirichlet. Having established the QRL to his own satisfaction, however, he promptly goes on to use it in his investigation of divisors of the forms $t^2 + cu^2$.

∝∝∝∝∝∝∝∝

§X. INVESTIGATION OF THE LINEAR FORMS THAT AGREE WITH DIVISORS OF THE FORMULA $t^2 + cu^2$.

We examine first the case where c is a prime number ...

(198) ... given c, one can determine *a priori* all the linear forms $4cz + b$ that are admissible as divisors of the formula $t^2 + cu^2$, and, on the other hand, one can also determine all the quadratic forms $py^2 + 2qyz + rz^2$, that have the same divisors. It follows that every prime number contained in one of the linear forms $4cz + b$ is also contained in one of the quadratic forms $py^2 + 2qyz + rz^2$. A very fruitful proposition, and from the development for the different values of c, it furnishes a multitude of interesting theorems about prime numbers.

∝∝∝∝∝∝∝∝

Here Legendre is extolling that with the quadratic reciprocity law he can determine with certainty all the linear forms of divisors of quadratic forms $t^2 + cu^2$, i.e., the law enables him to solve any divisor problem, which Lagrange could not do in general. Legendre ends with tables for divisors, and proofs of representation theorems about the forms of prime numbers, similar to what we saw in restricted situations with Lagrange. He concludes [153, p. 307]:

∞∞∞∞∞∞∞∞

(232) ... Lagrange is the first who has opened the road to investigations of
these kinds of theorems. (See the Memoirs of Berlin, 1775.) But the methods
this great geometer has used are only applicable to a very few cases of the prime
numbers $4n + 1$; and this difficulty can only be completely resolved with the aid
of the law of reciprocity that I have given for the first time in the Memoirs of the
Academy of Sciences of Paris, in 1785.

∞∞∞∞∞∞∞∞

Let us witness how easily Legendre obtains some of the results that Lagrange either achieved only with his trick for primes of the form $4q + 3$, or
could not achieve in that way for the form $4q + 1$. For instance, consider again
Fermat's claim that every prime of the form $3n + 1$ is the sum of a square and
the triple of another square, i.e., is represented by $x^2 + 3y^2$.

Lagrange's theory of quadratic forms provides an excellent descent, showing that any nontrivial prime divisor of $x^2 + 3y^2$ must have the same form
(Exercise 4.34). It remains only to solve the divisor problem, i.e., to find
arithmetic progressions that contain precisely all the primes that are nontrivial divisors of $x^2 + 3y^2$. So we wish to know the odd primes p for which -3 is a
quadratic residue, i.e., $\left(\frac{-3}{p}\right) = 1$. Using the quadratic reciprocity law, the multiplicativity of the Legendre symbol (from the introduction), and the quadratic
character of -1 (from the introduction, proved above by Euler), we calculate:

$$\left(\frac{-3}{p}\right) = \left(\frac{-1}{p}\right)\left(\frac{3}{p}\right) = \begin{cases} 1 \cdot \left(\frac{p}{3}\right) & \text{if } p = 4n + 1 \\ (-1) \cdot -\left(\frac{p}{3}\right) & \text{if } p = 4n + 3 \end{cases}$$
$$= \left(\frac{p}{3}\right).$$

Thus we conclude that $\left(\frac{-3}{p}\right) = \left(\frac{p}{3}\right) = 1$ if and only if $p \equiv 1 \bmod 3$. So the divisor problem is solved: The nontrivial divisors of numbers of the form $x^2 + 3y^2$
are precisely those primes of the form $3n + 1$. Combined with descent, this
proves Fermat's claim completely, whereas Lagrange, as we saw above, could
manage it only for primes of the form $12n + 7$.

As another illustration of what Legendre could easily do with his version
of quadratic reciprocity, recall that when discussing Lagrange we used the
fact that the nontrivial prime divisors of $x^2 + 5y^2$ are precisely those in the
arithmetic progressions $20n + 1$, $20n + 3$, $20n + 7$, $20n + 9$. In Exercise 4.41
the reader may verify this à la Legendre, find divisors of some other quadratic
forms as well, and begin to see a connection to Euler's version of reciprocity.

By tracing the odyssey of the theory of quadratic forms and the QRL
over the last two millennia, one cannot help but be impressed by the pre-
cariousness of its journey and by the small number of mathematicians who
took the trouble to nurse it along. A testimony to the genius of these pioneers
is their intuition in singling out aspects of the yet almost nonexistent theory

whose pursuit would later prove to be of the utmost importance. We have now reached the time, at the turn of the nineteenth century, when number theory is to metamorphose from ugly duckling to queen of the mathematical sciences, at the hands of Carl Friedrich Gauss, one of the great scientific geniuses of all time. The next section will contain a proof of the quadratic reciprocity law, and display how Euler's, Legendre's, and Gauss's versions are really all about one and the same phenomenon.

Exercise 4.38. Rewrite the statements and proofs of Legendre's results in (132), (133) in the language of congruences.

Exercise 4.39. Rewrite the statements and proofs of Legendre's (134) in the language of congruences.

Exercise 4.40. Explain in the language of congruences how Legendre's (134) yields Euler's criterion (stated with the QRL in the introduction).

Exercise 4.41. Use Legendre's statement of the QRL to prove that the non-trivial prime divisors of $x^2 + 5y^2$ are precisely those in the arithmetic progressions $20n + 1$, $20n + 3$, $20n + 7$, $20n + 9$. Then find the arithmetic progressions for nontrivial prime divisors of $x^2 + 7y^2$ and $x^2 + 13y^2$. Euler discovered that the periodicity $4N$ was important for the progressions containing divisors of $x^2 + Ny^2$. Explain how this can be seen in your results, and why it is occurring.

4.5 Gauss Proves the "Fundamental Theorem"

Outwardly, Carl Friedrich Gauss's life was rather simple. Born in 1777 in Braunschweig, Germany, to working-class parents, he distinguished himself early on as extremely gifted. Gauss is said to have taught himself to calculate before he knew how to talk, and to have corrected an error in his father's wage calculations at age three. He received a ducal stipend for study from age fourteen, and in 1799 he ended his career as a student with a doctorate in mathematics, having studied in Braunschweig, Göttingen, and the University of Helmstedt. Before the age of twenty-five he was famous for his work in mathematics as well as astronomy, and mathematical ideas spewed forth almost faster than he could write them down, as his diary demonstrates. The first monument to this incredible burst of creativity was the *Disquisitiones Arithmeticae* [80], from which we are about to read some excerpts, a work so impressive that it gained him recognition at the time as the prince of mathematicians. After its publication his attention shifted elsewhere, mostly to astronomy. Looking for a secure position without too many onerous duties (such as teaching basic mathematics) he decided on a career in astronomy, and in 1807 became director of the Göttingen observatory, where he remained for the rest of his life, with only few travels. Periodically his interests would return to number theory for short periods, but his great contribution to that field remained the *Disquisitiones*. His other mathematical contributions cover a wide range of areas, such as algebra, analysis, geometry, and probability and statistics (beginning with his invention of the method of least squares). Gauss was equally distinguished as a natural scientist, making fundamental contributions to astronomy, geodesy, geomagnetism, mechanics, and physics.

Photo 4.7. Gauss.

While Gauss carried on an extensive scientific exchange with his peers in the natural sciences, he seemed to have largely isolated himself from the mathematical community. In part the lack of good mathematicians in Germany at the time may have been responsible for this, in part it may have been Gauss's personality. Consequently, his mathematical influence extended only through his publications. Gauss had no students in mathematics and provided little guidance to the emerging mathematical community in Germany that was destined to make that country a mathematical world center later in the nineteenth century. For more details on Gauss's life and work, the reader is encouraged to consult [26, 92].

Returning to the QRL, Gauss called it the "fundamental theorem" about primes and quadratic residues. In the *Disquisitiones* he gave two different proofs, adding four more in later years. The third published proof appeared in 1808 [82]. It relies on what now is commonly called Gauss's lemma, and is very closely related to the proof of Eisenstein's we will present shortly. In 1805, after having left number theory for researches in astronomy, Gauss was lured back to it by several letters from a certain Monsieur LeBlanc in Paris, who had been inspired by the *Disquisitiones*, and who later turned out to be a pseudonym for Sophie Germain, a French number theorist who made significant contributions to the proof of Fermat's last theorem [24, 150, 151]. In correspondence with his astronomer colleague Wilhelm Olbers, Gauss gives this account:

> Through various circumstances — partly through several letters from LeBlanc in Paris, who has studied my *Disq. Arith.* with a true passion, has completely mastered them, and has sent me occasional very respectable communications about them ... I have been tempted into resuming my beloved arithmetic investigations [201, p. 268].

DISQVISITIONES

ARITHMETICAE

AVCTORE

D. CAROLO FRIDERICO GAVSS

LIPSIAE

IN COMMISSIS APVD GERH. FLEISCHER, Jun.

1801.

Photo 4.8. Title page of *Disquisitiones Arithmeticae*.

Gauss enthusiastically continued his correspondence with Germain after learning her true identity. In a letter of April 30, 1807 [83, vol. 10, pp. 70–74], he included to her without proof a result (Gauss's lemma) from which he says one can derive special cases of the quadratic reciprocity law. In a May 12, 1807, letter to Olbers, Gauss says:

> Recently I replied to a letter of hers and shared some Arithmetic with her, and this led me to undertake an inquiry again; only two days later I made a very pleasant discovery. It is a new, very neat, and short proof of the fundamental theorem of art. 131. [83, vol. 10, p. 566].

The proof Gauss is referring to, based on the above lemma in his letter to Germain, is the above-mentioned third published proof of the quadratic reciprocity law, where he says he has finally found "the simplest and most natural way to its proof" (see also [148, 149]). We shall return to this proof shortly, when we prepare for reading Eisenstein's geometric proof of the law.

Here, first, is Gauss's original statement of the QRL, Art. 131, translated from [81]. He formulates the general result after presenting numerical examples, indicative of his general method of discovering new results.

∞∞∞∞∞∞∞∞∞

Carl Friedrich Gauss, from
Disquisitiones Arithmeticae

131.

We will soon prove that what we have found here through induction is true in general. But before we undertake this labor, it will be necessary to list everything that follows from that theorem, under the assumption that it is correct. We state the Theorem itself as follows:

If p is a prime of the form $4n+1$ respectively $4n+3$, then $+p$, respectively $-p$, is a quadratic residue or nonresidue of any prime number which, taken positively, is a quadratic residue or nonresidue of p.

Since almost everything one can say about quadratic residues follows from this theorem, the name "Fundamental Theorem," which we shall use subsequently, will not be inappropriate.

∞∞∞∞∞∞∞∞

We mentioned earlier that Gauss seems to have believed that it was Legendre who first discovered this result, but points out problems with Legendre's proof. Even though he was well aware of Euler's number-theoretic work, he seems to have judged Euler as falling short of the actual discovery of the QRL. Here is what Gauss has to say about his view of his predecessors' contributions.

∞⋉∞⋉∞⋉∞⋉∞⋉∞

On the Works of Others Concerning These Investigations.
151.

The Fundamental Theorem, which surely has to be counted among the most elegant discoveries in this subject, has not yet been stated by anybody in this simple form. This is all the more curious, since certain other theorems that follow from it and which would easily lead back to it were already known to Euler. He had known that there are certain forms which contain all prime divisors of numbers of the form $x^2 - A$, and that there are others that contain all prime nondivisors of numbers of the same form, in a mutually exclusive way. And he had discovered a method to find those forms. But all his efforts to arrive at a proof have always been in vain, and have only served to make this truth, found by induction, more probable ...

After Euler, Legendre concerned himself with the same topic in his excellent treatise *Researches in Indeterminate Analysis, Hist. de l'Ac. des Sc. 1785, p. 465 ff.* There (page 465) he arrives at the theorem, which, as concerns the topic itself, is identical to the Fundamental Theorem, namely that, if p and q are two positive prime numbers, then the smallest positive remainders of the powers $p^{\frac{q-1}{2}}$ and $q^{\frac{p-1}{2}}$ modulo q, respectively p, are either both $+1$ or -1, if either p or q is of the form $4n + 1$, but that, if both p and q are of the form $4n + 3$, then the one smallest remainder is $+1$ and the other is -1. From this and Art. 106 it follows that the relationship (with the meaning adopted in Art. 146) of p to q and of q to p is the same if either p or q is of the form $4n + 1$, but opposite if both p and q are of the form $4n + 3$. This theorem is contained among the results of Article 131; it also follows from Theorems 1, 3, 9 of Article 133. Conversely, the Fundamental Theorem can be derived from it. Legendre has also attempted a proof, about which we will say more in the following section, since it is very imaginative. But since in it he has assumed much without proof (as he himself admits on p. 520: *We have only assumed etc.*), which in parts has not been proven by anybody up to now, and which, in parts, at least in our opinion, cannot be proven without the Fundamental Theorem, it seems that the path taken by him cannot lead to success. Therefore our proof must be considered to be the first. — Incidentally, we will provide farther below two more proofs of this most important theorem, which are completely different from the former and from each other.

∞⋉∞⋉∞⋉∞⋉∞⋉∞

We have at this point four separate statements that we claim are the quadratic reciprocity law or at least antecedents to it. It is time to clarify this state of affairs, and then finally see a proof. Euler's original claims in his 1744 paper were reconciled with those of his 1772 paper in Exercises 4.27 and 4.28. Gauss's claim just above, that Legendre's law is equivalent to his own "fundamental theorem" as stated in §131, is easy to prove from what we already know (Exercise 4.42). To complete reconciling the four versions, let us then show that Gauss's statement of the fundamental theorem is equivalent to Euler's statement in the *Conclusion* of his second paper [140].

Henceforth in this section let s and p be odd primes. First recall from our discussion following Euler's *Conclusion* that his claims about $+s$ and $-s$ are equivalent, so we can restrict our attention to $+s$. Moreover, his first column lists four mutually exclusive linear forms that an odd prime p (the "divisor") may have. Then clearly his *Conclusion* can be phrased thus:

Euler's *Conclusion.* s is a quadratic residue modulo p if and only if either p or $-p$ is an [odd] square modulo $4s$.

Notice that although Euler states this using odd squares only, the word "odd" is redundant, since p being odd, the only squares it could be congruent to modulo $4s$ are the odd ones. It will be convenient to leave out the the word "odd" in what follows.

On the other hand, turning to Gauss, since

$$(-1)^{\frac{p-1}{2}} = \begin{cases} +1 \text{ if } p = 4n + 1, \\ -1 \text{ if } p = 4n + 3, \end{cases}$$

we see that Gauss's statement can be phrased as follows:

Gauss's *Fundamental Theorem.* s is a quadratic residue modulo p if and only if $(-1)^{\frac{p-1}{2}} p$ is a square modulo s.

So to prove that these two claims are equivalent, we need only verify the following, for s and p odd primes:

Euler and Gauss Reconciled. p or $-p$ is a square modulo $4s$ if and only if $(-1)^{\frac{p-1}{2}} p$ is a square modulo s.

Proof:
p or $-p$ is a square modulo $4s$ if and only if   (Exercise 4.43)
$(-1)^{\frac{p-1}{2}} p$ is a square modulo $4s$ if and only if (Exercise 4.44)
$(-1)^{\frac{p-1}{2}} p$ is a square modulo s.

Thus at last we have seen how Euler's claims about prime divisors of numbers of a certain quadratic form were precursors to the reciprocity statements of Legendre and Gauss, all describing a single amazing phenomenon. We are finally ready for a proof of the quadratic reciprocity law.

Exercise 4.42. Show that Legendre's formulation of his law of reciprocity is equivalent to the fundamental theorem stated by Gauss in his §131. Hint: Use the multiplicativity of the Legendre symbol, and the quadratic character of negative one, which came from Euler's second paper.

Exercise 4.43. Prove that for odd primes s and p, p or $-p$ is a square modulo $4s \Leftrightarrow (-1)^{\frac{p-1}{2}} p$ is a square modulo $4s$. Hint: Modulo $4s$, an odd square is always congruent to $1 \bmod 4$.

Exercise 4.44. Prove that for odd primes s and p, $(-1)^{\frac{p-1}{2}} p$ is a square modulo $4s \Leftrightarrow (-1)^{\frac{p-1}{2}} p$ is a square modulo s. Hint: Modulo s, any square can be arranged to be an odd square. And an odd square is always congruent to $1 \bmod 4$, as is $(-1)^{\frac{p-1}{2}} p$. You may also need the fundamental theorem of arithmetic.

4.6 Eisenstein's Geometric Proof

We are all familiar with the cliché of the starving artist producing great masterworks in some cold attic, only to die at a young age, under tragic circumstances, and sometimes virtually unknown. Names like van Gogh remind us that this image has its roots in bitter reality. Mathematics too has its share of such tragedies. The untimely deaths of Niels Henrik Abel (1802–1829), Evariste Galois (1811–1832), and Gotthold Eisenstein (1823–1852) deprived the world of many mathematical masterworks these pioneers would probably have produced during a longer life. Even the few years allowed them left a lasting mark on mathematics.

Already early in his childhood Eisenstein showed exceptional talent for mathematics. By the time he enrolled as a student at the University of Berlin in 1843, his mathematical knowledge was far advanced, including a thorough study of Gauss's *Disquisitiones*. His soaring creativity led to a flood of works and published articles. During 1844, he contributed no fewer than twenty-five articles for Volumes 27 and 28 of Crelle's Journal, one of the foremost among the few scientific journals at the time, including the proof that we are about to study. Even the usually rather aloof Gauss took note when Eisenstein paid him a visit in the summer of 1844, recognizing that the young man was one of the few mathematicians in Europe who had anything to tell Gauss that he didn't already know. Fame came quickly to Eisenstein, but it was not accompanied by much financial support. His uncertain economic situation, combined with his lifelong poor health, took a serious toll on him, compounded by the

Photo 4.9. Eisenstein.

effects of political troubles he had gotten himself into. By 1848 Eisenstein was in extremely poor health, suffering from tuberculosis, mostly unable to perform his teaching duties at the university. Despite the support of Gauss and the famous scientist Alexander von Humboldt, who in 1847 recommended him for a professorship at the University of Heidelberg, Eisenstein was unable to obtain significant financial support for his studies. When he died of tuberculosis in 1852, depressed, alienated from his family, and without any close friends, he had created in a few short years a mathematical legacy that is still an important part of several subjects, especially number theory and the theory of elliptic functions.

We now turn to Eisenstein's proof of the QRL, published in [59] (Figure 4.3). Its essence is similar to the ideas in Gauss's "third" proof, discussed earlier. But Eisenstein introduces some rather ingenious improvements resulting in great elegance and economy. (See [148, 149] for a detailed comparison of the two proofs (Exercise 4.45).) Some annotations follow our presentation of Eisenstein's proof.

Geometrischer Beweis des Fundamentaltheorems für die quadratischen Reste.

Es sei p eine positive ungerade Primzahl, a der Complexus aller *geraden* Zahlen $<p$ und >0, also $a = 2, 4, \ldots 6, p-1$; q sei irgend eine durch den Modul p nicht theilbare ganze Zahl. Bezeichnet man durch r das allgemeine Glied der Reste der Vielfachen qa nach dem mod. p, so werden offenbar die Zahlen der Reihe, deren allgemeines Glied $(-1)^r \cdot r$ ist, mit den Zahlen der Reihe a bis auf Vielfache von p übereinstimmen; also wird man die beiden Congruenzen haben:

$$q^{\frac{1}{2}(p-1)} \Pi a \equiv \Pi r \pmod{p}, \quad \text{und} \quad \Pi a \equiv (-1)^{\Sigma r} \Pi r \pmod{p},$$

woraus folgt

$$q^{\frac{1}{2}(p-1)} \equiv (-1)^{\Sigma r} \pmod{p}, \quad \text{also} \quad \left(\frac{q}{p}\right) = (-1)^{\Sigma r}.$$

Bedeutet $E\left(\frac{qa}{p}\right)$ die gröfste in dem Bruche $\frac{qa}{p}$ steckende ganze Zahl, so ist offenbar $\Sigma qa = p\Sigma E\left(\frac{qa}{p}\right) + \Sigma r$; und da alle a *gerade* sind, und $p \equiv 1 \pmod 2$ ist, so folgt hieraus $\Sigma r \equiv \Sigma E\left(\frac{qa}{p}\right) \pmod 2$; also ist auch

$$\left(\frac{q}{p}\right) = (-1)^{\Sigma E\left(\frac{qa}{p}\right)}.$$

Wenn $q = 2$ ist, so giebt diese Formel sogleich den Werth von $\left(\frac{2}{p}\right)$: ist dagegen q ungerade, also $q-1$ gerade, so findet man durch eine leichte Transformation

$$\Sigma E\left(\frac{qa}{p}\right) \equiv -E\left(\frac{a}{p}\right) + E\left(\frac{2a}{p}\right) - E\left(\frac{3a}{p}\right) + \ldots \pm E\left(\frac{\frac{1}{2}(q-1)a}{p}\right)$$

$$\equiv E\left(\frac{a}{p}\right) + E\left(\frac{2a}{p}\right) + E\left(\frac{3a}{p}\right) + \ldots + E\left(\frac{\frac{1}{2}(q-1)a}{p}\right) \pmod 2.$$

Wird letztere Summe durch μ bezeichnet, so hat man auch $\left(\frac{q}{p}\right) = (-1)^{\mu}$.

Man stelle sich jetzt in der Ebene ein rechtwinkliges Coordinatensystem (x, y) und die ganze Ebene durch Parallelen mit den Axen in den Abständen $=1$ von einander in lauter Quadrate von den Dimensionen $=1$ getheilt vor.

Fig. 4.3. Eisenstein's *Geometrischer Beweis*.

∞∞∞∞∞∞∞∞∞

Gotthold Eisenstein, from
Geometric
Proof of the Fundamental Theorem for Quadratic Residues

Let p be a positive odd prime number, let a denote the elements of the set of all *even* numbers $< p$ and > 0, that is, $a = 2, 4, 6, \ldots, p-1$. Let q be any integer not divisible by the modulus p. If r denotes the general term of the remainders of the multiples qa modulo p, then obviously the numbers in the sequence of terms $(-1)^r \cdot r$ will agree with the numbers in the sequence a, up to multiples of p. Thus, one will have the following two congruences:

$$q^{\frac{1}{2}(p-1)} \prod a \equiv \prod r \pmod{p}, \quad \text{and} \quad \prod a \equiv (-1)^{\sum r} \prod r \pmod{p},$$

from which it follows that

$$q^{\frac{1}{2}(p-1)} \equiv (-1)^{\sum r} \pmod{p}, \quad \text{thus} \quad \left(\frac{q}{p}\right) = (-1)^{\sum r}.$$

If $E\left(\frac{qa}{p}\right)$ denotes the largest integer less than or equal to the fraction $\frac{qa}{p}$, then clearly $\sum qa = p \sum E\left(\frac{qa}{p}\right) + \sum r$. And since all the numbers a are *even*, and $p \equiv 1 \pmod{2}$, it follows from this that $\sum r \equiv \sum E\left(\frac{qa}{p}\right) \pmod{2}$. Thus we also have the equality

$$\left(\frac{q}{p}\right) = (-1)^{\sum E\left(\frac{qa}{p}\right)}.$$

When $q = 2$, this formula readily gives the value of $\left(\frac{2}{p}\right)$ (Exercise 4.46). If, on the other hand, q is odd, that is, $q - 1$ is even, then one finds by way of an easy transformation that

$$\sum E\left(\frac{qa}{p}\right) \equiv -E\left(\frac{q}{p}\right) + E\left(\frac{2q}{p}\right) - E\left(\frac{3q}{p}\right) + \cdots \pm E\left(\frac{\frac{1}{2}(p-1)q}{p}\right)$$

$$\equiv E\left(\frac{q}{p}\right) + E\left(\frac{2q}{p}\right) + E\left(\frac{3q}{p}\right) + \cdots + E\left(\frac{\frac{1}{2}(p-1)q}{p}\right)$$

$$\pmod{2}.$$

If we denote the latter sum by μ, then one also has $\left(\frac{q}{p}\right) = (-1)^{\mu}$.

Now imagine a right-angled coordinate system (x, y) in the plane, and the whole plane divided into squares of dimension $= 1$ by parallels to the axes at distances $= 1$. We call *lattice points* all those corners of squares which do not lie on the coordinate axes (Figures 4.4, 4.5).

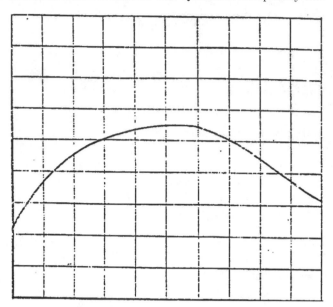

Fig. 4.4. Lattice points and a curve.

If one now chooses a point on any of the vertical parallels corresponding to the ordinate y, then $E(y)$ expresses the number of lattice points that lie between this point and the horizontal axis. And if one now chooses a point on any of the horizontal parallels corresponding to abscissa x, then $E(x)$ will express the number of lattice points that lie between this point and the vertical axis. If one draws any curve in the plane, whose equation is $y = \phi(x)$ (Figure 4.4), the sum

$$E\phi(1) + E\phi(2) + E\phi(3) + E\phi(4) + etc.$$

will therefore give the number of lattice points that lie between this curve and the x-axis, counting those lattice points which might by chance lie on the curve itself.

To return to our subject, let AB (Figure 4.5) denote the straight line whose equation is $y = \frac{q}{p}x$, where p and q are now both assumed to be positive odd *prime numbers*. Let $AD = FB = p$, $AF = DB = q$, $AC = EG = \frac{1}{2}(p-1)$, $AE = CG = \frac{1}{2}(q-1)$. Let μ denote the number of lattice points between AB and AD, up to and including the ordinate CG (which are marked with $*$ in the figure). Then it follows from the above that $\left(\frac{q}{p}\right) = (-1)^\mu$. Since the equation of our straight line can also be expressed as $x = \frac{p}{q}y$, one obtains in the same manner that $\left(\frac{p}{q}\right) = (-1)^\nu$, where ν denotes the number of lattice points between AB and AF, up to and including the abscissa EG, indicated by small zeros (∘) in the figure. But all the lattice points indicated by $*$ together with all lattice points indicated by ∘, that is, all lattice points to the *right* and all lattice points to the *left* of AB, exhaust *all* lattice points of the rectangle $AEGC$, whose number is

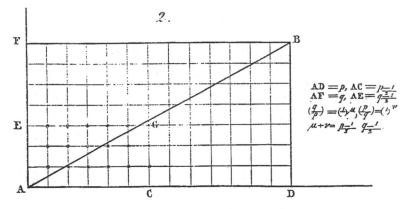

Fig. 4.5. Eisenstein's rectangle.

$\frac{1}{2}(p-1)\frac{1}{2}(q-1)$. Therefore it follows that $\mu + \nu = \frac{1}{2}(p-1)\frac{1}{2}(q-1)$, and

$$\left(\frac{q}{p}\right)\left(\frac{p}{q}\right) = (-1)^{\mu+\nu} = (-1)^{\frac{1}{2}(p-1)\frac{1}{2}(q-1)},$$

which was to be proven.

Incidentally, the above transformation $\sum E\left(\frac{qa}{p}\right) \equiv \mu \pmod{2}$ can also be proven by a very easy geometric observation. Namely, note that $\sum E\left(\frac{qa}{p}\right)$ is nothing other than the number of lattice points which lie on the *even* ordinates (corresponding to the abscissas $x = 2, 4, 6, \ldots, p-1$) between AB and AD up to BD. Furthermore, each ordinate, from the axis AD up to FB exclusive, contains $q-1$ lattice points, an even number. Finally, the two triangles BAD and ABF are congruent, and the former is placed in relation to BF and BD exactly the same way as the latter is placed in relation to AD and AF. We leave the details to the reader.

The first order of business for Eisenstein is to establish a representation

$$\left(\frac{q}{p}\right) = (-1)^{\sum r},$$

which reduces the problem to determining the parity of the exponent. This is done by first multiplying all even numbers a less than p by q and taking their remainders under division by p. His claim then is that the sequence of numbers $(-1)^r r$ gives the same list of remainders as the original list of a's. This can be seen by observing first that each of the numbers $(-1)^r r$ has an even least positive remainder under division by p. Furthermore, if there were any duplication among the remainders, e.g.,

$$(-1)^{qa}qa \equiv (-1)^{qa'}qa' \pmod{p},$$

then $a = \pm a'$ (mod p). Since the remainders are distinct, it follows that the only possibility is that of opposing signs, whence $a + a' \equiv 0$ (mod p), which cannot occur because $0 < a + a' < 2p$ and $a + a'$ is even.

From this result he now derives several congruences that imply the relevant expression for $\left(\frac{q}{p}\right)$. For this he uses Euler's criterion (from our introduction, with proof in our Legendre source) that

$$\left(\frac{q}{p}\right) \equiv q^{\frac{p-1}{2}} \text{ (mod } p).$$

To determine the parity of the exponent $\sum r$, he transforms it into

$$\sum r \equiv \sum E\left(\frac{qa}{p}\right) \text{ (mod 2),}$$

since all we care about is its parity. One of the most beautiful aspects of the proof is the representation now of the right-hand sum as the number of lattice points in a certain geometric figure, then using the figure to further transform the exponent. Redrawing Eisenstein's Figure 4.5 as Figure 4.6, we see that the exponent $\sum E\left(\frac{qa}{p}\right)$ is indeed just the number of lattice points with even abscissas lying in the interior of triangle ABD (note that no lattice points lie on the line AB).

Now consider an even abscissa $a > p/2$. Because the number of lattice points on each abscissa in the interior of rectangle $ADBF$ is $q - 1$, which is even, the number $E\left(\frac{qa}{p}\right)$ of lattice points on the abscissa below AB has the same parity as the number of lattice points above AB. This, in turn, is the same as the number of points lying below AB on the odd abscissa $p - a$. The one-to-one correspondence between even abscissas in triangle BHJ and odd abscissas in AHK now implies that

$$\sum E\left(\frac{qa}{p}\right) \equiv \mu \text{ (mod 2),}$$

where μ is the number of points inside triangle AHK, and thus

$$\left(\frac{q}{p}\right) = (-1)^{\mu}.$$

This is the essence of Eisenstein's geometric transformation, and the rest of his argument is straightforward (Exercise 4.47).

Most readers will now either be amazed at the beauty and simplicity of this proof, or they will say that if the proof is that simple, then the result cannot be all that deep. A discussion of this issue might begin with the section "Is the Quadratic Reciprocity theorem a Deep theorem?" in [207].

Eisenstein concludes his article with a brief and lamentably cryptic remark, pointing to a great generalization of his geometric method.

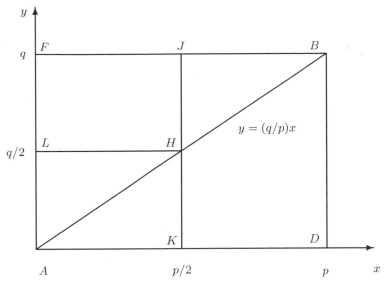

Fig. 4.6. Eisenstein's counting argument.

∞∞∞∞∞∞∞∞∞

Remark. There are figures for which one can determine the number of lattice points contained in them through simple formulas. Consider for instance a circle, whose center is at the origin and whose radius is $\sqrt{m}$. Then the number S of lattice points enclosed by this circle, counting those on the axes, is given by the following formula:

$$S = 1 + 4 \left(E(m) - E\left(\frac{1}{3}m\right) + E\left(\frac{1}{5}m\right) - E\left(\frac{1}{7}m\right) + \cdots \right),$$

where the sum is extended until it terminates. As is easy to see, this equation expresses a relationship between the number of lattice points of a circle and the number of lattice points in a segment enclosed by two hyperbolas. If one sets $m = \infty$ in the formula

$$\frac{1}{m}S = \frac{1}{m} + 4 \left(\frac{1}{m}E(m) - \frac{1}{m}E\left(\frac{1}{3}m\right) + \frac{1}{m}E\left(\frac{1}{5}m\right) - \text{etc.} \right),$$

then the left side is transformed into π, while the right side is transformed into $4(1 - \frac{1}{3} + \frac{1}{5} - \text{etc.})$. Thus, one obtains *Leibnitz's* formula for π. There are similar formulas for the number of lattice point of segments enclosed by a system of ellipses or hyperbolas, and similar relations occur in space and in situations with more than three dimensions. We will return to this important subject, which is closely related to properties of higher forms, on another occasion. Berlin, July 1844.

∞∞∞∞∞∞∞∞

While Eisenstein never returned to the topic of his *Remark*, it was taken up by Arthur Cayley (1821–1895). In the article *Eisenstein's Geometrical Proof of the Fundamental Theorem for Quadratic Residues* [32], he gives an English translation of Eisenstein's article above, and then takes up the topic of the *Remark*.

∞∞∞∞∞∞∞

Arthur Cayley, from
Eisenstein's Geometrical
Proof of the Fundamental Theorem for Quadratic Residues

(*Addition by the Translator.*) Eisenstein is now, alas! dead; too soon for the complete development of his various and profound researches in elliptic functions and the theory of numbers; and the promise at the conclusion of the foregoing memoir has not, I believe, been fulfilled. The formula in the Remark must, I think, have been established by geometrical considerations, and would have served to give the number of decompositions of a number into the sum of two squares; but, as I do not perceive how this is to be done, I shall follow a reverse course, and establish the theorem from considerations founded on the theory of numbers.

∞∞∞∞∞∞∞

Cayley goes on to give a proof for the formula of Eisenstein for the number of lattice points inside a circle. The reader is invited to study his argument.

While Eisenstein's geometric approach to the QRL seems rather ad hoc (and probably was), it adumbrated an extremely fruitful approach to certain aspects of number theory and was developed into a powerful tool by later mathematicians, principally Hermann Minkowski (1864–1909) (see his book *Geometry of Numbers* [171]), in connection with his research in the theory of quadratic forms. A discussion of this theory is beyond the scope of this book; see [200, Chapter 9] for details.

Exercise 4.45. Compare Eisenstein's proof with Gauss's third proof, an English translation of which can be found in [220].

Exercise 4.46. Prove the second part of the supplementary theorem stated in the introduction, on the quadratic character of 2, i.e., calculate $\left(\frac{2}{p}\right)$. Use Eisenstein's formula $\left(\frac{q}{p}\right) = (-1)^{\sum E\left(\frac{qa}{p}\right)}$.

Exercise 4.47. Prepare a write-up of Eisenstein's proof in your own words, including all statements that he uses without proof.

4.7 Gauss Composes Quadratic Forms: The Class Group

As mentioned earlier, the *Disquisitiones Arithmeticae* also contains Gauss's treatment of the theory of quadratic forms. Most importantly, he defines a *composition* of forms, in which any two forms with the same determinant combine to produce another form with the same determinant. This leads to the concept of a *group*, a modern algebraic structure formalized in the nineteenth century, of which Gauss's construction of composing forms is one of the first. Essentially to have a group means not only that one has a composition law, but that inverses and an identity also exist for this "multiplication" of objects. Our final original source gives a taste of this direction via the introduction from this part of the *Disquisitiones* [81] and Gauss's justification for developing the subject from scratch rather than building on the work of Lagrange and Legendre.

Carl Friedrich Gauss, from
Disquisitiones Arithmeticae

On Forms and Indeterminate Equations of Degree Two.
Subject of the Investigation;
Definition of the Forms and Terminology.
153.

In this section we will treat in particular the functions of two indeterminates x and y of the following form

$$ax^2 + 2bxy + cy^2,$$

where a, b, c are given integers. We will call these **forms** of degree two, or simply forms. This investigation ends with the solution of the famous problem to find all solutions of any given indeterminate equation of degree two in two unknowns, whether these unknowns are assumed to take on integer or only rational values. This problem has already been solved by Lagrange, and much about the nature of forms has been found by this great geometer, as well as by Euler, or first proven after it had been discovered by Fermat. But a thorough investigation of these forms has produced so much that is new, that it seemed worth the effort to treat the whole subject anew. This is all the more true since the discoveries of these men are disbursed in many places and are not widely known, according to our experience. Furthermore, the treatment of this subject is our own, and much of our contributions would hardly be comprehensible without a new presentation of the earlier discoveries. But there should be no doubt that much worth knowing in this direction has remained hidden, on which others can apply their strength. Incidentally, we will always indicate the history of the most distinguished properties at appropriate places.

We will denote the form $ax^2 + 2bxy + cy^2$ by (a, b, c), as long as the indeterminates x, y are not relevant. Thus, this expression will denote generally a sum of three parts, namely the product of the given number a by the square of an arbitrary indeterminate, double the product of the number b with this indeterminate and another indeterminate, and, finally, the product of the number c with the square of this second indeterminate. For instance, $(1, 0, 2)$ represents the sum of a square and double a square. Even though the forms (a, b, c) and (c, b, a) denote the same thing if we only consider the parts themselves, we will nonetheless distinguish them, when we also consider the order of the parts. We will therefore distinguish them carefully subsequently, and the advantage for us will become clear.

Representation of Numbers; the Determinant.
154.

We say that a given number is **represented** by a given form, if one can assign values to the indeterminates of the form so that the form takes the given number as value. We have the following:

Theorem. If the number M can be represented by the form (a, b, c) in such a way that the values of the indeterminates giving the representation are relatively prime, then $b^2 - ac$ is a quadratic residue of M.

Proof. If the values of the indeterminates are m, n, so that

$$am^2 + 2bmn + cn^2 = M,$$

and if one takes two numbers μ, ν such that $\mu m + \nu n = 1$ (Art. 40),[22] then one can show easily by expansion that

$$(am^2 + 2bmn + cn^2)(a\nu^2 - 2b\mu\nu + c\mu^2) =$$
$$[\mu(mb + nc) - \nu(ma + nb)]^2 - (b^2 - ac)(m\mu + n\nu)^2,$$

or

$$M(a\nu^2 - 2b\mu\nu + c\mu^2) = [\mu(mb + nc) - \nu(ma + nb)]^2 - (b^2 - ac).$$

Therefore:

$$b^2 - ac \equiv [\mu(mb + nc) - \nu(ma + nb)]^2 \pmod{M},$$

that is, $b^2 - ac$ is a quadratic residue modulo M.

We will call the number $b^2 - ac$, whose nature largely determines the properties of the form (a, b, c), as we shall see below, the **determinant** of this form.

---

[22] See the Appendix.

∞∞∞∞∞∞∞∞∞

To understand this choice of terminology, observe that if one uses matrix notation, then one can write the form (a, b, c) as

$$ax^2 + 2bxy + cy^2 = (x, y) \begin{pmatrix} a & b \\ b & c \end{pmatrix} \begin{pmatrix} x \\ y \end{pmatrix},$$

and $b^2 - ac$ is none other than the negative of the determinant of this matrix. By way of comparison, note that Lagrange's K for quadratic forms is always -4 times Gauss's determinant, and that in Lagrange's analysis of possible divisors of forms $t^2 \pm au^2$ in his section (16), he finds that the middle coefficient of a divisor form is always even, which Gauss simply takes as a starting point.

∞∞∞∞∞∞∞∞∞

A Form Containing Another
One or Being Contained in Another; Transformation, Proper and Improper.
157.

If a form F with indeterminates x, y can be transformed into another one F'' whose indeterminates are x', y', via a substitution of the form

$$x = \alpha x' + \beta y', \quad y = \gamma x' + \delta y',$$

where α, β, γ, δ are integers, then we say that the former **contains** the latter, or that the latter is contained in the former. If F is the form

$$ax^2 + 2bxy + cy^2,$$

and F' is

$$a'x'^2 + 2b'x'y' + c'y'^2,$$

then one has the following equations:

$$a' = a\alpha^2 + 2b\alpha\gamma + c\gamma^2,$$
$$b' = a\alpha\beta + b(\alpha\delta + \beta\gamma) + c\gamma\delta,$$
$$c' = a\beta^2 + 2b\beta\delta + c\delta^2.$$

If one multiplies the second equation by itself, the first by the third, and subtracts, then one obtains after cancellation:

$$b'^2 - a'c' = (b^2 - ac)(\alpha\delta - \beta\gamma)^2.$$

From this it follows that the determinant of the form F' is divisible by the determinant of the form F, and the quotient is a square. Hence, the two determinants obviously have the same sign. If in addition the form F' can be transformed into F by a similar transformation, that is, if F' is contained in F and F is contained in F', then the determinants of these forms become equal to each other, and $(\alpha\delta - \beta\gamma)^2 = 1$. In this case we call the forms **equivalent**. Consequently, *equality of the determinants is necessary for equivalence of forms*, even though the latter does not follow from the former alone.

∞∞∞∞∞∞∞

Gauss thus defines an equivalence relation on forms (Exercise 4.48), and the equivalence classes can become the objects of study. Lagrange's results above on transformation of forms (Exercise 4.49) show that for a given fixed determinant there are only finitely many equivalence classes (Exercise 4.50).

Gauss goes on to define a way of composing forms with the same determinant.[23] How to compose forms is not at all obvious, and for us to delve into it would be beyond the scope of our chapter, and take us into the heart of modern algebraic number theory. But it is presaged by something we observed already in the introduction, that the representation problem is "multiplicative," in the sense that if m and n are of the form $x^2 + ay^2$, then so is mn, using the ancient identity of Brahmagupta (see again Exercise 4.8). The more general idea is that if one has two positive definite forms with the same determinant, but not necessarily in the same equivalence class, then there is another form with the same determinant, called their *composition*, which represents all products of all pairs of numbers represented by the original forms (note that this composition should thus be commutative). Legendre had actually begun systematically composing forms in this way [245, pp. 204, 332ff], but ran into difficulties with making the composition uniquely defined. Gauss saw that this could be solved by refining the notion of equivalence to *proper equivalence*, in which he requires $\alpha\delta - \beta\gamma = 1$ above, not just $(\alpha\delta - \beta\gamma)^2 = 1$ as for ordinary equivalence. Then composition is not only well defined, it also actually respects proper equivalence, so that composition becomes an operation on the proper equivalence classes themselves, thereby defining a group structure on this finite set, today called the *class group*. The structure of this group is linked very closely to the determinant. For a detailed discussion of the influence of Gauss's work on the development of group theory see [251, Part I.3]. And to pursue the theory of quadratic forms further, in all the richness they provide to new research in mathematics even today, the reader may read the excellent books [41, 96]. In the final Exercise 4.51 we guide the reader on a small journey into composing forms and the group structure this creates, building on our earlier analysis after our Lagrange excerpt of the divisor, descent, and representation problems for the form $x^2 + 5y^2$.

We have included this short final section in order to show the dramatic transformation that has taken place, beginning with Fermat's results on sums of squares, and ending with an abstract theory of forms, equivalence classes,

---

[23] One often needs to restrict to *primitive* forms, i.e., those with relatively prime coefficients, which goes along with our very early notion of considering only non-trivial divisors of forms.

and the rise of modern algebra in the form of group theory. Such a development is quite characteristic of many subjects in mathematics. Increasing abstraction allows deeper understanding of the objects of study and their interrelationships.

Exercise 4.48. Show that Gauss's definition of equivalence of forms gives an equivalence relation.

Exercise 4.49. Examples of Containment.

1. Illustrate the linear transformation of Lagrange's Theorem I as an example of Gauss's containment. Write out the corresponding matrix:

$$\begin{pmatrix} t \\ u \end{pmatrix} = \begin{pmatrix} m_{11} & m_{12} \\ m_{21} & m_{22} \end{pmatrix} \begin{pmatrix} s \\ x \end{pmatrix}.$$

2. Illustrate the linear transformation of Lagrange's Theorem II as an example of Gauss's equivalence. Write out the corresponding matrix. Find its inverse.
3. Prove that if two quadratic forms are equivalent, then the forms take the same values, that is, they have the same ranges.

Exercise 4.50. Show that Lagrange's Theorem III and its corollaries imply that there are only finitely many equivalence classes of forms with a fixed positive determinant according to Gauss's definition.

Exercise 4.51. We are interested in studying the class group for determinant -5. Details we leave out may be found in Cox [41, p. 24ff]. Lagrange's excerpted analysis above in his (18.II.V) shows that every form of determinant -5 is equivalent to one of $p^2 + 5q^2$, $2p^2 \pm 2pq + 3q^2$. But in fact $2p^2 + 2pq + 3q^2$ is equivalent to $2p^2 - 2pq + 3q^2$ (replace p by $\;p$), and even properly equivalent (replace p by $p - q$). This leaves $p^2 + 5q^2$ and $2p^2 + 2pq + 3q^2$. In fact they are not equivalent, although we will not show that here. Everything we have said is also true for proper equivalence, so the class group for -5 has only two elements. Let us explore the composition of its forms to see how it yields the group multiplication. We have already seen via Exercise 4.8 how $p^2 + 5q^2$ composes with itself, via Brahmagupta's identity:

$$(x^2 + 5y^2)(z^2 + 5w^2) = (xz \pm 5yw)^2 + 5(xw \mp yz)^2,$$

showing that products of numbers of form $p^2 + 5q^2$ are again always of the same form. Since $p^2 + 5q^2$ appears to be acting as an identity element ought to for multiplication, we expect its equivalence class to be the identity element of the class group. Now we suggest to the reader identities to find that represent the compositions for the remaining multiplications possible in the two-element class group.

(a) Find an identity in x, y of the form

$$\left(2x^2 + 2xy + 3y^2\right)\left(2z^2 + 2zw + 3w^2\right) = (???)^2 + 5\left(xw - yz\right)^2,$$

showing that the square in the class group of the equivalence class containing $2p^2 + 2pq + 3q^2$ is the equivalence class containing $p^2 + 5q^2$, i.e., the identity element.

(b) Find an identity in x, y, z, w of the form

$$\left(x^2 + 5y^2\right)\left(2z^2 + 2zw + 3w^2\right)$$
$$= 2\left(p\left(x, y, z, w\right)\right)^2 + 2p\left(x, y, z, w\right)q\left(x, y, z, w\right) + 3\left(q\left(x, y, z, w\right)\right)^2,$$

where p and q are polynomials, showing that the equivalence class containing $x^2 + 5y^2$ really is acting as an identity element in the class group. Hint: Use what Lagrange showed in his section (17), and Brahmagupta's identity again.

(c) Write down a multiplication table for the class group.

(d) After the Lagrange source we determined that primes $20n + 1$, $20n + 9$ are precisely those of the form $x^2 + 5y^2$, and primes $20n + 3$, $20n + 7$ are precisely those of the form $2x^2 + 2xy + 3y^2$. Try multiplying these linear forms together in their various combinations, and see whether the results fit with the product structure we have just determined in the class group.

4.8 Appendix on Congruence Arithmetic

This appendix contains a short introduction to congruence arithmetic and some number-theoretic results needed in this chapter. We leave it to the reader to create examples as needed, and to prove some basic results. For a more extensive treatment the reader is encouraged to consult a basic number theory text, e.g., [27]. N.B: In this appendix all numbers are integers.

Recall that a natural number greater than one is defined to be *prime* if it has no positive divisors other than itself and one. An extremely important result, whose proof the reader may find elsewhere, or create a proof for (Exercise 4.52), is the following:

Theorem 4.1. *(Fundamental Theorem of Arithmetic) Any integer n greater than 1 has a unique prime factorization*

$$n = p_1^{a_1} p_2^{a_2} \cdots p_m^{a_m},$$

where the p_i are distinct primes, $p_1 < p_2 < \cdots < p_m$, and the a_i are positive.

The uniqueness is the deep part of this result, and is intimately related to a key feature of primes (Exercise 4.53):

Theorem 4.2. *(Euclid's Lemma) Given positive numbers m, n, and given p a prime, if p is a divisor of mn, then p is a divisor of either m or n, or both.*

For numbers a and b and a positive n, we say that a *is congruent to b modulo n*, denoted by

$$a \equiv b \bmod n,$$

if n divides $a - b$, or, equivalently, if there is a k such that $a = kn + b$.

The notion of congruence was first treated systematically by Gauss in his *Disquisitiones Arithmeticae* [80], although it can already be found implicitly in the work of Euler, Lagrange, and Legendre. The notation above is due to Gauss.

Theorem 4.3. *Let n be positive, and let a, b, c be numbers. Then*
  1. $a \equiv a \mod n$;
  2. If $a \equiv b \mod n$, then $b \equiv a \mod n$;
  3. If $a \equiv b \mod n$ and $b \equiv c \mod n$, then $a \equiv c \mod n$.

Proof. Exercise 4.54.

The three conditions of Theorem 4.3 show that congruence is an *equivalence relation* (Exercise 4.55).

For any a, the *equivalence* (or *residue*) *class* of a modulo n consists of all numbers that are congruent to a modulo n. There are clearly n such equivalence classes. Using the *division algorithm*, we can find numbers q, r such that

$$a = qn + r \quad \text{where} \quad 0 \le r < n.$$

So a is in the residue class of some nonnegative r less than n; and since q and r are uniquely determined by the algorithm, a is in exactly one such class. The number r is called the *remainder* of a modulo q.

We will use the notation $n|a$ to indicate that n divides a; that is, there is a number q such that $a = qn$.

Theorem 4.4. *Let n be positive, and let a, b, c, d be numbers. Then*
  1. If $a \equiv b \mod n$ and $c \equiv d \mod n$, then $(a + c) \equiv (b + d) \mod n$.
  2. If $a \equiv b \mod n$ and $c \equiv d \mod n$, then $ac \equiv bd \mod n$.

Proof. For part 1, we know that $n|(a - b)$ and $n|(c - d)$; so

$$n|[(a - b) + (c - d)] \quad \text{and} \quad n|[(a + c) - (b + d)].$$

That is, $(a + c) \equiv (b + d) \mod n$. In part 2, since $n|(a - b)$ and $n|(c - d)$, we have $n|[(a - b)c + (c - d)b]$, so $n|(ac - bd)$; thus $ac = bd \mod n$.

Example. Theorem 4.4 essentially says that addition and multiplication make sense modulo n. For instance, suppose we are working modulo 4, and we wish to add 1247 and 10118. Suppose we add first: $1247 + 10118 = 11365$, and then find the residue of the result modulo 4: $11365 = 2841(4) + 1$. So $1247 + 10118 \equiv 1 \mod 4$. If we instead find the residues of 1247 and 10118 first, and then add, we will get the same result: $1247 \equiv 3 \mod 4$ and $10118 \equiv 2 \mod 4$. So $1247 + 10118 \equiv 3 + 2 = 5 \equiv 1 \mod 4$. Multiplication works similarly: $1247(10118) = 12617146 = (3154286)4 + 2$; so $1247(10118) \equiv 2 \mod 4$. We also have $1245(10118) \equiv 3(2) = 6 \equiv 2 \mod 4$. In many cases, this property of multiplication modulo n can greatly simplify computation. Suppose we wish to find $56^{99} \mod 5$. Clearly we would prefer not to have to compute 56^{99}! We get around it by observing that $56 \equiv 1 \bmod 5$; so $56^{99} \equiv 1^{99} \equiv 1 \bmod 5$. To compute $2^{99} \bmod 7$, note that $2^{99} = (2^3)^{33} = 8^{33}$; and since $8 \equiv 1 \bmod 7$, we get $2^{99} \equiv 1^{33} \equiv 1 \bmod 7$.

Part 2 of Theorem 4.4 implies that if we multiply both sides of a congruence by the same number, the congruence is preserved. Unfortunately, division is not as nice, as the next example shows.

Example. Note that $9 \equiv 15 \bmod 6$; i.e., $3 \cdot 3 \equiv 3 \cdot 5 \bmod 6$. If we attempt to "divide both sides by 3," we discover that $3 \not\equiv 5 \bmod 6$. If we look at the congruence $36 \equiv 60 \bmod 8$, we see that it says that $3 \cdot 12 \equiv 3 \cdot 20 \bmod 8$; and in this case, we do have $12 \equiv 20 \bmod 8$.

To explain the difference between these two cases, and to obtain a result appealed to in our sources by both Lagrange and Gauss, we will examine what we can learn about greatest common divisors from the Euclidean algorithm. The greatest common divisor of m and n will be denoted by $\gcd(m, n)$.

Given $a_1 > a_2 > 0$, the *Euclidean algorithm* iterates the division algorithm as follows:

$$a_1 = q_1 a_2 + a_3;$$
$$a_2 = q_2 a_3 + a_4;$$
$$\cdots$$
$$a_{n-2} = q_{n-2} a_{n-1} + a_n;$$
$$a_{n-1} = q_{n-1} a_n + 0,$$

where $a_1 > a_2 > a_3 > \cdots > a_n > 0$. From these equations one can see that a_n is the greatest common divisor of a_1 and a_2, and moreover the equations clearly allow one to write

$$\gcd(a_1, a_2) = sa_1 + ta_2,$$

for some numbers s and t (Exercises 4.56 and 4.57). This is often called Bezout's equation. Both Lagrange and Gauss used this in our source excerpts.

Now we can address the cancellation question in modular arithmetic.

Theorem 4.5. *Suppose that n is positive and a and n are relatively prime. Then there is a number s such that $sa \equiv 1 \bmod n$. We call s a reciprocal of a modulo n.*

Proof. Bezout's equation yields $1 = sa + tn$ for some s, t.

Theorem 4.6. *Suppose that n is positive and a, b, c are such that $ab \equiv ac \bmod n$. Then if a and n are relatively prime, $b \equiv c \bmod n$.*

Proof. Let s be a reciprocal of a modulo n. Multiplying $ab \equiv ac \bmod n$ by s, we have $sab \equiv sac \bmod n$, so that $b \equiv c \bmod n$.

Alternative proof. Since $ab \equiv ac \bmod n$, we have $n | (a(b - c))$. Since a and n are relatively prime, we have $n|(b-c)$ by the uniqueness in the fundamental theorem of arithmetic, or Bezout's equation (Exercise 4.58). So $b \equiv c \bmod n$.

For an analysis of the connection between the Euclidean algorithm, which underlies the Bezout equation approach, and the fundamental theorem of arithmetic, see [186].

Finally, we prove the wonderful result of Fermat's that underlies so much number theory, and which Legendre used in his section (134) above to prove Euler's criterion.

Theorem 4.7. *(Fermat's Little Theorem) If p is a prime and a is not divisible by p, then*

$$a^{p-1} \equiv 1 \pmod{p}.$$

Proof. Consider the sequence of multiples

$$a, \ 2a, \ 3a, \ \ldots, \ (p-1)a.$$

These are all distinct and nonzero modulo p, because a has a reciprocal modulo p, and $1, 2, 3, \ldots, p-1$ are distinct and nonzero modulo p. Thus, modulo p, the two lists are identical, up to ordering, so that

$$a \cdot 2a \cdot 3a \cdots (p-1)a \equiv 1 \cdot 2 \cdot 3 \cdots (p-1) \pmod{p}.$$

Therefore

$$a^{p-1}(p-1)! \equiv (p-1)! \pmod{p}.$$

Since $(p-1)!$ is not divisible by p (Exercise 4.59), we can cancel it on both sides and are left with our result.

This is a very clever proof. A more traditional proof might proceed by mathematical induction (Exercise 4.60).

Exercise 4.52. Prove the fundamental theorem of arithmetic.

Exercise 4.53. Prove Euclid's lemma.

Exercise 4.54. Prove Theorem 4.3.

Exercise 4.55. Show that the relation defined in Theorem 4.3 is an equivalence relation.

Exercise 4.56. Explain how, given any $a_1 > a_2 > 0$, the Euclidean algorithm allows one to write $\gcd(a_1, a_2) = sa_1 + ta_2$, for some numbers s and t.

Exercise 4.57. Prove Lagrange's claim that given any three numbers A, B, C, with B, C relatively prime, the equation $A = Bt + Cu$ always has a solution in numbers t, u.

Exercise 4.58. Prove that if a and n are relatively prime, and if $n|(ad)$, then $n|d$. Hint: Base your proof either on the uniqueness in the fundamental theorem of arithmetic, or on Bezout's equation.

Exercise 4.59. Prove that if p is prime, then $(p-1)!$ is not divisible by p. Hint: Use the fundamental theorem of arithmetic.

Exercise 4.60. Prove Fermat's little theorem by mathematical induction on a. Use congruences throughout and the binomial theorem to expand at the inductive step. Then use the fundamental theorem of arithmetic to prove that the vast majority of the binomial coefficients are divisible by p

References

1. R.H. Abraham and C.D. Shaw, *Dynamics — the Geometry of Behavior; Part 3: Global Behavior*, Aerial Press, Santa Cruz, Calif., 1984.
2. P.S. Addison, *Fractals and Chaos: An Illustrated Course*, Institute of Physics Publishing, Bristol and Philadelphia, 1997.
3. D. Alexander, *Newton's Method — or Is It?*, Focus (Oct. 1996), 32–33.
4. H.G. Apostle, *Aristotle's Categories and Propositions*, Peripatetic Press, Grennel, Iowa, 1980.
5. Apollonius, *Apollonius of Perga*, T. Heath, (ed.), W. Heffer and Sons, Cambridge, 1961.
6. T. Apostol, *Introduction to Analytic Number Theory*, Springer-Verlag, New York, 1976.
7. Archimedes, *The Works of Archimedes, Including the Method*, T.L. Heath (ed.), Cambridge University Press, 1897, 1912, reprinted by Dover, New York, 1990, and in Great Books of the Western World (ed. R. Hutchins), vol. 11, Encyclopaedia Britannica, Chicago, 1952.
8. I.K. Argyros and Ferenc Szidarovszky, *The Theory and Applications of Iteration Methods*, CRC Press, Boca Raton, FL, 1993.
9. M. Atiyah, *The Work of Simon Donaldson*, Proceedings of the International Congress of Mathematicians, 1986, A.M. Gleason (ed.), American Mathematical Society, Providence, RI, 1987.
10. R. Ayoub, *Euler and the Zeta Function*, American Mathematical Monthly **81** (1974), 1067–1086.
11. P.G. Bachmann, *Niedere Zahlentheorie*, Teubner, Leipzig, 1902–1910; republished by Chelsea, New York, 1968.
12. I. Bashmakova, *Diophantus and Diophantine Equations*, Mathematical Association of America, Washington, D.C., 1997.
13. S. Batterson, *Stephen Smale: The Mathematician Who Broke the Dimension Barrier*, American Mathematical Society, Providence, RI, 2000.
14. G. Berkeley, *The Analyst, Or A Discourse Addressed to an Infidel Mathematician. Wherein It is examined whether the Object, Principles, and Inferences of the modern Analysis are more distinctly conceived, or more evidently deduced, than Religious Mysteries and Points of Faith. "First cast out the beam out of thine own Eye; and then shalt thou see clearly to cast out the mote out of thy brother's Eye."* in *The Works*, G. Bell and Sons, 1898. vol. 3, 1–51.

15. J. Bernoulli, *Die Werke von Jakob Bernoulli*, Naturforschende Gesellschaft in Basel, Birkhäuser Verlag, Basel, 1975.

16. J. Bernoulli *The Art of Conjecturing: Together with "Letter to a Friend on Sets in Court Tennis"* (transl. E.D. Sylla), Johns Hopkins University Press, Baltimore, 2006.

17. R. Bonola, *Non-Euclidean Geometry*, Dover, New York, 1955.

18. C.B. Boyer, U.C. Merzbach, *A History of Mathematics* (second ed.), John Wiley & Sons, New York, 1989.

19. C. Boyer, *Pascal's Formula For the Sums of Powers of the Integers*, Scripta Mathematica **9** (1943), 237–244.

20. A. Bréard, E-mail, Feb. 12, 2000.

21. D. Bressoud, *A Radical Approach to Real Analysis*, Math. Assoc. of America, Washington, D.C., 1994.

22. E.J. Britten, *Britten's Old Clocks and Watches and Their Makers*, 8th ed., C. Clutton, G.H. Baillie, C.A. Ilbert, (eds.), Dutton and Co., New York, 1973.

23. R.A. Brualdi, *Introductory Combinatorics.* Second edition. North-Holland Publishing Co., New York, 1992.

24. L.L. Bucciarelli, N. Dworsky, *Sophie Germain: An Essay in the History of the Theory of Elasticity*, D. Reidel Publishing Co., Boston, 1980.

25. D.A. Buell, *Binary Quadratic Forms*, Springer Verlag, New York, 1989.

26. W. K. Bühler, *Gauss, A Biographical Study*, Springer Verlag, New York, 1981.

27. D. Burton, *Elementary Number Theory* (third ed.), Wm. C. Brown Publishers, Dubuque, Iowa, 1994.

28. F. Cajori, *An Introduction to the Modern Theory of Equations*, Macmillan, New York, 1943 (first published 1904).

29. F. Cajori, *A History of the Arithmetical Methods of Approximation to the Roots of Numerical Equations of One Unknown Quantity*, Colorado College Publication, General Series, nos. 51–52, Science Series **12** (1910), Colorado Springs, pp. 171–287.

30. R. Calinger (ed.), *Vita Mathematica: Historical Research and Integration with Teaching*, Math. Assoc. of Amer., Washington, D.C., 1996.

31. G. Cardano, *The Great Art, or The Rules of Algebra*, translated and edited by T. Richard Witmer, M.I.T. Press, 1968.

32. A. Cayley, *Eisenstein's Geometric Proof of the Fundamental Theorem for Quadratic Residues*, Quarterly Mathematical Journal **1** (1857), 186–191; also in *The Collected Mathematical Papers of Arthur Cayley*, vol. III, Cambridge University Press, Cambridge, 1890.

33. F. Cesari, *Fractal Explorer: Mandelbrot and Julia sets*, On the Internet: www.geocities.com/fabioc, 2001.

34. J.-L. Chabert, Evelyne Barbin, Michel Guillemot, Anne Michel-Pajus, Jacques Borowczyk, Ahmed Djebar and Jean-Claude Martzloff, *Histoire d'Algorithmes — Du Caillou à la Puce.* Berlin, Paris, 1994. Translated into English by Chris Weeks, *A History of Algorithms — From the Pebble to the Microchip*, Springer-Verlag, Berlin, 1999.

35. A.B. Chace, Ludlow Bull, Henry Parker Manning, *The Rhind Mathematical Papyrus.* Vol. I, Free Translation and Commentary, 1927. Vol. II, Photographs, Transcription, Transliteration, Literal Translation, 1929. Mathematical Association of America, Oberlin, Ohio, U.S.A.

36. W. Cheney and D. Kincaid, *Numerical Mathematics and Computing*, 2nd ed., Brooks/Cole, Pacific Grove, California, 1985.

37. C. Christensen, *Newton's Method for Resolving Affected Equations*. College Mathematics Journal **27** (Nov. 1996), 330–340.
38. R.V. Churchill, James W. Brown and Roger F. Verhey, *Complex Variables and Applications*, 3rd ed., McGraw-Hill, New York, 1974.
39. F.M. Clarke, *Thomas Simpson and His Times*, Waverly Press, New York, 1929.
40. *Convergence: Where Mathematics, History and Teaching Interact*, http://convergence.mathdl.org/jsp/index.jsp.
41. D. Cox, *Primes of the Form $x^2 + ny^2$*, Wiley, New York, 1989.
42. M. Davis, *Hilbert's tenth problem is unsolvable*, Amer. Math. Monthly **80** (1973), 233–269.
43. M. Davis, R. Hersh, *Hilbert's 10th Problem*, Scientific American **229** (1973), 84-91.
44. M. Davis, Y. Matijasevič, J. Robinson, *Hilbert's Tenth Problem. Diophantine Equations: Positive Aspects of a Negative Solution*, in Mathematical Developments Arising from Hilbert Problems, pp. 323–378 in Proceedings of Symposia in Pure Mathematics, vol. 28, Amer. Math. Soc., 1976. See §1: Diophantine Representation of the Set of Primes, pp. 328–331.
45. R. Dedekind, *Was Sind und Was Sollen die Zahlen?* 10th ed., Vieweg, Braunschweig, 1969.
46. R.L. Devaney, *Film and Video as a Tool in Mathematical Research*, Math. Intelligencer, **11** (1989), 33–38.
47. L.E. Dickson, *New First Course in the Theory of Equations*, John Wiley & Sons, New York, 1939.
48. E.J. Dijksterhuis, *Archimedes*, Princeton University Press, Princeton, New Jersey, 1987.
49. Diophantus, *Diophanti Alexandrini Arithmeticorum libri sex, et de numeris multangulis liber unus*, Paris, 1621.
50. M.P. do Carmo, *Riemannian Geometry*, Flaherty, F. (trans.), Birkhäuser, Boston, 1992.
51. P. Dombrowski, *150 Years After Gauss' "Disquisitiones generales circa superficies curvas,"* Astérisque **62** (1979), 99–153.
52. W. Dunham, *A "Great Theorems" Course in Mathematics*, Amer. Math. Monthly **93** (1986), 808–811.
53. W. Dunham, *Journey Through Genius*, John Wiley, New York, 1980.
54. W. Dunham, *Euler: The Master of Us All*, Mathematical Association of America, Washington D.C., 1999.
55. C. Dupin, *Développements de Géométrie, Avec des Applications à la Stabilité des Vaisseaux, aux Déblais et Remblais, au Défilement, à l'Optique, etc.*, Mme. Ve. Courcier, Paris, 1813.
56. H.M. Edwards, *Riemann's Zeta Function*, Academic Press, New York, 1974.
57. H.M. Edwards, *Euler and Quadratic Reciprocity*, Mathematics Magazine **56** (1983), 285–291.
58. A. Einstein, *Relativity, the Special and General Theory*, Lawson, R.W. (trans.), Henry Holt and Co., New York, 1920.
59. G. Eisenstein, *Geometrischer Beweis des Fundamentaltheorems für die Quadratischen Reste*, Journal für die Reine und Angewandte Mathematik (Crelle's Journal) **28** (1844), 246–249; also in *Mathematische Werke* (2nd ed.), Chelsea Publ. Co., New York, 1989, 64–166.
60. G. Eisenstein, *Mathematische Werke* (2nd ed.), Chelsea Publ. Co., New York, 1989.

61. Euclid, *The Thirteen Books of Euclid's Elements*, T.L. Heath (ed.), Dover, New York, 1956.

62. L. Euler, *Introduction to Analysis of the Infinite* (transl. John D. Blanton), Springer-Verlag, New York, 1988.

63. *Encyclopædia Britannica*, Chicago, 1986.

64. L. Euler, *Recherches sur la courbure des surfaces*, Mémoires de l'académie des sciences de Berlin **16** (1767), 119–143.

65. L. Euler, *Correspondance Mathématique et Physique*, P.-H. Fuss (ed.), The Sources of Science **35**, Johnson Reprint Corporation, New York, 1968; initially published by the Imperial Academy of Sciences, St. Petersburg, 1843.

66. L. Euler, *Opera Omnia*, series I, B.G. Teubner, Leipzig and Berlin, Lausanne, 1911– .

67. L. Euler, *Vollständige Anleitung zur Differenzial-Rechnung* (transl. Johann Michelsen), Berlin, 1790, reprint of the 1798 edition by LTR-Verlag, Wiesbaden, 1981.

68. L. Euler, *Foundations of Differential Calculus* (transl. John D. Blanton), Springer Verlag, New York, 2000.

69. L. Euler, Excerpts on the Euler-Maclaurin summation formula, from *Institutiones Calculi Differentialis* (transl. David Pengelley), http://www.math.nmsu.edu/~davidp/institutionescalcdiff.dvi, New Mexico State University, 2000.

70. P. Fatou, *Sur les Équations Fonctionnelles*, Bull. Soc. Math. de France **47** (1919), 161–270; **48** (1920), 33–95, 208–314.

71. J. Fauvel, J. Van Maanen, *History in Mathematics Education*, Kluwer, Boston, 2000.

72. J. Fauvel, R. Flood, M. Shortland & R. Wilson (eds.), *Let Newton Be!* Oxford University Press, Oxford, 1988.

73. P. de Fermat, *Oeuvres*, P. Tannery (ed.), Paris, 1891–1922.

74. J. Fourier, *Analyse des Équations Determinées, Première Partie*, Firmin Didot Frères, Paris, 1830.

75. D.H. Fowler, *Ratio in Early Greek Mathematics*, Bull. Amer. Math. Soc. (N.S.) **1** (1979), pp. 807–846.

76. A.T. Fuller, *Horner versus Holdred: An Episode in the History of Root Computation*, Historia Mathematica **26** (1999), 29–51.

77. L. Gaal, *Classical Galois Theory with Examples*, Chelsea Pub. Co., New York, 1979.

78. Galileo Galilei, *Two New Sciences, Including Centers of Gravity & Force of Percussion*, Stillman Drake (trans.), University of Wisconsin Press, Madison, 1974.

79. E. Galois, *Ecrits et Mémoires Mathématiques d'Evariste Galois* (R. Bourgne and J.-P. Azra, eds.) Gauthiers-Villars, Paris, 1962.

80. C.F. Gauss, *Disquisitiones Arithmeticae* (English transl., A.A. Clarke), Yale University Press, New Haven, 1966.

81. C.F. Gauss, Untersuchungen über Höhere Arithmetik (German transl. of Disquisitiones Arithmeticae, H. Maser), Springer, Berlin, 1889, reprinted by Chelsea Pub. Co., Bronx, N.Y., 1965.

82. C.F. Gauss, *Commentationes Societatis Regiae Scientiarum Gottingensis*, **16**, 1808, Göttingen; also in *Werke*, Göttingen, 1876, v. 2, pp. 1–8.

83. C.F. Gauss, *Werke*, Teubner, Leipzig, 1863–1929.

84. C.F. Gauss, *Disquisitiones generales circa superficies curvas*, Gauss Werke, Königliche Gesellschaft der Wissenschaften, Göttingen, v. IV, 1880.

85. C.F. Gauss, *Neue allgemeine Untersuchungen über die krummen Flächen*, Gauss Werke, Königliche Gesellschaft der Wissenschaften, Göttingen, B.G. Teubner, Leipzig, v. VIII, 1900.

86. C.F. Gauss, *General Investigations of Curved Surfaces*, Hiltebeitel, A., Morehead, J. (trans.), The Princeton University Library, Princeton, 1902.

87. C.F. Gauss, *General Investigations of Curved Surfaces*, Hiltebeitel, A., Morehead, J. (trans.), Raven Press, Hewlett, New York, 1965.

88. S. Germain, *Recherches sur la théorie des surfaces élastiques*, Mme. V. Courcier, Paris, 1821.

89. S. Germain, *Remarques sur la nature, les bornes et l'étendue de la question des surfaces élastiques, et équation générale de ces surfaces*, Huzard-Courcier, Paris, 1826.

90. M. Gerstenhaber, *The 152nd Proof of the Law of Quadratic Reciprocity*, American Mathematical Monthly **70** (1963), 397–398.

91. P. Gidal, *Andy Warhol: Films and Paintings*, Studio Vista Ltd., London, 1971.

92. C.C. Gillispie, F.L. Holmes (eds.), *Dictionary of Scientific Biography*, Scribner, New York, 1970.

93. D. Gjertsen, *The Newton Handbook*, Routledge, New York, 1987, 606.

94. A.M. Gleason, *Angle Trisection, the Heptagon, and the Triskaidecagon*, American Mathematical Monthly **95** (1988), pp. 185–194.

95. J. Gleick, *Chaos: Making a New Science*, Viking, New York, 1987.

96. J.R. Goldman, *The Queen of Mathematics, A Historically Motivated Guide to Number Theory*, A. K. Peters, Wellesley, MA, 1998.

97. H.H. Goldstine, *A History of Numerical Analysis from the 16th through the 19th Century*, Springer-Verlag, New York, 1977.

98. H.W. Gould, *Explicit Formulas for Bernoulli Numbers*, American Mathematical Monthly **79** (1972), 44–51.

99. S.J. Gould, *Leonardo's Mountain of Clams and the Diet of Worms: Essays on Natural History*, Three Rivers Press, New York, 1998.

100. *Great Books of the Western World*, Mortimer Adler (ed.), Encyclopædia Britannica, Inc., Chicago, 1991.

101. M.J. Greenberg, *Euclidean and Non-Euclidean Geometries*, W.H. Freeman and Company, New York, 1980.

102. E. Grosswald, *Representations of Integers as Sums of Squares*, Springer Verlag, New York, 1985.

103. R.K. Guy, *Unsolved Problems in Number Theory*, 2nd edition, Springer-Verlag, New York, 1994.

104. E. Hairer and G. Wanner, *Analysis by its History*, Springer Verlag, New York, 1996.

105. G. Hämmerlin and K. Hoffmann, *Numerical mathematics*. Translated from the German by Larry Schumaker. Undergraduate Texts in Mathematics. Readings in Mathematics. Springer-Verlag, New York, 1991.

106. G.H. Hardy, *Divergent Series*, Chelsea Publishing, New York, 1991.

107. G.H. Hardy, *A Mathematician's Apology*, Cambridge University Press, New York, 1982.

108. P. Hazard, *The European Mind, 1680–1715*, May, J.L. (trans.), World Publishing Co., Cleveland, OH, 1963.

109. T.L. Heath, *Diophantus of Alexandria, A Study in the History of Greek Algebra*, Cambridge University Press, Cambridge, 1885.
110. T.L. Heath, *Apollonius of Perga*, Barnes and Noble (reprint), 1961.
111. T.L. Heath, *The Thirteen Books of Euclid's Elements*, 3 vols., Cambridge University Press, 1926.
112. T.L. Heath, *A History of Greek Mathematics*, 2 vols., Clarendon Press, Oxford, 1921.
113. T. L. Heath, *A Manual of Greek Mathematics*, Dover, New York, 1963.
114. H. von Helmholtz, *Über die Tatsachen die der Geometrie zu Grunde liegen*, Gött. Nachrichten **9** (1868), 618–639.
115. I.N. Herstein, *Topics in Algebra*, John Wiley & Sons, Inc., New York, 1975.
116. F.B. Hildebrand, *Introduction to Numerical Analysis*, 2nd edition, McGraw-Hill, New York, 1974.
117. J.E. Hofmann, *Leibniz in Paris, 1672–1676; His Growth to Mathematical Maturity*, Cambridge University Press, Cambridge, 1974.
118. S. Hollingdale, *Makers of Mathematics*, Penguin Books, London, 1989, New York, 1994.
119. G. de L'Hospital, *Analyse des infiniment petits pour l'intelligence des lignes courbes*, François Montalant, Paris, 1696.
120. HP Museum of Calculators: `www.hpmuseum.org/srinst.htm`
121. C. Huygens, *The Pendulum Clock or Geometrical Demonstrations Concerning the Motion of Pendula as Applied to Clocks*, R.J. Blackwell (trans.), The Iowa State University Press, Ames, Iowa, 1986.
122. *International Study Group on the History and Pedagogy of Mathematics Newsletter,* `http://www.hpm-americas.org/`.
123. K. Itô, ed., *Encyclopedic Dictionary of Mathematics*, 2nd ed., MIT Press, Cambridge, Massachusetts, 1993.
124. E. Jahnke, F. Emde, *Tables of Functions with Formulae and Curves*, 4th edn., Dover, New York.
125. G.G. Joseph, *The Crest of the Peacock: The Non-European Roots of Mathematics*, Tauris, London, 1991; Princeton University Press, 2000.
126. G. Julia, *Mémoire sur l'Iteration des Fonctions Rationelles*. J. de Math. Ser. 7. **4** (1918), 47–245.
127. S.C. Kak, *The Brahmagupta Algorithm for Square Rooting, Ganita Bhāratī*, Bull. Ind. Soc. Hist. Math. **11** (1989), 27–29.
128. K. Kanim, *Proof Without Words: How Did Archimedes Sum Squares in the Sand?*, Mathematics Magazine **74** (2001), 314–315.
129. I. Kant, *Critique of Pure Reason*, J.M.D. Meiklejohn, (trans.), Prometheus Books, Buffalo, New York, 1990.
130. J.-M. Kantor, *Hilbert's problems and their sequels*, Math. Intelligencer **18** (1996), 21–30.
131. D.S. Kasir, *The Algebra of Omar Khayyam*, Teachers College, Columbia University, 1931.
132. V. Katz (ed.), *Using History to Teach Mathematics: An International Perspective*, Mathematical Assoc. of America, Washington DC, 2000.
133. V.J. Katz, *A History of Mathematics: An Introduction* (second ed.), Addison-Wesley, New York, 1998.
134. F. Klein, *Famous Problems of Elementary Geometry*, orig. pub. 1895 (English trans. Atheneum Press 1897).

135. M. Kline, *Mathematical Thought from Ancient to Modern Times*, Oxford University Press, New York, 1972.

136. D. Klyve, L. Stemkoski, *The Euler Archive*, http://www.math.dartmouth.edu/~euler/ or http://www.eulerarchive.org.

137. K. Knopp, *Theory and Application of Infinite Series*, Dover Publications, New York, 1990.

138. D. Knuth, *The Art of Computer Programming*, 2nd ed., vol. 1, Addison-Wesley, Reading, Mass., 1973.

139. N. Kollerstrom, *Thomas Simpson and "Newton's Method of Approximation": An Enduring Myth*, British J. Hist. Sci. **25**, no. 86, part 3 (1992), 347–354.

140. L. Kronecker, *Zur Geschichte des Reciprocitätsgesetzes*, Berliner Monatsberichte 1875, 267–275; in *Gesammelte Werke*, vol. II, 1–10, Chelsea Publ. Co., New York, 1968.

141. J.L. Lagrange, *Oeuvres*, Gauthiers-Villars, Paris, 1869.

142. J. Lagrange, *Réflexions sur la Résolution Algébriques des Equations*, in Oeuvres de Lagrange. v. 3, Gauthier-Villars, Paris, 1869.

143. J. Lagrange, *Traité de la Résolution des Équations Numériques de Tous les Degres, avec des Notes sur Plusieurs Points de la Théorie des Équations Algébriques*, 1798, revised in 1808, 3rd ed., Bachelier, Paris, 1826, in *OEuvres de Lagrange*, J.-A. Serret, ed., Gauthier-Villars, Paris, 1867–1892, VIII.

144. R. Laubenbacher, D. Pengelley, *Teaching with Original Historical Sources in Mathematics*, http://www.math.nmsu.edu/~history/.

145. R. Laubenbacher, D. Pengelley, and M. Siddoway, *Recovering Motivation in Mathematics: Teaching with Original Sources*, Undergraduate Mathematics Education Trends **6**:4 (Sept. 1994).

146. R. Laubenbacher and D. Pengelley, *Great Problems of Mathematics: A course Based on Original Sources*, Amer. Math. Monthly **99** (1992), 313–317.

147. R. Laubenbacher and D. Pengelley, *Mathematical Masterpieces: Teaching with Original Sources*, Vita Mathematica: Historical Research and Integration with Teaching (R. Calinger, ed.), Math. Assoc. of Amer., Washington, D.C., 1996, 257–260.

148. R. Laubenbacher, D. Pengelley, *Gauss, Eisenstein, and the "Third" Proof of the Quadratic Reciprocity Theorem: Ein kleines Schauspiel*, The Mathematical Intelligencer **16** (1994), 67–72.

149. R. Laubenbacher, D. Pengelley, *Eisenstein's misunderstood geometric proof of the Quadratic Reciprocity Theorem*, College Mathematics Journal **25** (1994) 29–34.

150. R. Laubenbacher, D. Pengelley, *Mathematical Expeditions: Chronicles by the Explorers*, Springer, New York, 1999.

151. R. Laubenbacher, D. Pengelley, *The Number-Theoretic Work of Sophie Germain*, in preparation.

152. R. Laubenbacher, G. McGrath and D. Pengelley, *Lagrange and the Solution of Numerical Equations*, Historia Mathematica **28** (2001), 220–231.

153. A.-M. Legendre, *Théorie des Nombres* (3rd ed.), A. Hermann, Paris, 1830.

154. G.W. Leibniz, *Mathematische Schriften*, C.I. Gerhardt (ed.), Georg Olms Verlagsbuchhandlung, Hildesheim, v. VII, 1962.

155. F. Lemmermeyer, *Reciprocity Laws: From Euler to Eisenstein*, Springer-Verlag, Berlin, 2000.

156. Leonardo Fibonacci, *The Book of Squares* (transl. L.E. Sigler), Academic Press, Boston, 1987.

157. U. Libbrecht, *Chinese Mathematics in the Thirteenth Century: The Shu-shu chiu-chang of Ch'in Chiu-shao*, MIT Press, Cambridge, 1973, 177–193.

158. S. Lie, G.W. Scheffers, *Vorlesungen über continuierliche Gruppen mit geometrischen und anderen Anwendungen*, B.G. Teubner, Leipzig, 1893.

159. S. Lie, G.W. Scheffers, *Geometrie der Berührungstransformationen*, B.G. Teubner, Leipzig, 1896.

160. J. Lodder, *Curvature in the Calculus Curriculum*, The American Mathematical Monthly **110** (2003), 593–605.

161. C.C. MacDuffee, *Theory of Equations*, Wiley, NewYork, 1954.

162. P.A. Macmahon, *Combinatory Analysis*, Cambridge University Press, 1915–1916.

163. M. Mahoney, *The Mathematical Career of Pierre de Fermat*, 2nd edition, Princeton University Press, Princeton, New Jersey, 1994.

164. B.B. Mandelbrot, *The Fractal Geometry of Nature*, Freeman, New York, 1982.

165. G.E. Martin, *Geometric Constructions*, Springer, New York, 1998.

166. J. McCleary, *Geometry from a Differentiable Viewpoint*, Cambridge University Press, New York, 1996.

167. J.-B. Meusnier, *Mémoire sur la courbure des surfaces*, Mémoires de mathématique et physique **10** (1785), 477–510.

168. J. Milnor, *The Work of M.H. Freedman*, Proceedings of the International Congress of Mathematicians, 1986, A.M. Gleason (ed.), American Mathematical Society, Providence, RI, 1987.

169. J.W. Milnor and J.D. Stasheff, *Characteristic Classes*, Princeton University Press, Princeton, New Jersey, 1974.

170. F. Minding, *Wie sich entscheiden lässt ob zwei gegebene krumme Flächen auf einander abwickelbar sind oder nicht*, Journal für die Reine und Angewandte Mathematik **19** (1839), 370–387.

171. H. Minkowski, *Geometrie der Zahlen*, Teubner, Leipzig, 1910; reprinted by Johnson Reprint Co., 1968.

172. H. Minkowski, *Space and Time*, The Principle of Relativity: A collection of papers on the special and general theory of relativity, Dover, New York, 1952.

173. M. Monastyrsky, *Riemann, Topology, and Physics*, J. King, V. King (trans.), R.O. Wells (ed.), Birkhäuser, Boston, 1987.

174. J. Needham, *Science and Civilization in China*, vol. 3: Mathematics and the Sciences of the Heavens and the Earth, Cambridge Univ. Press, Cambridge, 1959.

175. J. Needham, *The Grand Titration*, Allen & Unwin, London, 1969.

176. O. Neugebauer, *The Exact Sciences in Antiquity*, Princeton Univ. Press, 1952.

177. J.R. Neuman, *The World of Mathematics*, Simon and Schuster, New York, 1956.

178. I. Newton, *The Mathematical Papers of Isaac Newton*, D.T. Whiteside (ed.), Cambridge University Press, Cambridge, v. III, 1969.

179. I. Newton, *Of Analysis by Equations of an infinite Number of Terms*, in *The Mathematical Works of Isaac Newton*, D.T. Whiteside (ed.), Johnson Reprint Corp., New York, 1964.

180. Nicomachus of Gerasa, *Introduction to Arithmetic*, in Great Books of the Western World (ed. R. Hutchins), vol. 11, Encyclopaedia Britannica, Chicago, 1952.

181. R. Osserman, *Poetry of the Universe: A Mathematical Exploration of the Cosmos*, Anchor Books, New York, 1995.

182. B. Pascal, *Oeuvres*, L. Brunschvieg (ed.), Paris, 1908–14; Kraus Reprint, Vaduz, Liechtenstein, 1976.

183. T.E. Peet, *The Rhind Mathematical Papyrus, British Museum 10057 and 10058*. Introduction, Transcription, Translation and Commentary, Hodder & Stoughton, London, 1923.

184. H.O. Peitgen and P.H. Richter, *The Beauty of Fractals: Images of Complex Dynamical Systems*, Springer-Verlag, Berlin, 1986.

185. H. O. Peitgen, D. Saupe and F. v. Haesler, *Cayley's Problem and Julia Sets*, Math. Intelligencer **6** (1984), 11–20.

186. D. Pengelley, F. Richman, *Did Euclid need the Euclidean algorithm to prove unique factorization?*, American Mathematical Monthly **113** (2006), 196–205.

187. D. Pengelley, *A graduate course on the role of history in teaching mathematics*, in Study the Masters: The Abel-Fauvel Conference, 2002 (ed. Otto Bekken et al), pp. 53–61, National Center for Mathematics Education, University of Gothenburg, Sweden, 2003.

188. Z. Pogoda, L.M. Sokołowski, *Does Mathematics Distinguish Certain Dimensions of Space? Part II*, The American Mathematical Monthly **104**, 5 (1998), 456–463.

189. R. Rashed (ed.), *Diophantus, Lés Arithmétiques. Livres IV–VII, Zweisprachige Ausg. (Oeuvres de Diophante, vol. III et IV)*, Les Belles Lettres, Paris, 1984.

190. R. Rashed, *The Development of Arabic Mathematics: Between Arithmetic and Algebra*, Kluwer, Dordrecht, 1994. Original title: *Entre Arithmétique et Algèbre, Recherches sur l'Histoire des Mathématiques Arabes*, Société d'Edition Les Belles Lettres, Paris, 1984.

191. R. Rashed, A. Djebbar, *L'oeuvre algébrique d'Al-Khayyām*, University of Aleppo, Aleppo, 1981.

192. K. Reich, *Die Geschichte der Differentialgeometrie von Gauss bis Riemann (1828–1868)*, Archive For History of Exact Sciences **11** (1973), 273–382.

193. B. Riemann, *Über die Hypothesen, welche der Geometrie zu Grunde liegen*, Abhandlungen der Königlichen Gesellschaft der Wissenschaften zu Göttingen **13** (1868), 133–152.

194. B. Riemann, *The Collected Works of Bernhard Riemann*, second edition, Weber, H. (ed.), Dover, New York, 1953.

195. R. Rivest, A. Shamir, L.N. Adleman, *A Method for Obtaining Digital Signatures and Public-key Cryptosystems*, Communications of the ACM **21** (1978), 120–126.

196. S. Rockey, *A Bibliography of Collected Works and Correspondence of Mathematicians*, Cornell University Mathematics Library, Ithaca, New York, 1996–; http://www.library.cornell.edu/math/collectedworks.php.

197. O. Rodrigues, *Recherches sur la théorie analytique des lignes et des rayons de courbure des surfaces, et sur la transformation d'une classe d'intégrales doubles, qui ont un rapport direct avec les formules de cette théorie*, Correspondance sur l'École Polytechnique **3** (1815), 162.

198. B.A. Rosenfeld, *A History of Non-Euclidean Geometry: Evolution of the Concept of Geometric Space*, Springer Verlag, New York, 1988.

199. T. Sakai, *Riemannian geometry*, American Mathematical Society, Translations of Mathematical Monographs, v. 149, Providence, RI, 1996.

200. W. Scharlau, H. Opolka, *From Fermat to Minkowski*, Springer Verlag, New York, 1995.

201. C. Schilling (ed.), *Wilhelm Olbers: Sein Leben und Seine Werke*, v. 2, Berlin, Springer Verlag, 1900.

202. I. Schneider, *Die Situation der mathematischen Wissenschaften vor und zu Beginn der wissenschaftlichen Laufbahn von Gauß*, in *Carl Friedrich Gauß (1777–1855)*, I. Schneider (ed.), Minerva Publikation, Munich, 1981.

203. M.R. Schroeder, *Number Theory in Science and Communication* (2nd ed.), Springer Series in Information Sciences **7**, Springer Verlag, New York, 1990.

204. G. Schuppener, *Geschichte der Zeta-Funktion von Oresme bis Poisson*, Deutsche Hochschulschriften 533, Hänsel-Hohenhausen, Egelsbach, Germany, 1994.

205. G.T. Seaborg & E.G. Valens, *Elements of the Universe*, New York, Dutton, 1958.

206. J. Sesiano (ed.), *Books IV to VII of Diophantus' Arithmetica in the Arabic Translation Attributed to Qusta Ibn Luqa*, Springer Verlag, New York, 1982.

207. D. Shanks, *Solved and Unsolved Problems in Number Theory*, Chelsea Publ. Co., New York, 1985.

208. Y.A. Shashkin, *Fixed Points*, American Mathematical Society, Providence, 1991.

209. *Honors*, SIAM News **21** (1988), 5.

210. K. Shen, *Seki Takakazu and Li Shanlan's Formulae on the Sum of Powers and Factorials of Natural Numbers. (Japanese)* Sugakushi Kenkyu **115** (1987), 21–36.

211. G. Simmons, *Calculus Gems: Brief Lives and Memorable Mathematics*, McGraw-Hill, New York, 1992.

212. G.F. Simmons, *Calculus with Analytic Geometry*, McGraw Hill, Inc., New York, 1985.

213. T.J. Simpson, *A New Treatise on Fluxions: Wherein the Direct and Inverse Method Are Demonstrated after a New, Clear and Concise Manner, with Their Applications to Physics and Astronomy*, London, 1737.

214. T.J. Simpson, *Essays on Several Curious and Useful Subjects, in Speculative and Mix'd Mathematicks. Illustrated by a Variety of Examples*, J. Nourse, London, 1740. (Reprinted in vol. 29 of Readex Microprint: *Landmarks of Science*, ed. by I. Bernard Cohen, et al., 1972.)

215. M.-K. Siu, *The ABCD of Using History of Mathematics in the Classroom*, in Using History To Teach Mathematics: An International Perspective (ed. V. Katz), pp. 3–9, Mathematical Assoc. of America, Washington DC, 2000; archived at http://www.math.nmsu.edu/~history/.

216. S. Smale, *The Fundamental Theorem of Algebra and Complexity Theory*, Bulletin Amer. Math. Soc. **4** (1981), 1–36.

217. S. Smale, *On the Steps of Moscow University*, Math. Intelligencer **6** (1984), 21–27.

218. S. Smale, *On the Efficiency of Algorithms of Analysis*, Bulletin Amer. Math. Soc. **13** (1985), 87–121.

219. S. Smale, *Steve Smale*, www.math.berkeley.edu/~smale, 1996.

220. D.E. Smith, *A Source Book in Mathematics*, Dover, New York, 1959.

221. D. Sobel, *Longitude: The True Story of a Lone Genius Who Solved the Greatest Scientific Problem of His Time*, Walker and Co., New York, 1995.

222. M. Spivak, *A Comprehensive Introduction to Differential Geometry*, Publish or Perish, Houston, v. I, 1979.

223. Ibid., v. II.

224. Ibid., v. III.
225. P. Stäckel, *Friedrich Ludwig Wachter, ein Beitrag zur Geschichte der Nichteuklidischen Geometrie*, Mathematische Annalen **54** (1901), 49–85.
226. S. Stahl, *Real Analysis: A Historical Approach*, Wiley, New York, 1999.
227. J. Stewart, *Calculus*, 5th ed., Brooks/Cole, Belmont, CA, 2003.
228. J. Stillwell, *Sources of Hyperbolic Geometry*, American Mathematical Society, Providence, Rhode Island, 1996.
229. D.R. Stinson, *Cryptography: Theory and Practice*, CRC Press, Boca Raton, 1995.
230. D.J. Struik, *Outline of a history of differential geometry*, Isis **19** (1933), 92–120.
231. Ibid., Isis **20** (1934), 161–191.
232. D.J. Struik, *A Source Book in Mathematics, 1200–1800*, Cambridge, Harvard University Press, 1969; Princeton University Press, Princeton, 1986.
233. F. Swetz, J. Fauvel, O. Bekken, B. Johansson, V. Katz, (eds.), *Learn from the Masters!* Math. Assoc. of Amer., Washington, D.C., 1995.
234. F. Swetz (ed.), *From Five Fingers to Infinity: A Journey Through the History of Mathematics*, Open Court, Chicago, 1994.
235. F. Swetz, *The Evolution of Mathematics in Ancient China*, Mathematics Magazine **52** (1979), 10–19.
236. R. Taton, *L'Oeuvre Scientifique de Monge*, Presses Universitaires de France, Paris, 1951.
237. R. Thom, *Sur les Travaux de Stephen Smale*, Proceedings of the International Congress of Mathematicians, Izdatel'stvo Mir, Moscow, 1968.
238. J.V. Uspensky, *Theory of Equations*, McGraw-Hill, New York, 1948.
239. B.L. van der Waerden, *Modern Algebra*, Ungar, New York, 1949.
240. B.L. van der Waerden, *Science Awakening*, Noordhoff, Groningen, 1954.
241. François Viète, *The Analytic Art* (translated by T. Richard Witmer), Kent State University Press, Kent, Ohio, 1983.
242. N.Y. Vilenkin, *Successive Approximation*, Macmillan, New York, 1964.
243. F.W. Warner, *Foundations of Differentiable Manifolds and Lie Groups*, Springer-Verlag, New York, 1983.
244. J.R. Weeks, *The Shape of Space: How to Visualize Surfaces and Three-Dimensional Manifolds*, M. Dekker, New York, 1985.
245. A. Weil, *Number Theory: An Approach Through History: From Hammurapi to Legendre*, Birkhäuser, Boston, 1983.
246. H. Wilf, *Generatingfunctionology*, Academic Press, Boston, 1990.
247. J. J. Winter & W. 'Arafat, *The Algebra of 'Umar Khayyam*, Journal of the Royal Asiatic Society **41** (1950), 27–78.
248. M.J. Winter, *The WebElements Periodic Table*, http://www.webelements.com, 2004.
249. H.E. Wolfe, *Introduction to Non-Euclidean Geometry*, Holt, Rinehart and Winston, New York, 1945.
250. Wolfram Research Inc., http://mathworld.wolfram.com/PrimeFormulas.html.
251. H. Wussing, *The Genesis of the Abstract Group Concept*, The MIT Press, Cambridge, MA, 1984.
252. B.F. Wyman, *What Is a Reciprocity Law?*, American Mathematical Monthly **79** (1972), 571–586.
253. B. H. Yandell, *The honors class: Hilbert's problems and their solvers*, A.K. Peters, Natick, Mass., 2002.

254. R.C. Yates, *The Trisection Problem*, Baton Rouge, The Franklin Press, 1942.
255. L. Yau & A. Ben-Israel, *The Newton and Halley Methods for Complex Roots*, American Mathematical Monthly **105** (1998), 806–818.
256. J.G. Yoder, *Unrolling Time: Christiaan Huygens and the mathematization of nature*, Cambridge University Press, Cambridge, 1988.
257. K. Yosida, *A Brief Biography on Takakazu Seki (1642?–1708)*, Math. Intelligencer **3** (1980/81), no. 3, 121–122.
258. R.M. Young, *Excursions in Calculus: An Interplay of the Continuous and the Discrete*, Mathematical Association of America, Washington, D.C., 1992.
259. T.J. Ypma, *Historical Development of the Newton-Raphson Method*, SIAM Review **37**, no. 4 (Dec. 1995), 531–551.

Credits

Front cover: Clockwise from top left, same as Photo 1.3, Figure 3.5, Photo 2.4, Photo 3.1, Photo 2.3, Photo 4.9.

Every effort has been made to obtain permission for the reproduction of the postage stamps featured in this book. Special thanks go to the postal authorities from each country whose stamp is represented here:

Greece [Philatelic Service, Hellenic Post], for the stamp of Archimedes.

France [La Poste, Service National des Timbres-poste et de la Philatelie], for the stamp of Lagrange.

Germany [Deutschepost], for the stamp of Gauss.

Figure 1.5 from R.M. Young, *Excursions in Calculus: An Interplay of the Continuous and the Discrete*, Mathematical Association of America, Washington, D.C., 1992; by permission.

Figure 1.6 from B. Pascal, *Oeuvres; publiées suivant l'ordre chronologique, avec documents complémentaires, introductions et notes, par Léon Brunschvicq et Pierre Boutroux*, Paris, Hachette, 1904–14. Kraus Reprint Ltd., Vaduz [Liechtenstein], vol. 3, 1965–78.

Figure 1.7 courtesy of Kathe Kanim; by permission.

Photo 1.2 from D.E. Smith, *Portraits of Eminent Mathematicians: with Brief Biographical Sketches, Part II*, Scripta Mathematica, New York, 1938–1946.

Figure 1.10 from C.B. Boyer, U.C. Merzbach, *A History of Mathematics* (second ed.), John Wiley & Sons, New York, 1989.

Photo 1.3 from Jakob Bernoulli: Wahrscheinlichkeitsrechnung, Ostwalds Klassiker der exakten Wissenschaften Band 107, 2. Auflage, Verlag Harri Deutsch 1999, 2002; by permission.

Figure 2.3 from A.B. Chace, L. Bull, H.P. Manning, *The Rhind Mathematical Papyrus*. Vol. II, Photographs, Transcription, Transliteration, Literal Translation, 1929. Mathematical Association of America, Oberlin, Ohio, U.S.A.; by permission.

Figure 2.4 from O. Neugebauer, *Mathematische Keilschrifttexte*, Quellen und Studien, A 3, Springer-Verlag, Berlin, 1935; with kind permission of Springer Science and Business Media. Also in B.L. Van der Waerden, *Science Awakening*, P. Noordhoff Ltd., Groningen, Holland, 1954.

Photo 2.1 from R. Rashed, *L'oeuvre algébrique d'Al-Khayyām*, University of Aleppo, Aleppo, 1982.

Photo 2.2 from D.E. Smith, *A Source Book in Mathematics*, Dover Publications, New York, 1959.

Photo 2.3 from U. Libbrecht, *Chinese Mathematics in the Thirteenth Century: The Shu-shu chiu-chang of Ch'in Chiu-shao*, MIT Press, Cambridge, 1973; by permission.

Photo 2.4 from D.E. Smith, *A Source Book in Mathematics*, Dover Publications, New York, 1959.

Photo 2.5 from T.J. Simpson, *Essays on Several Curious and Useful Subjects, in Speculative and Mix'd Mathematicks. Illustrated by a Variety of Examples*, J. Nourse, London, 1740.

Photo 2.6 courtesy of Steve Smale.

Figures 2.17 and 2.18 from R.H. Abraham and C.D. Shaw, *Dynamics — the Geometry of Behavior; Part 3: Global Behavior*, Aerial Press, Santa Cruz, Calif., 1984; by permission.

Photo 3.1 from C. Boyer, U. Merzbach, *A History of Mathematics*, Second Edition, John Wiley & Sons, New York, 1989; by permission.

Figure 3.2 from C. Huygens, *Horologium oscillatorium*, 1673.

Figure 3.5 from J.G. Yoder, *Unrolling Time: Christiaan Huygens and the Mathematization of Nature*, Cambridge University Press, Cambridge, 1988.

Photo 3.2 by permission, Deutsches Museum.

Figure 3.12 M.C. Escher's "High and Low" © 2005 The M.C. Escher Company-Holland. All rights reserved. www.mcescher.com.

Photo 4.1 from A. Weil, *Number Theory: An Approach Through History: From Hammurapi to Legendre*, Birkhäuser, Boston, 1983; with kind permission of Springer Science and Business Media.

Photo 4.2 from D.E. Smith, *Portraits of Eminent Mathematicians: with Brief Biographical Sketches, Part II*, Scripta Mathematica, New York, 1938–1946.

Photo 4.3 from H. Meschkowski, *Denkweisen Grosser Mathematiker*, Friedrich Vieweg & Sohn, Braunschweig, 1990.

Photo 4.5 from D.J. Struik, *A Concise History of Mathematics*, Fourth Revised Edition, Dover, New York, 1987.

Photo 4.8 from K. Biermann, *Carl F. Gauss: "Der Fürst der Mathematiker" in Briefen und Gesprächen*, Verlag C. H. Beck, München, 1990; by permission from Universitätsbibliothek Leipzig.

Photo 4.9 from G. Eisenstein, *Mathematische Werke, vol. 1*, Second Edition, Chelsea Publishing Company, New York, 1989; by permission from American Mathematical Society.

Name Index

Āryabhaṭa, 7
A'h-mosè, 88, 89
Abel, Niels Henrik, 101
al Ṭūsī, Sharaf al-Din, 96
al-Karajī, Abū Bakr, 7, 8, 17, 32, 38
al-Ṭūsī, 142
Apollonius, 94, 161
Archimedes, 2–6, 8, 18–25, 32, 39
Aristotle, 229

Beltrami, Eugenio, 224
Berkeley, George, 134
Bernoulli, Daniel, 13, 15
Bernoulli, Jakob, 1, 2, 12, 28, 39, 41–50,
    52, 56, 58, 64
Bernoulli, Johann, 1, 42, 185, 251
Bianchi, Luigi, 224
Bolyai, János, 160, 165, 196, 197,
    216
Bolyai, Wolfgang, 197
Bombelli, Rafael, 127
Bulliald, Ismael, 48
Büttner, J. G., 5

Cajori, Florian, 103
Cardano, Girolamo, 96–101, 108, 141
Cauchy, Augustin-Louis, 146
Cayley, Arthur, 300
Chevalier de Méré, 11
Ch'in, see Qin
Clairaut, Alexis, 163, 187
Clarke, Francis, 132
Clausen, 64, 69

Clifford, William, 216

d'Alembert, Jean-Baptist, 262
Darboux, Gaston, 224
DeBranges, Louis, 148
Debreau, Gerard, 148
del Ferro, Scipione, 97
Digby, Kenelm, 233, 276
Dijksterhuis, E.J., 20
Diophantus, 231
Dirichlet, Lejeune, 240, 284
Dirichlet, Peter Gustav Lejeune, 214,
    215
Donaldson, Simon, 167
Dupin, Charles, 187, 188, 193, 194
Döbereiner, Johann W., 248

Einstein, Albert, 160, 167, 224
Eisenstein, Gotthold, 246, 292–300
Empedocles, 229
Escher, Maurits Cornelis, 221
Euclid, 94, 160, 230
Euler, Leonhard, 1–4, 13–15, 49–82,
    161, 163–166, 187–195, 197–199,
    203, 223, 225, 235, 237–240,
    251–262, 272, 273, 290

Faber, Georg, 52
Fatou, Pierre, 152
Faulhaber, Johann, 9, 48
Fauvel, John, 125
Fermat, Pierre de, 9, 10, 26–32, 45, 231,
    233–236, 238, 239, 272

Ferrari, Lodovico, 97, 101
Fibonacci, 233
Freedman, Michael, 167

Galois, Evariste, 101, 103
Gauss, Carl Friedrich, 5, 80, 159, 161,
    164, 165, 166, 196–204, 214, 220,
    223–225, 239, 244, 245, 286–292,
    301–305
Germain, Sophie, 164, 198, 199, 287, 289
Goldbach, Christian, 13, 235, 253
Gray, Thomas, 132
Gregory, James, 1

Hadamard, Jacques, 80
Hardy, Godfrey H., 247
Harrison, John, 162
Helmholtz, Hermann von, 216
Herbart, Johann Friedrich, 218
Hilbert, David, 231
Hirsch, Morris W., 148
Holdred, Theophilus, 102
Horner, William G., 102, 129
Hui, Yang, 111
Huygens, Christiaan, 161, 167, 168,
    174–181, 234

ibn al-Haytham, Abū ʿAlī al-Ḥasan,
    8–10, 17, 32, 38

Jacobi, Carl Gustav Jacob, 214
Julia, Gaston, 152

Kanim, Kathe, 21
Kant, Immanuel, 160
Karp, Richard, 148, 149
Khayyam, Omar, see ʿUmar
Kummer, Ernst, 64

l'Hospital, Guillaume, 185
Lagrange, Joseph-Louis, 103, 238, 239,
    248, 262–279, 285, 303
Lambert, Johann, 60
Lebesgue, Henri, 151
Leeuwenhoek, Anton van, 169
Legendre, Adrien-Marie, 80, 214, 240,
    241, 245, 248, 279–286, 290
Leibniz, Gottfried Wilhelm, 1, 133, 135,
    162, 176, 185

Leucippus, 229
Libbrecht, Ulrich, 110
Lie, Sophus, 224, 226
Lobachevsky, Nikolai Ivanovich, 160,
    196, 216, 224
Lorenz, Edward, 153

MacDuffee, Cyrus C., 141
Matiyasevič, Juri, 231
Menaechmus, 106
Mendeleyev, Dmitry I., 229, 248
Mengoli, Pietro, 2
Mercator, Nicolaus, 48
Mersenne, Marin, 26, 28, 30, 233
Meusnier, Jean-Baptiste, 166, 193–195
Meyer, Julius L., 229
Minding, Ferdinand, 164
Minkowski, Hermann, 160, 167, 224, 300
Monge, Gaspard, 188, 195

Needham, Joseph, 110, 123
Newton, Isaac, 1, 102, 125–133, 155,
    161, 163, 168, 181–185
Nicomachus, 16

Otero, Daniel E., 43
Oughtred, William, 127

Pascal, Blaise, 10–12, 26, 28–41, 43, 49,
    56
Ping, Wang, 110
Plutarch, 18
Poincaré, Henri, 224
Poussin, Charles-Jean de la Vallée, 80
Ptolemy, 225

Qin, Jiu Shao, 101, 110–125, 129

Raphson, Joseph, 129
Riemann, Georg Friedrich Bernhard,
    77, 94, 159, 166, 167, 196, 214–227
Roberval, Gilles Persone de, 9
Rodrigues, Oline, 164, 198
Ruffini, Paolo, 102

Seki, Takakazu, 49
Shi-Jie, Zhu, 111
Simpson, Thomas J., 102, 132–139

Smale, Stephen, 85, 103, 142, 147–152, 157
Steven, Simon, 127

Tartaglia, Niccolò, 97
Thom, René, 147

'Umar ibn Ibrahīm al-Khayyāmī, Abu'l Fatḥ, 92–96, 98, 106

Viète, François, 90, 102, 127, 142

Voltaire, 125
von Staudt, Christian, 64, 69

Wallis, John, 48
Warhol, Andy, 220
Weber, Wilhelm, 216
Weyl, Hermann, 223
Wingate, Edmund, 127

Ye, Li, 111

Subject Index

absolute error, 143
age of exploration, 162
age of reason, 132
al-jabr, 92
al-muqabala, 92
Algebrista y Sangrador, 92
algorithm
  fixed-point, 144, 155
  linear, 142
  quadratic, 143
  robust, 142
algorithms, nature of, 86, 87, 140–141
analytic continuation, 77
approximate zero, 150, 151, 156
approximately
  asymptotically, 65
approximation, 17, 38, 70, 71, 73, 75,
    101, 104, 113, 114, 117, 141, 143,
    155
approximations, 3, 51, 72
arithmetic progression, 40
arithmetical triangle, 10–12, 29, 31, 39,
    40, 48
attractor basins, 85
auxiliary sphere, 164, 165, 202, 204,
    205, 208, 227

Babylon, 90, 106, 107, 139
Basel problem, 2–4, 13, 14, 51, 70, 71
basic elements, 229
Bernoulli family, 42
Bernoulli number, 4, 12, 13, 15–17, 43,
    49, 52, 62–65, 69, 70. See number

Bezout's equation, 308, 309
Bieberbach conjecture, 148, 149
binomial coefficient, 10, 12, 29, 31, 32
biquadratic equation or polynomial, 97,
    101 See equation
biquadratic residue, 244
bisection method, 148, 155
brachistrochrone problem, 42
Brahmagupta's identity, 250, 305

Cardano's formula, 107
center of curvature, 170, 181, 183–186.
    See curvature
China, 110–112
class group, 301, 304
clock, 169–181
combination number, 29
combinatorial coefficients, 10
composition, 301
composition of forms, 304
congruence, quadratic, 243
containment, 303, 305
convergence
  cubic, 143
  global, 103
  linear, 146, 150, 156
  local, 103
  quadratic, 84, 103, 139, 143, 146, 150,
    153
  rate of, 137, 141, 142, 146, 156
  speed of, 142, 149, 150
coordinate chart, 223, 226, 227
correcting terms, 38

Cramer's rule, 138
cross section, 189–191
cryptography
    public key, 248
cube root, 101
cubic equation or polynomial, 92, 96,
        127. *See* equation
cuneiform tablet, 91
curvature, 147, 161
    center of, 170, 181, 183–186
    constant, 166
    defining conditions of, 183–184
    Gaussian, 161, 165, 166, 197, 198,
        212–214, 223
    integral, 198, 202, 208, 209
    maximum, 164, 187, 193–195, 203, 213
    mean, 164, 198, 199
    minimum, 164, 187, 193–195, 203, 213
    negative, 165
    radius of, 162, 163, 170, 174, 176,
        179, 180, 181, 183, 185, 186, 188,
        189, 193
    total, 165, 198
cycloid, 169, 170, 178, 179

descent method, 238, 239, 273, 277
desk calculator, 140
determinant, 269, 302
Diophantus, 249
discriminant, 108. *See* determinant
distance-preserving map, 200, 201, 224
divisor plus descent, 238, 271, 277
divisor problem, 236, 238, 273, 276, 277,
        280, 284

Egypt, 87
Eisenstein's rectangle, 297
electronic computer, 140
envelope, 161
equation
    biquadratic, 97, 101
    cubic, 92, 96, 127
    fifth-degree, 101
    fourth-degree, 101, 116, 120, 124
    linear, 87
    quadratic, 90, 106, 243
equivalence of forms, 304, 305
Eratosthenes
    sieve of, 230, 249

Euclid's lemma, 306
Euler
    product formula, 76, 81
Euler's conclusion, 258–261, 291
Euler's constant, γ, 71, 75
Euler's criterion, 240, 241, 280, 283, 298
Euler–Maclaurin summation formula, 3.
        See summation formula
Eureka, 20
evolute, 161–163, 170, 174–176, 178–180
exhaustion
    method of, 6, 24, 25
extrapolation
    linear, 130, 141
    quadratic, 130, 143

factorial function, 76
false position
    method of, 128
Fermat, 253, 271
Fermat's last theorem, 231
Fermat's little theorem, 282, 309, 310
Fibonacci
    number, 155
figurate number, 9 *See* number
first circle, 23
fixed point, 142, 144–146, 155, 156
fixed-point method, 103
fluxion, 133, 135, 138, 139, 163, 182, 184
forms of degree two (= quadratic form),
        301
fourth-degree polynomial, 101, 116, 120,
        124. *See* equation
fractal, 85, 104, 152, 156, 157
    with basins, 84, 152
    with iterations, 152
functional equation, 77
fundamental theorem of arithmetic,
        229, 306
Fundamental theorem of Gauss,
        Quadratic Reciprocity Law, 286,
        287, 289, 291

gamma function, 76
Gauss map, 198, 212, 213, 227
Gauss's
    third proof, 287
Gaussian curvature, 161, 165, 166, 197,
        198, 212–214, 223. *See* curvature

generating function, 58, 59, 69
genus theory, 277
geodesic, 161, 166, 216, 223
geodesic triangle, 165
geometrical binomial, 99
geometry
  elliptic, 161
  Euclidean, 160, 161, 165
  foundation, 159, 160
  hyperbolic, 160, 161, 196, 197, 211,
    224
  non-Euclidean, 159, 165, 196, 216, 224
  Riemannian, 167
global behavior, 147
gnomon, 7
Greece, 18–21
greedy algorithm, 105
group, 301

higher parabola, 9. *See* parabola
higher reciprocity laws, 244
Hilbert's tenth problem, 231
Horner–Ruffini method, 102, 113,
    117–124, 131, 141, 142
Huguenots, 132
Huygens pendulum, 171
hyperbolic function, 60
hyperbolic logarithm, 59

induction, mathematical, 5, 8, 40, 45, 49
infinite polynomial, 51
infinitesimal, 180–182, 184, 198, 204, 205
initial guess, 83, 115, 116, 125, 130, 135,
    136, 138, 139, 144
integral curvature, 198, 202, 208, 209.
    See curvature
interpolation
  linear, 141
involute, 179
inward-pointing normal, 212, 227
Islamic mathematicians, 7–9
isochrone, 169
isolation of root, 115, 141
isometry, 196
iteration, 101, 114, 115, 130, 140, 143,
    144

lattice points, 295
Legendre symbol, 240, 241, 250, 251, 283

Leibniz's formula for π, 1, 299
Lie group, 226
linear algorithm, 142, 144, *See* algorithm
linear convergence, 146, 150, 156. *See*
    convergence
linear equation, 87. *See* equation
linear extrapolation, 130, 141. *See*
    extrapolation
local behavior, 142
longitude, 162
Lorenz attractor, 153

manifold, 159–161, 163, 166, 216–227
maximum radius of curvature, 164, 187,
    193–195, 203, 213. *See* curvature
mean curvature, 164, 198, 199. *See*
    curvature
mean proportion, 106
metric, 161, 163, 166, 200, 201, 216,
    217, 220, 223
minimal surface, 166, 199
minimum radius of curvature, 164, 187,
    193 195, 203, 213. *See* curvature
modified Newton's method, 150, 156
motion
  center of, 184
  of pendulum bob, 169–181
multiplicativity of the Legendre symbol,
    242
multiply extended quantity, 217, 218
Möbius band, 227

Newton's proportional method, 83, 102,
    127–130, 137–139, 141, 143, 154
Nicomachus of Gerasa, 7
normal
  inward-pointing, 197
  outward-pointing, 197
number
  Bernoulli, 4, 12, 13, 15–17, 43, 49, 52,
    62–65, 69, 70
  combination, 10, 12, 31
  figurate, 10, 12, 27–29, 31, 33, 45, 48
  polygonal, 5, 16
  pyramidal, 28, 45
  rectangular, 4
  square, 4, 16
  triangular, 4, 5, 10, 16, 26, 28, 45
numerical methods, 140–142

orientation, 205
orientation-preserving, 198
orientation-reversing, 198, 212
osculating circle, 162–164, 176, 179,
    189, 191
osculatory radius, 192, 193
outward-pointing normal, 227

$P = NP$ problem, 149
parabola
    area under a higher, 39, 48
paraboloid, 5, 8
parallel postulate, 159–161, 165, 166,
    216, 217, 223
Pascal's equation, 38, 40, 55, 69
pendulum
    clock, 162, 169, 170
    Huygens's, 178, 184
    isochronous, 162, 172, 173, 177, 180,
       181
    simple, 169–171, 173, 176, 177, 180
periodic table, 229
Persia, 92
pi, π, 2, 70, 75, 81
piling up, 27, 28
Poincaré conjecture, 167, 224
polynomial formula, 9, 43, 69
positive definite form, 273, 278
prime divisor, 255, 257
prime divisors, trivial, 236
prime number
    infinitely many, 229, 230
prime number theorem, 80
probability, 10–12, 31, 43, 151
product expansion
    infinite, 68
progression, 33, 34, 36, 47, 48
    arithmetic, 5, 10, 16, 20, 24, 38,
       235–237, 240, 253, 256, 257, 284
    natural, 36, 37
proportion, 96, 106, 129, 130
pyramid
    collateral, 9, 10, 26, 27, 30
pyramidal number, 28, 45. See number
Pythagoreans, 2–5, 229

Qin's method, 117, 120–142
QRL, 237. See quadratic reciprocity law
QRL, different proofs, 247

quadratic algorithm, 143. See algorithm
quadratic character of negative one,
    259
quadratic convergence, 84, 103, 139,
    143, 146, 150, 153. See convergence
quadratic equation, 90, 106, 243. See
    equation
quadratic extrapolation, 130, 143. See
    extrapolation
quadratic form, 235, 237, 239, 251, 253,
    261, 271–273, 280, 284, 301, 304
quadratic reciprocity law, 237, 279, 284
quadratic reciprocity law
    four separate statements of, 290
quadratic residue, 236, 238, 244, 245,
    248, 250, 257–260, 276, 283, 295,
    302
    key properties, 257

radius of curvature, 162, 163, 170, 174,
    176, 179–181, 183, 185, 186, 188,
    189, 193. See curvature
rate of convergence, 137, 141, 142, 146,
    156. See convergence
real projective space, 227
recursion relation, 27, 29
recursive formula, 13, 17, 62, 64, 68
Regula falsi, 128
relative error, 143
remainder, 258
Renaissance, 97
representability problem, 238
representation problem, 235, 273, 277,
    280, 304
residue, 258. See also remainder and
    quadratic residue
restoration, 126
Rhind mathematical papyrus, 87–89
Riemann
    hypothesis, 81
Riemann integral, 215
Riemann sum, 25
Riemann zeta function, 76, 77–80, 215.
    See zeta function
Roberval, 17, 38
robust algorithm, 142. See algorithm

saddle surface, 197, 212
self-similarity, 85

series,
    alternating, 1
    asymptotic, 74
    divergent, 72–74
    geometric, 1
    harmonic, 71
    infinite, 51, 52
    Leibniz's, 1, 13, 18
    of reciprocal even powers, 52
    of reciprocal squares, 73
    power, 1, 15, 51, 59, 62, 69
    Stirling's, 75
    Taylor, 54, 68
similar figure, 21
Simpson's
    method fails, 85
Simpson's fluxional method, 83–85,
        102, 103, 105, 123, 131, 133–139,
        141–143, 145, 146, 148, 153
slide rule, 140
Smale horseshoe, 153
Smale's theorem, 147, 151, 152
Song dynasty, 110
space-time continuum, 224
speed of convergence, 142, 149, 150.
        See convergence
spiral, 5, 18, 20, 21, 23, 24
    Archimedes, 16
square root, 101, 113, 114, 124
squaring, 9
Stirling's approximation, 75
strange attractor, 153
sum
    discrete, 5
    infinite, 51
    of cubes, 7, 8, 16
    of first powers, 25
    of fourth powers, 8, 17, 69
    of gnomons, 7
    of powers, 3, 6, 9, 11, 14, 15, 26, 28,
        30–32, 36, 37, 43, 45–48, 55, 63,
        64, 66
    of reciprocal powers, 65, 70

of reciprocal squares, 14, 15, 18, 51,
        70, 75, 81
of squares, 7, 16, 20–22, 25, 233, 239
of kth powers, 12
Riemann, 6
summation
    continuous, 12
    discrete, 12
summation formula, 5, 21
    Euler–Maclaurin, 4, 13–16, 18, 51, 52,
        56, 66, 71, 74, 75, 77
Supplementary theorem, 242
synthetic division, 113, 118–120, 122,
        128

table of logarithms, 140
tautochrone, 169, 176, 178
theorema egregium, 164–166, 196–200,
        211
thing, 94, 100
third proof, 293
torus, 212, 225
total curvature, 165, 198. See curvature
touch, 204
tractable method, 148, 149
trapezoid rule, 18
triangle
    collateral, 9, 10, 26
triangular number, 4, 5, 10, 16, 26, 28,
        45. See number
triangulo-triangle
    collateral, 26
trisection, 109
Turing machine, 149
    nondeterministic, 149
twelfth consequence, 29, 30, 39, 40

'Umar's method, 107
unit fractions, 87

Wallis's theorem, 41

zeta function, 76–80, 215

Undergraduate Texts in Mathematics

Abbott: Understanding Analysis.
Anglin: Mathematics: A Concise History and Philosophy.
Readings in Mathematics.
Anglin/Lambek: The Heritage of Thales.
Readings in Mathematics.
Apostol: Introduction to Analytic Number Theory. Second edition.
Armstrong: Basic Topology.
Armstrong: Groups and Symmetry.
Axler: Linear Algebra Done Right. Second edition.
Beardon: Limits: A New Approach to Real Analysis.
Bak/Newman: Complex Analysis. Second edition.
Banchoff/Wermer: Linear Algebra Through Geometry. Second edition.
Beck/Robins: Computing the Continuous Discretely.
Berberian: A First Course in Real Analysis.
Bix: Conics and Cubics: A Concrete Introduction to Algebraic Curves. Second edition.
Brémaud: An Introduction to Probabilistic Modeling.
Bressoud: Factorization and Primality Testing.
Bressoud: Second Year Calculus.
Readings in Mathematics.
Brickman: Mathematical Introduction to Linear Programming and Game Theory.
Browder: Mathematical Analysis: An Introduction.
Buchmann: Introduction to Cryptography.
Buskes/van Rooij: Topological Spaces: From Distance to Neighborhood.
Callahan: The Geometry of Spacetime: An Introduction to Special and General Relativity.
Carter/van Brunt: The Lebesgue–Stieltjes Integral: A Practical Introduction.
Cederberg: A Course in Modern Geometries. Second edition.
Chambert-Loir: A Field Guide to Algebra.
Childs: A Concrete Introduction to Higher Algebra. Second edition.
Chung/AitSahlia: Elementary Probability Theory: With Stochastic Processes and an Introduction to Mathematical Finance. Fourth edition.
Cox/Little/O'Shea: Ideals, Varieties, and Algorithms. Second edition.
Croom: Basic Concepts of Algebraic Topology.

Cull/Flahive/Robson: Difference Equations: From Rabbits to Chaos.
Curtis: Linear Algebra: An Introductory Approach. Fourth edition.
Daepp/Gorkin: Reading, Writing, and Proving: A Closer Look at Mathematics.
Devlin: The Joy of Sets: Fundamentals of Contemporary Set Theory. Second edition.
Dixmier: General Topology.
Driver: Why Math?
Ebbinghaus/Flum/Thomas: Mathematical Logic. Second edition.
Edgar: Measure, Topology, and Fractal Geometry.
Elaydi: An Introduction to Difference Equations. Third edition.
Erdős/Surányi: Topics in the Theory of Numbers.
Estep: Practical Analysis in One Variable.
Exner: An Accompaniment to Higher Mathematics.
Exner: Inside Calculus.
Fine/Rosenberger: The Fundamental Theory of Algebra.
Fischer: Intermediate Real Analysis.
Flanigan/Kazdan: Calculus Two: Linear and Nonlinear Functions. Second edition.
Fleming: Functions of Several Variables. Second edition.
Foulds: Combinatorial Optimization for Undergraduates.
Foulds: Optimization Techniques: An Introduction.
Franklin: Methods of Mathematical Economics.
Frazier: An Introduction to Wavelets Through Linear Algebra.
Gamelin: Complex Analysis.
Ghorpade/Limaye: A Course in Calculus and Real Analysis.
Gordon: Discrete Probability.
Hairer/Wanner: Analysis by Its History.
Readings in Mathematics.
Halmos: Finite-Dimensional Vector Spaces. Second edition.
Halmos: Naive Set Theory.
Hämmerlin/Hoffmann: Numerical Mathematics.
Readings in Mathematics.
Harris/Hirst/Mossinghoff: Combinatorics and Graph Theory.
Hartshorne: Geometry: Euclid and Beyond.

Undergraduate Texts in Mathematics

Hijab: Introduction to Calculus and Classical Analysis. Second edition.

Hilton/Holton/Pedersen: Mathematical Reflections: In a Room with Many Mirrors.

Hilton/Holton/Pedersen: Mathematical Vistas: From a Room with Many Windows.

Iooss/Joseph: Elementary Stability and Bifurcation Theory. Second Edition.

Irving: Integers, Polynomials, and Rings: A Course in Algebra.

Isaac: The Pleasures of Probability. Readings in Mathematics.

James: Topological and Uniform Spaces.

Jänich: Linear Algebra.

Jänich: Topology.

Jänich: Vector Analysis.

Kemeny/Snell: Finite Markov Chains.

Kinsey: Topology of Surfaces.

Klambauer: Aspects of Calculus.

Knoebel/Laubenbacher/Lodder/Pengelley: Mathematical Masterpieces: Further Chronicles by the Explorers. *Readings in Mathematics.*

Lang: A First Course in Calculus. Fifth edition.

Lang: Calculus of Several Variables. Third edition.

Lang: Introduction to Linear Algebra. Second edition.

Lang: Linear Algebra. Third edition.

Lang: Short Calculus: The Original Edition of "A First Course in Calculus."

Lang: Undergraduate Algebra. Third edition.

Lang: Undergraduate Analysis.

Laubenbacher/Pengelley: Mathematical Expeditions.

Lax/Burstein/Lax: Calculus with Applications and Computing. Volume 1.

LeCuyer: College Mathematics with APL.

Lidl/Pilz: Applied Abstract Algebra. Second edition.

Logan: Applied Partial Differential Equations, Second edition.

Logan: A First Course in Differential Equations.

Lovász/Pelikán/Vesztergombi: Discrete Mathematics.

Macki-Strauss: Introduction to Optimal Control Theory.

Malitz: Introduction to Mathematical Logic.

Marsden/Weinstein: Calculus I, II, III. Second edition.

Martin: Counting: The Art of Enumerative Combinatorics.

Martin: The Foundations of Geometry and the Non-Euclidean Plane.

Martin: Geometric Constructions.

Martin: Transformation Geometry: An Introduction to Symmetry.

Millman/Parker: Geometry: A Metric Approach with Models. Second edition.

Moschovakis: Notes on Set Theory. Second edition.

Owen: A First Course in the Mathematical Foundations of Thermodynamics.

Palka: An Introduction to Complex Function Theory.

Pedrick: A First Course in Analysis.

Peressini/Sullivan/Uhl: The Mathematics of Nonlinear Programming.

Prenowitz/Jantosciak: Join Geometries.

Priestley: Calculus: A Liberal Art. Second edition.

Protter/Morrey: A First Course in Real Analysis. Second edition.

Protter/Morrey: Intermediate Calculus. Second edition.

Pugh: Real Mathematical Analysis.

Roman: An Introduction to Coding and Information Theory.

Roman: Introduction to the Mathematics of Finance: From Risk management to options Pricing.

Ross: Differential Equations: An Introduction with Mathematica®. Second Edition.

Ross: Elementary Analysis: The Theory of Calculus.

Samuel: Projective Geometry. *Readings in Mathematics.*

Saxe: Beginning Functional Analysis.

Scharlau/Opolka: From Fermat to Minkowski.

Schiff: The Laplace Transform: Theory and Applications.

Sethuraman: Rings, Fields, and Vector Spaces: An Approach to Geometric Constructability.

Shores: Applied Linear Algebra and Matrix Analysis.

Sigler: Algebra.

Silverman/Tate: Rational Points on Elliptic Curves.

Simmonds: A Brief on Tensor Analysis. Second edition.

Singer: Geometry: Plane and Fancy.

Singer: Linearity, Symmetry, and Prediction in the Hydrogen Atom.

Singer/Thorpe: Lecture Notes on Elementary Topology and Geometry.

Smith: Linear Algebra. Third edition.

Undergraduate Texts in Mathematics

Smith: Primer of Modern Analysis.
Second edition.

Stanton/White: Constructive Combinatorics.

Stillwell: Elements of Algebra: Geometry,
Numbers, Equations.

Stillwell: Elements of Number Theory.

Stillwell: The Four Pillars of Geometry.

Stillwell: Mathematics and Its History.
Second edition.

Stillwell: Numbers and Geometry.
Readings in Mathematics.

Strayer: Linear Programming and Its
Applications.

Toth: Glimpses of Algebra and Geometry.
Second Edition.
Readings in Mathematics.

Troutman: Variational Calculus and Optimal
Control. Second edition.

Valenza: Linear Algebra: An Introduction to
Abstract Mathematics.

Whyburn/Duda: Dynamic Topology.

Wilson: Much Ado About Calculus.

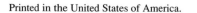Printed in the United States of America.